HYDROGEN TECHNOLOGY FOR ENERGY

HYDROGEN TECHNOLOGY FOR ENERGY

David A. Mathis

NOYES DATA CORPORATION
Park Ridge, New Jersey, U.S.A.
1976

Library of Congress Catalog Card Number: 76-17938
ISBN: 0-8155-0629-5
Printed in the United States

Published in the United States of America by
Noyes Data Corporation
Noyes Building, Park Ridge, New Jersey 07656

FOREWORD

This is the ninth volume of our Energy Technology Review Series. The technological information found here is based on government-sponsored studies (often performed by highly qualified personnel from industrial companies) and two U.S. patents.

In the United States we are fortunate in receiving direct help from the numerous surveys, together with active research and development programs that are being supported by the federal government.

Advanced composition and production methods developed by Noyes Data are employed to bring our new durably bound books to you in a minimum of time. Special techniques are used to close the gap between "manuscript" and "completed book." Industrial technology is progressing so rapidly that time-honored, conventional typesetting, binding and shipping methods are no longer suitable. We have bypassed the delays in the conventional book publishing cycle and provide the user with an effective and convenient means of reviewing up-to-date information in depth.

The last chapter constitutes a listing of 263 hydrogen technology experts, their affiliations and addresses, arranged by their specialties.

The table of contents is organized in such a way as to serve as a subject index and provides easy access to the information contained in this book. Bibliographic reference lists follow each chapter in order to provide the reader with easy access to further information on this timely topic. The Bibliography at the end of the book lists the highly important government reports and a source of their purchase.

CONTENTS AND SUBJECT INDEX

INTRODUCTION 1

THE HYDROGEN ECONOMY 3
Primary Energy Sources 5
Solar Energy 5
Wind Energy 7
Fossil Energy 9
Nuclear Energy 10
Miscellaneous 11
The Hydrogen Energy Conversion System 13
Hydrogen Production 15
Transportation and Storage of Hydrogen 16
Utilization of Hydrogen 17
Advantages and Disadvantages of Hydrogen as a Fuel 18
References 18

GASEOUS HYDROGEN 21
Comparison of Hydrogen Storage Methods 21
Existing Hydrogen Handling Systems 23
Economics of Transmission 24
A Hydrogen Pipeline System 24
Hydrogen Compatibility Problems 27
Types of Hydrogen Embrittlement 27
Embrittlement Mechanisms 29
Pressure Vessels and Pipelines for High Pressure Hydrogen 31
Aerojet Pressure Vessels 31
Los Alamos Hydrogen Pipeline 32
Fracture Mechanics Calculations 33
Compressor Requirements for Hydrogen Pipeline Transmission 36
Description of Existing Natural Gas Transmission System 36
Types of Pipeline Compressors 37
Reciprocating and Centrifugal Compressors 38
Regenerative Compressors 40

Cost of Compression Equipment for Hydrogen Storage 46
References. 46

LIQUID HYDROGEN. 49
Technology of Liquefaction and Storage. 50
Liquefaction 51
Energy Requirements 52
Capital Costs. 52
Liquefaction Costs 53
Recovery of Liquefaction Energy 53
Para-Hydrogen. 54
Storage and Transfer 55
Representative Storage and Transfer Losses 55
Storage Dewars 57
Transfer Systems. 63
Storage and Transfer Problems 70
Safety 72
Fire and Explosion Hazards. 73
Operation 75
Liquid Hydrogen Pumping 77
Pump Types 77
Technical Problems. 81
The Role of Liquid Hydrogen in an Energy System 84
The Energy Pipe 84
Liquid Hydrogen-Cooled Electrical Transmission Cable 84
Superconducting Transmission Cable 86
Peak Shaving Facility 88
Small Town Usage of Liquid Hydrogen. 91
Comparison of Energy Transmission Costs for a Near-Urban Environment. . . 94
References. 96

SOLID HYDRIDES. 100
Mechanism of Hydride Storage 100
Technical Problems of Hydride Storage. 104
Heat Transfer 104
Deterioration. 105
Safety 105
Steel Problems. 105
Compressor 110
Metal Hydride Screening Experiments. 110
Iron-Titanium-Manganese Alloy for Hydrogen Storage 114
Modelling Studies of Fixed-Bed Metal Hydride Storage Systems. 116
Convection Bed Model 117
Conduction Bed Model 118
Results of Studies 119
High Efficiency Power Cycles Using Metal Hydride Compressors 120
Operation of an Iron-Titanium Bed as a Hydrogen Compressor. 121
High and Low Temperature Heat Sources. 123
Transmission of Hydrogen from Compressors to High Temperature Source. 126
Implications of High Efficiency Power Cycle 127
Metal Hydrides for Ground Transport. 127
References. 133

USAGE OF HYDROGEN FOR FUEL AND ENERGY STORAGE 135
Hydrogen Utilization Devices 135
Water-Modified H_2-O_2 (Aphodid) Burner 135
Hydrogen Fueled Internal Combustion Engine 137
Catalytic Burners. 141
Flame Burners. 141
Fuel Cells 143
Other Devices 145
H_2-O_2 Combustion Powered Steam-MHD Central Power Systems 145
Steam MHD System Performance 146
Capital and Operating Costs for Steam MHD Systems. 148
Comparison of Power Systems 149
Advantages and Disadvantages. 151
Operating Experiences with Hydrogen Fueled Reciprocating Engines. 152
Hydrogen/Air Engines. 153
Hydrogen/Oxygen Engines 163
Mixed-Fuel Hydrogen Engines. 165
Energy Storage for Utilities-Brookhaven Optimization Model Studies. 166
Elements and Cost of Hydrogen Storage Electric Peaking Plants. 166
Cases Considered. 168
Conclusions. 170
Energy Storage for Utilities–PSE&G Maplewood Test System. 171
Description of System. 171
Conceptual Design of Metal Hydride Bed 175
Overall System Design Considerations. 178
Economic Perspective 179
Fuel for Transportation Systems. 180
Ground (Surface) Transportation 180
Water Transportation 188
Air (and Space) Transportation. 189
Liquid Hydrogen System for Automotive Transportation. 193
Vehicle Fuel Storage 193
Hydrogen Production and Distribution System. 197
Safety Considerations 201
Metal Hydride Storage System for Vehicles. 205
Description of System. 205
Safety Considerations 209
Weight, Economy and Range. 210
Hydride-Dehydride Power System Useful for Refrigeration 211
References. 216

NONTECHNICAL ASPECTS OF A HYDROGEN ECONOMY. 221
Safety Implications. 221
Legal Aspects 221
Political Aspects 224
Assessment of Safety Factors 226
Legal Implications. 227
Energy Law. 227
Regulatory Law. 229
Environmental Law. 231
International Law 234
Environmental Implications 236
Socioeconomic Implications 238

Social Costs and Values 238
Energy and the Social Future 240
Economic Costs 240
Hydrogen and Society 241
Political Implications 242
Politics and Energy 243
International Politics 244
Domestic Politics 246
References 247

HYDROGEN TECHNOLOGY EXPERTS 248
Hydrogen Properties 248
Heat Transfer 250
Hydrogen Production 251
Hydrogen Liquefaction Plants in the United States 251
Hydrogen LIquefaction and Purification 252
Ortho-Para Conversion 253
Slush Hydrogen 253
Gel Hydrogen 254
Metallic Hydrogen 254
Electrolysis Systems 255
Thermochemical Production of Hydrogen 255
Materials of Construction 256
Metals 256
Polymeric Composites 258
Insulations 258
Engineering Systems and Subsystems 259
Large Storage Tanks and Vessels (Field-Erected) 259
Vacuum-Jacketed Transfer Lines 260
Heat Exchangers 261
Liquid Hydrogen Pumps 261
Refrigerators 263
Compressors (Nonlubricated) 263
Hydrogen Disposal Systems 264
Transportation Systems for Hydrogen 264
Commercial Operations 264
NASA Operations 265
Regulatory Agencies 265
Advisory Agencies 266
Instrumentation for Hydrogen Systems 266
Temperature Measuring Instruments 267
Pressure Measuring Instruments 268
Flow Measuring Instruments 268
Liquid Level and Quantity Measuring Instruments 269
Combustible Gas Detectors 270
Hydrogen Applications 270
Rocket Propulsion (LH_2-LO_2) 271
Hydrogen Fueled Aircraft 272
Hydrogen Superconducting Systems 273
Metal Hydrides 273
Hydrogen Fuel Cells 274
Internal Combustion Engines 275
Hydrogen Fuel, General 276

Safety and Hazards 277
Storage and Handling of Hydrogen 277
Electrical Codes and Equipment 278
Hydrogen Fires, Explosions and Spills. 278
Hydrogen Safety and Accident Investigations 280

BIBLIOGRAPHY 283

INTRODUCTION

It has been estimated that the total energy requirements in the United States may double from the year 1970 to the year 2000. At the present time, the United States is still very heavily dependent upon petroleum energy in which it is no longer self-sufficient. As the world's supply of hydrocarbon resources rapidly dwindles, new sources of energy must be found.

Some of the primary energy sources proposed for future use, e.g., nuclear or solar energy, can be utilized to produce hydrogen from water by electrolysis or thermochemical decomposition. The world's oceans, therefore, become vast reservoirs of a potential hydrogen source and the hydrogen itself becomes a means by which energy from the primary sources may be delivered to the consumer.

Hydrogen is attractive as a fuel of the future because it is abundant, relatively inexpensive and ecologically clean. Hydrogen gas can be burned in air, combining with oxygen to form water vapor which eventually returns to the earth's water supply where it is again available as a source of fuel. The use of hydrogen as a universal fuel necessitates the development of methods for storing and handling it.

This book is primarily concerned with the technical aspects of storage and transmission systems for a proposed "hydrogen economy." Therefore, the various methods of producing hydrogen are not described in depth because they are important to the hydrogen economy only insofar as the hydrogen must be produced in large quantities at reasonable cost.

The first chapter describes the hydrogen economy and suggests how it can be integrated into the energy system of the United States. The next three chapters are concerned with the technology of handling the various forms of hydrogen—gaseous, liquid and solid (in the form of metal hydrides).

The fifth chapter describes some of the work which has been done or is under way in utilizing hydrogen as a fuel or as an energy storage system. Another chapter delves into the safety, legal, political, socioeconomic and environmental implications of a hydrogen economy.

The final chapter provides a list of hydrogen technology experts and includes a brief description of each individual's expertise in the various aspects of hydrogen technology.

A hydrogen economy by the year 2000 would seem to be feasible and this book is intended to give the reader an overall view of the problems which can arise in implementing it.

THE HYDROGEN ECONOMY

The sources of information in this chapter were N-74-11728-735 and BNL-18224. For a complete bibliography of these reports please see page 283.

Hydrogen is a most promising future medium for energy transmission and storage. It may be extracted from fossil fuels, from water or both simultaneously. While fossil fuel supplies are diminishing and must be regarded as a limited resource, water may be considered inexhaustible.

In addition to the raw material sources consumed in hydrogen production, energy sources must be used to extract hydrogen from fossil supplies or from water. This energy can be obtained directly from fossil fuels but primary energy sources including nuclear, solar, wind, fossil, geothermal, or ocean temperature gradients must be used to obtain hydrogen from water.

The use of nuclear energy as a primary energy source has already proven to be economically attractive in electricity production, principally due to the relatively low cost of nuclear fuel. Therefore, development of the other energy resources, notably solar, wind, and geothermal, is not likely to displace nuclear energy in the short-term as a primary energy source.

Hydrogen is produced today in large quantities as a major chemical intermediate for use in the chemical process industries. Presently hydrogen production in the United States is about 3×10^{12} scf/yr. The major proportion of this intermediate is presently made from fossil fuel (natural gas or naphtha) by either steam reforming or partial oxidation processes.

If hydrogen is envisioned as a fuel, much larger amounts will be needed. In these cases, it is obvious that the primary raw material source for hydrogen must be water. Water can be split into its elements, hydrogen and oxygen, by a variety of processes. A commercially available process for producing hydrogen is the electrolysis of water. This process is practiced in areas of the world such as Canada, Norway and India where there is either an abundant supply of cheap hydroelectric power or no natural gas source. The main drawback to this method is that electricity must first be generated and then used; consequently, the generation cost of electrolytic hydrogen must always be more than that of electricity.

A more promising method of hydrogen production currently under consideration is a closed cycle thermochemical process for which energy to obtain the hydrogen from water is in the form of heat rather than work. These cycles (many different types of processes have been proposed) operate at maximum temperatures in the neighborhood of 1000°K with theoretical thermal efficiencies in the range of 20 to 50%. As yet, none of these cycles have been experimentally proven in their entirety; however, much research and development work is currently being done.

Since coal reserves represent a large fraction of our domestic fossil energy reserves, but are currently underutilized, this primary energy source may be used as a short and midterm source of energy. Even though coal resources could contribute a significant portion of the energy needs for at least a century, several important factors (developing technology, land reclamation, mine safety, possible water shortage, and shortage of personnel) could delay its large scale implementation until the late 1980s.

It is doubtful that coal will be used to produce hydrogen on a large scale for the hydrogen economy, because of factors favoring methane production. The photolytic decomposition of water, the direct thermal dissociation of water, and the production of hydrogen from organic waste by algae are not considered feasible in the near future because of low efficiency, high cost, or undeveloped technology.

If hydrogen is to become a widely used energy carrier, it must be produced economically and efficiently on a large scale for a long time period. Therefore hydrogen production must be linked to those primary energy sources which are currently available, as well as to those which might be available in the future. Figure 1.1 presents a summary of hydrogen production methods from a number of primary energy sources. There may be some primary energy sources and/or process paths which are not shown, but the chart does indicate the interrelationships between many possible sources of primary energy and the production of hydrogen fuel.

This chart shows that hydrogen can be produced by current primary energy sources (for example, gas, oil, coal, fission) as well as by those energy sources which might be available in the future (for example, solar, fusion, geothermal). Two important routes couple the several primary sources to hydrogen, thermochemical decomposition and electrolysis. Several primary sources could utilize either path (for example, nuclear or solar), while other sources only seem practical when coupled to one of these paths (for example, wind-electrolysis, geothermal-electrolysis).

An important characteristic of the future energy supply of the United States is that it will probably be based upon many primary sources. This trend is the result of the heavy demand on the limited domestic energy supplies. Economic and environmental constraints may cause energy to be extracted from many sources so that costs can be held down and the resulting pollutants (including reject heat) can be disposed of in many reservoirs. The section which follows will show the ways in which primary energy can be converted into hydrogen so that these economic and environmental costs can be minimized.

FIGURE 1.1: SUMMARY OF METHODS FOR HYDROGEN PRODUCTION

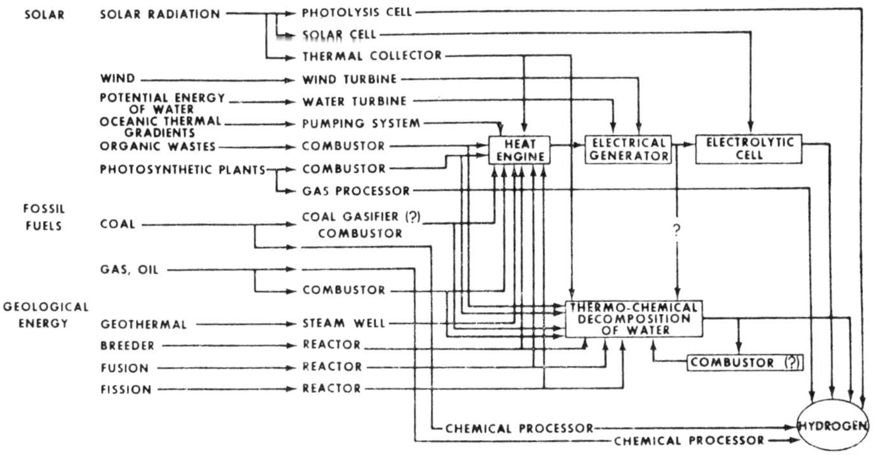

Source: N74-11728-735

PRIMARY ENERGY SOURCES

Solar Energy

The advantages of solar energy are well recognized. It is an available, nondepletable and nonpolluting energy source. Two major problems arise in an attempt to utilize this source. The first is caused by the low power density (or low flux) of sunshine, and the second is due to the intermittent nature of sunlight, that is, there is no direct sunlight at night or on cloudy days. The cost of utilizing solar energy is largely the cost of overcoming these two problems, which, until now, have prevented widespread solar energy use.

The amount of solar energy arriving at the outer edge of the earth's atmosphere is 2.0 calories per square centimeter minimum or 1,390 watts per square meter (1). Some of this is intercepted by the earth's atmosphere, and the amount of solar energy that finally impinges on the earth's surface depends upon the latitude, time of day, season of the year, and local weather conditions. These effects are illustrated in Figure 1.2. The top curve shows the solar radiation impinging on the desert floor near Yuma, Arizona, at the equinox (2). At noon, the energy flux is 920 w/m^2, with the average being 700 w/m^2 during the 12 hours of daylight. During the other 12 hours, the solar radiation is zero.

Some feeling for the effect of latitude and season can be obtained by considering conditions in Washington, D.C. On a clear summer day, the average intensity of the solar radiation striking the earth's surface during the 14 daylight hours is 478 w/m^2 (3). During the winter, this value drops to an average of 248 w/m^2 for 9.5 hours of daylight. The amount of energy impinging on one square meter in one day is the area under the curve in Figure 1.2. For the upper curve this quantity is 8.4 kwh, while the lower curve yields 2.36 kwh. The effects of geographic location and season are seen to be quite significant.

FIGURE 1.2: SOLAR ENERGY DENSITY AT TWO LOCATIONS

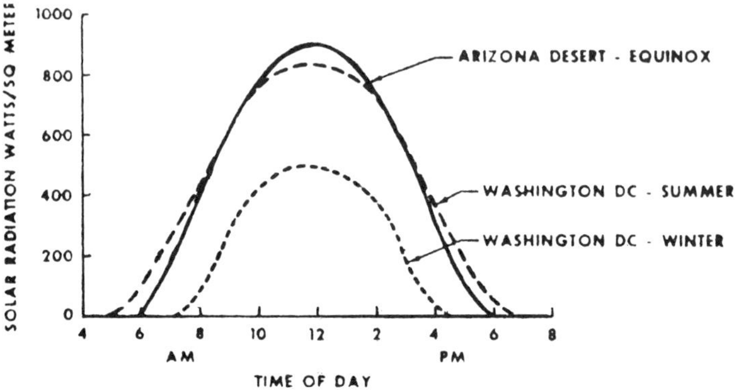

Source: N74-11728-735

Consider now the power generating capacity of the solar energy falling on a one square mile area. Both Meinel (4) and Ralph (5) have concluded that one square mile of radiation will produce electrical power at the average rate of about 67,000 kilowatts, assuming that the overall efficiency for converting solar energy to electrical energy is 10%.

In a similar calculation based on solar radiation in the Central Asian Soviet Socialist Republics, Vermishev (6) concluded that the solar energy collected over 1,000 square kilometers (360 square miles) would generate about 200,000 Mw. This represents the entire generating capacity of the USSR in 1964. Vermishev assumed an overall efficiency of 20%. However, Meinel estimates that our electrical needs in the year 2000 will be about 1,000,000 Mw which could be produced by solar energy falling on an area having dimensions of 75 miles by 75 miles, located in the Arizona desert.

These values indicate the problem caused by the low flux of solar energy. Large areas must be covered with solar collectors if significant quantities of solar energy are to be obtained.

The technical problems involved with solar energy collection are substantial. Although the energy is free, the cost of collecting it is high. As a result, the use of solar energy is a capital intensive enterprise. The cost of the energy depends largely on the capital costs of the very large area of solar collectors required. By contrast, in the electric power industry, the cost of fuel is a significant fraction of the cost of electric power, but the capital investment is comparatively lower.

Solar energy is an intermittent energy source. Its intensity depends upon the time of day, season of the year, latitude of the solar collector, and local weather conditions, as shown in Figure 1.2. The variations in power demand do not correspond to the variations in the solar energy supply. If solar energy is to be used during a 24 hour day, it is apparent that energy must be stored during the daylight hours and released at night. Energy must also be stored during sunny days for use during cloudy days. The problem has been considered for many years and a number of solutions have been proposed. Lof (7) has used beds of gravel to store heat during the day and release it at night to heat a home.

In the late 1940s, Telkes (8) suggested using the heat of fusion of various salts as an energy storage method, and this method is still discussed, but in a higher temperature range. Baranov (9), for example, suggests using the heat of fusion of metallic oxides for high temperature storage of solar energy, while Meinel (10) suggests using the heat of fusion of salt eutectics for energy storage at temperatures near 550°C.

Recently, various researchers have briefly suggested that energy can be stored by using solar energy to produce hydrogen. The hydrogen would then be transported or stored and burned as energy is needed. Hydrogen production assists in the energy storage and distribution problems since both are feasible.

Wind Energy

The energy of the winds is a form of solar energy. It is a renewable source of power. In the past, the exploitation of wind power was either in geographically special locations (farms on the plains and offshore oil platforms) or in the developing, often rural, nations. It was felt that wind power had no place in an industrialized, urban country like the United States. However, as other more conventional forms of primary energy become increasingly unobtainable or undesirable, the advantages of all forms of solar energy, including wind power, become apparent.

The environmental advantages of wind power are significant. Rather than being a dirty source, wind power is inherently clean. There is no fuel to be consumed and no resulting air, water, or thermal pollution. The energy which is extracted from the wind would otherwise appear in the atmosphere as thermal energy generated by viscous dissipation. A wind power system takes energy from the atmosphere at the wind turbine and delivers this energy at the point of use. Because the energy involved in the transfer is a small fraction of the total natural energy flow of the atmosphere, and because of the rapid mixing of the atmosphere, there are virtually no environmental effects.

The limitations on the exploitation of wind energy are economic. The diffused nature of wind energy requires that the wind turbines may need to be located at sites which are remote from the load center. The variable nature of wind energy requires that some fraction of extracted energy be stored so that the energy demand pattern can be matched. Because hydrogen appears to be an economical form in which to store and transmit energy, the combination of the two technologies—primary wind power and transfer of energy utilizing hydrogen—may result in a practical contribution to the total energy economy.

The technology of wind turbines is well established. Extensive research and prototype development were conducted in the 1940s and 1950s, in both the United States and Europe (11)(12). Most of this work was set aside because of the promise of low-cost nuclear energy. Current developments tend to make wind power more practical. The production of hydrogen from wind power allows the wind capacity to be counted toward the base capacity.

The size of the conventional base load plants can therefore be reduced. At the same time, rising fuel costs (both nuclear and fossil) make operating cost of conventional units more expensive. Wind power cost (for a particular installation) is constant. In addition, environmental impact and social costs of nuclear and fossil power tend to be proportional to total output. Thus any energy derived from the wind which reduces the output of conventional plants reduces the amount of these external costs of energy. These factors will serve to lower the cost of wind power, relative to fossil and nuclear power, in the future.

There is a trade-off between the total amount of wind energy available and the cost of energy produced. As less favorable sites are exploited, each new site will be less productive (assuming a constant wind turbine technology). Thus, the amount of wind energy available has economic limitations. Table 1.1 presents one proposal for the development of 1.536 x 10^{12} kwh/yr of electrical energy by wind power for the year 2000 (13). The hydrogen produced by these selected, favorable sites would represent 5 to 10% of the projected total energy demand of the year 2000.

TABLE 1.1: MAXIMUM ELECTRICAL ENERGY PRODUCTION FROM WIND POWER

Site	Annual Power Production	Maximum Possible by Year
Offshore, New England	159 x 10^9 kwh	1990
Offshore, New England	318 x 10^9 kwh	2000
Offshore, Eastern Seaboard, along the 100 meter contour, Ambrose shipping channel south to Charleston, S.C.	283 x 10^9 kwh	2000
Along the E-W Axis, Lake Superior (320 m)	35 x 10^9 kwh	2000
Along the N-S Axis, Lake Michigan (220 m)	29 x 10^9 kwh	2000
Along the N-S Axis, Lake Huron (160 m)	23 x 10^9 kwh	2000
Along the W-E Axis, Lake Erie (200 m)	23 x 10^9 kwh	2000
Along the W-E Axis, Lake Ontario (160 m)	23 x 10^9 kwh	2000
Through the Great Plains from Dallas, Texas North in a path 300 miles wide W-E, and 1,300 miles long, S to N. Wind Stations to be clustered in groups of 165, at least 60 miles between groups (sparse coverage)	210 x 10^9 kwh	2000
Offshore the Texas Gulf Coast, along a length of 400 miles from the Mexican border, eastward, along the 100 meter contour	190 x 10^9 kwh	2000
Along the Aleutian Chain, 1,260 miles, on transects each 35 miles long, spaced at 60-mile intervals, between 100 meter contours. Hydrogen is to be liquefied and transported to California by tanker	402 x 10^9 kwh	2000

Estimated Total Production Possible: 1.536 x 10^{12} kwh by year 2000

Note: This electrical output was produced from fuel cells which were assumed to be 51% efficient. Thus the energy of the hydrogen produced is 3 x 10^{12} kwh per year.

Source: N74-11728-735

Although wind power will probably not be able to economically supply a majority of the future energy needs, it can provide a small and environmentally desirable contribution. At a recent National Science Foundation sponsored conference on solar energy, it was recommended that wind power development be included in future expanded programs to exploit solar energy.

Fossil Energy

Fossil fuels (natural gas, petroleum, coal, shale oil, tar sands) are nonrenewable sources of primary energy, because the rate of their consumption is orders of magnitude greater than their natural rates of formation. The result is an open cycle in which carbon atoms, bound in the fuel, are irreversibly converted to carbon dioxide with the liberation of energy. Ultimately such fuel will be totally depleted. As the limit of such a resource is approached, a prudent society will conserve the resource while developing alternative primary sources.

Table 1.2 gives the estimated domestic energy reserves of fossil fuels in terms of quadrillion (10^{15}) Btu or simply, Q. The second column gives resources identified and recoverable using today's technology. The total represents the ultimate amount estimated to be available. The accuracy of these estimates probably ranges from 20 to 50% for identified-recoverable resources and may be an order of magnitude in error for the total resource available (14).

TABLE 1.2: ESTIMATED DOMESTIC FOSSIL FUEL RESERVES

Resource	Identified and Recoverable, Q	Total Q	Date of Estimate
Natural gas	299[a]	6,810	1970
Petroleum	302[b]	16,800	1970
Coal	5,200-10,000[c]	83,200	1972
Oil shale	928-3,480[a]	151,000	1972

[a]1,032 Btu/scf.
[b]5,800,000 Btu/bbl.
[c]26,000,000 Btu/ton.

Source: N74-11728-735

Of the estimates available for natural gas and petroleum, the estimate presented here is the highest because it includes the largest proportion of favorable ground for exploration. Tar sands are excluded because they represent a small fraction of available domestic resources (they are being developed in Canada for the United States energy market).

Several observations may be made from this table. Ultimate oil shale reserves may amount to more than all others combined, but identified and recoverable reserves of oil shale are less than those for coal. Another important factor, in the favor of coal, is that its geographic distribution is more uniform than for oil shale. Compare with these quantities the total energy consumption of about 69 Q by the United States in 1971. By 1980 this is expected to increase to at least 96 Q, and by 1990 to 140 Q/yr (15).

Even though coal may account for some 80% of natural energy resources available today, it is currently being used to supply only about 18% of the total domestic energy consumption, primarily in the generation of electrical energy (16). In order to maintain a highly reliable domestic energy resource, it is clear that coal usage must be expanded considerably.

Nuclear Energy

Nuclear energy will play an important role in fulfilling society's future energy requirements. Long term forecasts indicate that a growing proportion of this energy will be supplied by fission reactors and, in the distant future, perhaps by thermonuclear fusion. Because energy from controlled thermonuclear fusion has not yet proven to be feasible (17), it is assumed that the primary energy will be obtained from nuclear fission.

At present, the United States has available both light water reactors (LWR) and high temperature gas-cooled reactors (HTGR). The major emphasis on reactor development has been on water-cooled reactors using the slow neutron induced fission of ^{235}U (18). The LWR may be classified in either of two categories, the pressurized water reactor (PWR) or the boiling water reactor (BWR). The PWR has a low operating temperature in the range of 530° to 630°K, but operating pressures up to 2,500 psi are necessary to prevent boiling. Heat is usually transferred through a heat exchanger to generate steam for electrical generation. The BWR power reactor allows water to boil, producing saturated steam at about the same temperature as the PWR and 1,000 psi.

In contrast, the high temperature gas-cooled reactor is cooled by helium gas and provides a potential supply of high temperature heat in the range 1000° to 1100°K. The hot helium gas is used to generate steam at temperatures and pressures typically used in modern steam turbine-electrical generation plants. The fuel cycle is based on thorium conversion to ^{233}U with ^{235}U as the startup material.

Presently, a variety of breeder reactors is being developed. The main reactor types include the liquid-metal cooled fast breeder reactor (LMFBR), the gas-cooled fast breeder reactor (GCBR), and in Germany, the pebble-bed breeder reactor (PBBR). The LMFBR converts ^{238}U to ^{239}Pu, thereby using up nearly all of the energy potential of uranium supplied by nature in the more abundant (99%) ^{238}U form. The GCBR is also a plutonium breeder reactor. The German PBBR is in the research stage with work being done to develop a ceramic pebble fuel.

Current fissionable uranium resources identified and recoverable with today's technology and costs are of the same order as oil and gas. Because of the slowdown in development of nuclear installations, there is little incentive for further exploration.

Even if the fast breeder reactor technology were implemented to reduce the requirements on existing enrichment plant capacity, by 1985 the enrichment capability will be the limiting factor in the growth of the nuclear industry. The problem of waste disposal for breeder reactors is currently under investigation.

Table 1.3 summarizes the important characteristics for the nuclear reactor types available in the near future.

TABLE 1.3: NUCLEAR REACTOR COMPARISON

Reactor	Operating Temperature (°K)	Efficiency, Thermal to Electrical	Installed Capital Cost ($/ kW elec) 1972 $ (20)	Installed Capital Cost ($/ kW elec) 1972 $ (18) (19)	Date Available
Light Water Reactor	600	0.33	220	300	Now
High Temperature Gas-Cooled Reactor	1150	0.40	220	300	Now
Liquid metal Fast Breeder Reactor	800	0.40	250	350	1980's
Gas-Cooled Fast Breeder Reactor	800	0.40	220	300	1980's

Source: N74-11728-735

Miscellaneous

As traditional energy supplies become more limited, increased attention is being devoted to "unconventional" sources of energy. Rather than thoroughly analyze every possible miscellaneous source of energy, three of these, oceanic thermal gradients, geothermal, and bioconversion of organic wastes, will serve as examples of the ways in which a hydrogen energy system can be tied to the exploitation of multiple, unconventional energy sources.

Oceanic Thermal Gradients: A common problem in exploiting solar energy is that it is a very diffuse form of energy. One proposed way to make its use more feasible is to take energy from sites where nature has already concentrated it. One such proposal is to use the thermal gradients of the ocean to drive a heat engine.

The most promising location near the United States for the exploitation of oceanic thermal gradients is off the Florida and Georgia coast in the Gulf Stream. In this region, cold water (5°C) from the melting polar caps flows under a region of less dense water which has been heated (25°C) by the tropical sun. The temperature gradient occurs over a depth of 1,000 to 3,000 feet.

Anderson and Anderson (21) calculated that the energy content of the flow of warm Gulf Stream water is sufficient to generate 182×10^{12} kwh of electricity per year. This is almost 75 times the projected 1980 domestic demand. The recent NSF/NASA Solar Energy Panel (13), following suggested designs by Heronemus (22), concluded that oceanic thermal gradients could supply at least 10% of the projected electrical demand by 2020.

Although two demonstration oceanic thermal power units have been built (Cuba in 1929 and Ivory Coast in 1956), economic estimates on the cost of large scale units are speculative at this time. The preliminary estimate by Heronemus of $200 to $400/kwh makes this method look promising.

If the energy content off the Florida-Georgia coast were developed on a large scale, this energy would have to be transmitted in large quantities and over long distances throughout the country. Conversion of this energy into hydrogen is attractive, since hydrogen can be transmitted economically. Because of the relatively low temperature source available from oceanic thermal gradients, it appears that the only practical way to produce hydrogen from this source of energy is through the heat engine-electrical-electrolysis path. The temperature difference would be used to drive a heat engine (probably Rankine cycle) and an electric generator.

The technical problems associated with exploitation of oceanic thermal gradients center around the complexities of handling very large volumes of seawater. Heat exchangers need to be large, efficient, and resistant to corrosion by seawater. Further studies are needed to determine the optimum working fluid of the heat engine (23). Oceanic thermal energy appears to be very attractive environmentally (24). By reinjecting discharge water at a depth corresponding to its temperature, thermal effects are minimized. If the plant is submerged, it poses minimal aesthetic degradation, reduces safety problems to shipping, and is not susceptible to damage by storms.

The NSF/NASA Solar Energy Panel recommends a 15 year research and development program leading to the development of a 400 Mw demonstration plant. Renewable power from oceanic thermal gradients, if technically developed, offers many benefits. The advantages of a hydrogen energy system complement the development of power from oceanic thermal gradients.

Geothermal Power: Geothermal sources of energy can be classified into three types; dry steam, hot water, and hot rock. These are listed in order of increasing size of reserves. This is also the order of increasing technical complexity.

Dry steam wells are currently producing 298 Mw of electrical power at the Geysers site near San Francisco. This site and similar ones in Nevada might be able to supply up to 7,000 Mw by 1985 (25). The technology to exploit such dry steam wells is developed.

Hot water (brine) resources are estimated to be twenty times as large as dry steam resources, although the technology for this exploitation is not yet developed. Hot water sources would require a heat engine operating between smaller temperature differences than traditional power turbines. A working fluid other than water would likely be used. Fouling of heat exchangers by the minerals of the hot water source may be a problem. One of the largest sources of hot water energy is in the Imperial Valley of southern California. The output of this region could be as high as 30,000 Mw (26).

The third source of geothermal energy is from hot rock found in certain geological regions. In these regions, there is insufficient water in the hot structure to provide hot water or steam. Proposed schemes involve sinking two wells; the first, 8,000 to 20,000 feet deep, this would be used to inject water. Water injected through the first well would flow through the porous structure of the hot rock. As the water heats up, its density will decrease and it will be carried upward by convection where it will be extracted through the second, more shallow well. The energy content of promising hot rock sources in the western United States has been estimated to be 10^{17} kwh.

Geothermal sources are not the perfectly clean source of power that many would like to believe. Steam and water from subsurface sources often carry corrosive and/or odorous impurities. However, the impact of these impurities can be minimized, making geothermal power environmentally desirable. Many of the minerals removed from the heat source have economic value, aiding the economics of this power scheme.

The relatively low temperature of any geothermal source (less than 700°C) suggests that the most likely path for the production of hydrogen from geothermal energy is through the heat engine-electricity-electrolysis path. Because two of the most promising geothermal locations are near major load centers (Imperial Valley near Los Angeles and Geysers near San Francisco), the economic advantage of the low transmission costs of hydrogen energy will not likely be an impetus to the development of geothermal-hydrogen systems in the near future.

When larger and more remote geothermal sources are developed through the exploitation of hot rock, the production of hydrogen could be the economical way to distribute the energy. Large quantities of electrical energy at 15 mills per kilowatt hour from hot rock, would permit the production of hydrogen for \$5.01 to \$4.66/10^6 Btu (using previous electrolysis data, 1973 figures).

Bioconversion of Organic Wastes: A biological method of producing hydrogen from organic wastes has been suggested utilizing the photosynthetic system. Blankenship and Winget (27) have proposed that hydrogen be produced by certain algae and/or bacteria, if they are placed in a substrate of organic waste material. The metabolism of these microorganisms depends upon the photosynthetic process and, thus, requires sunlight. However, since part of the energy input to the system is from the organic waste material substrate, the energy output in the form of hydrogen is greater than the energy input of direct solar energy. The area of solar energy collection can then be less than the area required for a purely solar energy process.

Calculations based on experimental results with the photosynthetic bacterium *R. Rubrum* show that a net volume of about 1,800 liters per hour of waste is needed. Spread over an area of about 180 square meters (1,800 ft^2) this would release enough energy to supply the needs of a typical home. The household sewage system can be used as a growth medium for the bacteria. Such a method would not only serve as an energy supply, but would also reduce solid waste pollution. The technical and economic feasibility of this method needs to be established and it consequently merits further consideration. Recommendations for the development of this energy source are made by the NSF/NASA Solar Energy Panel.

The bioconversion of organic wastes is only one method of utilizing the energy of waste products. A more complete picture of energy supply from wastes and appropriate processing methods is given by Maugh (28). Although this energy source appears environmentally attractive, Maugh maintains that such energy sources can supply only a few percent of our total energy demand in the future.

THE HYDROGEN ENERGY CONVERSION SYSTEM

If hydrogen were adopted as the main general-purpose synthetic fuel, the energy conversion system would probably have a configuration as illustrated in Figure 1.3 (29)(30). As the system grows, hydrogen would probably be produced by a variety of primary energy sources, initially by coal gasification and off-peak power from nuclear steam-electric generating plants, supplemented later by solar, geothermal, and fusion steam-electric generating plants. Perhaps in the more distant future, methods can be developed to produce hydrogen more directly from nuclear and solar heat by thermochemical reactions.

FIGURE 1.3: THE HYDROGEN ENERGY CONVERSION SYSTEM

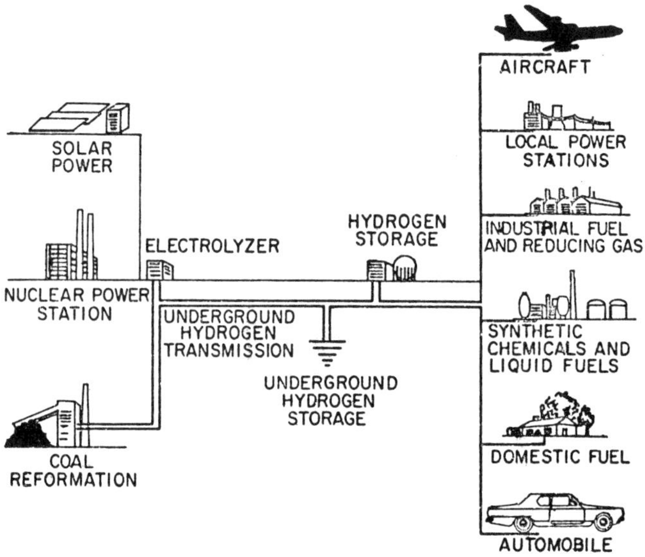

Source: BNL 18224

It is assumed that the primary energy sources will be located at sites remote from the load centers and the hydrogen will be transported via underground pipeline. The system will provide for hydrogen storage as high pressure gas in underground reservoirs, as liquid in cryogenic storage tanks, and as solids in metal hydride beds.

Most industrial, residential, and commercial fuel requirements can be met with hydrogen using existing technology. Of course, such technology is expected to improve over the present state-of-the-art. In urban areas a small fraction of the hydrogen would be used to produce electricity to satisfy those needs specifically requiring energy in electrical form (lighting, electric motors, etc.). Many household appliances now operating on electricity would be replaced by similar devices fueled directly by hydrogen. The hydrogen-electrical conversion would be accomplished in fuel cells or in gas turbines. Most of the hydrogen would be consumed directly as fuel in industrial, transportation or domestic applications.

Hydrogen Production

Production from Coal: For many decades, most American cities were supplied with gas manufactured from coal (variously called water gas, producer gas, etc.). Typically such gases contained 50 to 55% hydrogen, 25 to 30% methane, 4 to 7% carbon monoxide, plus lesser amounts of ethane, carbon dioxide, nitrogen and miscellaneous hydrocarbons. By more refined processes hydrogen of 90 to 98% purity can be manufactured from coal-steam reactions (31). Probably in the initial phases of system evolution, the cheapest method of production will be from coal. It should be noted, however, that for some applications the coal-produced gas will have to be further purified to avoid degradation of system components, for example, for metal hydride storage and fuel cell consumption.

Electrolysis of Water: Today in some parts of the world large quantities of hydrogen are produced by electrolysis of water. The installed capacity of such plants is estimated to be about 3 million pounds of H_2 per day (32). Electrolysis of water is a familiar process, but one that has received little research attention in recent years. In the electrolysis process, the various options and parameters which may be selected are: electrode materials, electrolyte, cell geometry, operating voltage, current density, temperature and pressure.

These factors all contribute to the final cost of the produced hydrogen. Optimization becomes a matter of trade-off between capital costs and operating costs. Operation at high current density decreases the capital costs, but at high current density electrical efficiency tends to drop thus increasing operating costs.

If electrode polarization effects can be suppressed, current densities can be increased without increasing cell voltage. This has the effect of decreasing the cell dimensions, and hence reducing capital costs. Teledyne Isotopes has been developing a cell which is capable of operating at higher temperatures and higher current densities. Test units have operated as high as 1,600 amp/ft^2. More advanced high temperature systems, which electrolyze water in the vapor phase with extremely high current density, have been reviewed by Gregory (33).

With such high performance design, cell efficiencies in the range of 95 to 97% appear technically feasible. Thus, with the development of new electrolysis technology the production costs of electrolytic hydrogen can be expected to decrease from today's costs.

Obviously, electrolysis of water yields oxygen as well as hydrogen. It is expected that there will be a reasonable market for the oxygen and this will help reduce production costs. Also, electrolysis plants can be arranged to separate deuterium which is expected to be in large demand sometime in the future as fuel for fusion reactors.

The production costs will be a major factor in determining at what point in history the hydrogen system will begin to compete successfully with natural gas and petroleum. Therefore, it is of great interest to estimate such costs; however, at the present time all cost factors are changing so rapidly that such estimates have limited validity. As an approximation, natural gas prices must increase four- or five-fold before electrolytic hydrogen becomes competitive. Because of these economic factors, hydrogen production by electrolysis is limited to countries which have plentiful supplies of low-cost electricity, plus a high demand for ammonia for fertilizers, and no low-cost hydrocarbon fuels. This combination of conditions does not exist today in the United States since hydrogen for ammonia production can be manufactured most economically from natural gas. However, the situation must be expected to change as natural gas becomes more expensive and scarce, and as more nuclear power stations come on line.

Nuclear power stations are designed to operate as base-load plants, that is, 24 hours a day, 7 days a week. In fact, because of the large capital investment, it is economically imperative to realize high plant capacity factors with the nuclear units. If an electrical system has several such plants with total capacity that exceeds the off-peak electrical demand, it is economically desirable to store the surplus power.

Many utilities aspire to build pumped storage facilities in which off-peak power can be used to pump water up-hill into a high reservoir. The energy thus stored can be used to generate electricity during peak demand periods. Unfortunately, sites for pumped storage are becoming more difficult and more costly to obtain. It is doubtful that most utilities can obtain adequate pumped storage capacity to balance their load. In such circumstances hydrogen production appears to offer an attractive application for off-peak power.

In the more distant future, nuclear plants might be built for the exclusive purpose of producing hydrogen. Such a plant would generate lower cost DC electricity and be optimized for electrolysis operation. In the long run, solar plants might become competitive for hydrogen production. Such plants, located at great distances from urban localities could transmit their energy output more economically as hydrogen rather than as high-voltage electricity.

Transportation and Storage of Hydrogen

Transportation: Hydrogen can be transported efficiently through underground pipelines using systems similar to the existing natural gas transmission networks. Several hydrogen pipelines of 50 mile lengths are already in industrial use (33). Cost studies of hydrogen pipeline transmission have been made at the Institute of Gas Technology by D.P. Gregory (30)(34).

Since hydrogen gas has only about one-third the energy content of a corresponding volume of natural gas, the volume through-put (STP) to deliver the same energy must be about three times greater for hydrogen. Thus hydrogen pipelines should be designed to operate at higher pressures than existing natural gas pipelines; however, because of the low density of hydrogen the pumping effort can be lower. To convert a pipeline from natural gas to hydrogen would require among other things modification to compressor systems and consideration of embrittlement in the pipes and fittings.

From Gregory's studies, it appears that a hydrogen transmission pipeline could move energy more economically than high voltage electric transmission lines, but at higher costs than natural gas. If we ever reach the point where electrical transmission must be predominantly underground, the cost advantage of hydrogen over electricity will become very large.

Storage: Economical hydrogen storage presents technological difficulties which require more research effort. High pressure gas tank storage is too costly to be competitive. Storage as a liquid in cryogenic tanks is technically practical and highly developed; however, the liquefaction process exacts an efficiency and economic penalty which will limit its use to special applications where liquid is needed for other reasons, for example, as aircraft fuel.

Gregory (30) has pointed out that underground storage in a proper geological formation is practical, just as some helium is now stored. Appropriate formations might include depleted oil or gas wells, certain types of aquifers, or specially mined caverns. Such storage is not without precedence since an aquifer near Paris stores 13 billion cubic feet of manufactured gas that contains 50% hydrogen (34).

One promising storage concept is the metal hydride bed. Certain metals and alloys readily absorb hydrogen exothermally to form hydrides. By heating the hydride bed, hydrogen gas can be released at a controlled rate. Such metal hydride systems are under intensive study at Brookhaven National Laboratory (BNL) by Wiswall and Reilly (35). A small prototype storage bed is under construction at BNL for testing by the Public Service Electric and Gas Company of New Jersey.

Utilization of Hydrogen

Local Distribution: There appear to be no technological barriers to the distribution of hydrogen to consumers via underground pipelines. Existing pipe networks are inadequate; however, Gregory (33) suggests that many systems could be upgraded by using the "pipe within a pipe" concept. Small diameter higher pressure tubing could be inserted inside existing larger diameter gas mains. The annulus surrounding the inner tube could be filled with a noncombustible low pressure gas such as nitrogen.

The Consuming Sectors: Hydrogen gas can serve as a general-purpose fuel for most industrial, commercial, and domestic needs, with only minor modification of existing equipment. Because of the higher flame velocity of hydrogen, conventional stove burners and furnace nozzles would have to be changed. In addition to meeting the needs for space and process heating, hydrogen-operated air-conditioning units and refrigerators are within existing technology.

In a hydrogen economy only a small fraction of the hydrogen supply would be used to generate electricity. This could be done in fuel cells or gas turbines at the substation level. Since reconversion to electricity is a relatively costly and inefficient process, the use of electricity would be limited to those devices specifically requiring electrical energy. Electrical space heating costs would be prohibitive.

The use of hydrogen to fuel aircraft and vehicles appears technically feasible, although much engineering effort is needed to solve the "on-board" tankage problem. Today's jet turbines and internal-combustion engines operate very efficiently on hydrogen with only minor modification.

It appears that engines specifically designed to operate on hydrogen could be simplified over present models. Obviously, the use of hydrogen fuel would greatly reduce our current air pollution problems from urban traffic. Some preliminary design concepts for hydrogen propelled jet aircraft indicate that performance and range could be improved over existing fuel systems.

The hydrogen would be carried as liquid in wing tanks. Design of the fuel tanks appear to present no unsolvable technical difficulties. For automotive applications, some preliminary design work has been done at BNL on metal-hydride fuel storage. In fact, at BNL Reilly and Wiswall are operating a small Wankel engine on hydrogen which is stored in a small Ti-Fe hydride bed. Only trivial modifications to the engine were necessary. Proposals for development of prototype tanks and road demonstration in modified current production model automobiles have been submitted to the National Science Foundation and the U.S. Department of Transportation.

Advantages and Disadvantages of Hydrogen as a Fuel

Hydrogen has many attractive features as the principal synthetic fuel:

It can be manufactured by decomposition of water;

All of the future primary energy sources are suitable for hydrogen production;

When combusted with air the only products are water vapor and oxides of nitrogen (the NO_x can be controlled and suppressed by selection of combustion conditions);

The heat of combustion per unit weight is larger than for any other practical fuel;

It can be transported cheaply by pipeline;

It is an excellent general purpose fuel, probably better than any other practical fuel (this is partly because in air mixtures the range of flammability is so broad);

Spills dissipate rapidly under conditions of good ventilation; and

It is less toxic than other fuels.

On the other hand, hydrogen has some drawbacks:

It is a gas at room temperature;

Compact storage requires high pressures, liquefaction, or chemical combination;

The temperature of liquefaction is extremely low (20.3°K or -423.0°F) and the liquid has relatively low density;

The heat of combustion per unit volume at STP is relatively small due to the low density of the gas;

In air mixtures the range of flammability is very broad (Note: this is both an advantage and a disadvantage depending on context);

In air mixtures the maximum flame velocity is quite high; and

In air mixtures the ignition energy is very small.

REFERENCES

(1) Baum, W.A., "Meteorology and Utilization of Solar Energy," *Solar Energy Research,* F. Daniels and J.A. Duffie, editors, University of Wisconsin Press, page 15 (1955).

(2) Meinel, A.B. and M.P. Meinel, "Briefings Before the Task Force on Energy of the U.S. House of Representatives," volume III, U.S. Government Printing Office, Washington, D.C. (1972).
(3) Cherry, W.R., "The Generation of Pollution-Free Electrical Power from Solar Energy," *J. of Engineering for Power*, page 78 (April 1972).
(4) Meinel, A.B. and M.P. Meinel, "Is It Time for a New Look at Solar Energy," *Bull. of Atomic Scientists,* volume 27, page 32 (1971).
(5) Ralph, E.L., "Large Scale Solar Electric Power Generation," *Solar Energy,* volume 14, page 11 (1972).
(6) Vermishev, K. Kh., "Some Problems in the Long Range Forecasting of Power Generation and Solar Energy Utilization," *Applied Solar Energy,* volume 6, page 1 (1972).
(7) Lof, G.O.G., et al, "Solar Energy Utilization for House Heating," Report No. PB 25375, Department of Commerce, Washington, D.C. (1946).
(8) Telkes, M., "Solar Heat Storage," in *Solar Energy Research,* F. Daniels and J.A. Duffie, editors, University of Wisconsin Press, page 57 (1955).
(9) Baranov, V.I., C.F. Muchnik and S.N. Trushevskii, "Investigation of High Temperature Solar Energy Absorbers and Thermal Storage Devices," *Semiconductor Solar Energy Converters,* V.A. Baum, editor, Consultant's Bureau, New York, page 183 (1969).
(10) Meinel, A.B. and M.P. Meinel, "Physics Looks at Solar Energy," *Physics Today,* volume 52, page 44 (1972).
(11) McCaull, J., "Windmills," *Environment,* volume 15, n 1, pages 6 to 17 (January 1973).
(12) Putnam, P.C., *Power from the Winds,* Van Nostrand (1948).
(13) NSF/NASA Solar Energy Panel, *An Assessment of Solar Energy as a National Energy Resource,* University of Maryland (December 1972).
(14) Theobald, P.K., et al, "Energy Resources of the United States," Geological Survey Circular 650, U.S.G.S., Washington, D.C. (1972).
(15) "The Potential for Energy Conservation," a staff study for the Office of Emergency Preparedness, Office of the President, Washington, D.C. (October 1972).
(16) "Clean Energy from Coal–A National Priority," 1973 Annual Report, Calendar Year 1972, Office of Coal Research, Washington, D.C. (1973).
(17) Post, R.F., "The Uncertain Certainty," *Bull. At. Sci.,* volume 27, pages 42 to 48 (Oct. 1971).
(18) Macmillan, D.P. and J.D. Balcomb, "Nuclear Reactors for High Temperature Process Heat–A Survey of Reactor Types and Temperature Regimes," LASL Paper LA-UR-73-877. Los Alamos Scientific Laboratory, Los Alamos, New Mexico, Paper presented at the 166th National A.C.S. Meeting (August 27, 1973).
(19) McTague, P.J., G.J. Davidson, R.M. Bredin and A.A. Herman, "The Evolution of Nuclear Plant Costs," *Nuclear News,* volume 15-2, pages 31 to 35 (February 1972).
(20) Leeth, G.G., "Nuclear Power Plants for Hydrogen Production," G.E. Publication 72-TMP-52, General Electric Company–TEMPO, Santa Barbara, California (November 1972).
(21) Anderson, J.H. and J.H. Anderson, Jr., "Thermal Power from Sea Water," *Mechanical Engineering,* volume 88, n 4 (April 1966).
(22) Heronemus, W.E., "The United States Energy Crisis: Some Proposed Gentile Solutions," ASME/IEEE Joint Meeting, West Springfield, Massachusetts (January 12, 1972).
(23) Walters, S., "Thermal Sea Power," *Mechanical Engineering,* volume 93, n 10 (Oct. 1971).
(24) Matz, W.D., "Ocean Temperature Gradients: Solar Power from the Sea," *Science,* volume 180, n 4092 (June 22, 1973).
(25) Briefings before the Task Force for Energy, Committee on Science and Astronautics, U.S. House of Representatives, 92nd Congress, Serial U, Washington, D.C., page 126 (1972).
(26) Weinberg, A.M., "Long-Range Approaches for Resolving the Energy Crisis," *Mechanical Engineering,* volume 95, n 6 (June 1973).
(27) Blankenship, D.T. and G.D. Winget, "Hydrogen Fuel: Production by Bioconversion," 8th Intersociety Energy Conversion Engineering Conference, University of Pennsylvania, Philadelphia, Pennsylvania (August 1973).
(28) Maugh, T.H., "Fuel from Wastes: A Minor Energy Source," *Science,* volume 176 (November 10, 1972).

(29) Winsche, W.E., K.C. Hoffman and F.J. Salzano, "The Potential Role of Hydrogen in the Nation's Energy Economy," *Science,* volume 180, pages 1325 to 1332 (June 29, 1973).
(30) Gregory, D.P., "The Hydrogen Economy," *Sci. Amer.* 228, No. 1, 13-21 (January 1973).
(31) Grisenthwaite, A.T., *Trans. Inst. Chem. Eng.* (London), 34, 235-46 (1956).
(32) Williams, C.F., *Hydrogen Production Through Electrolysis of Water,* Teledyne Isotopes, Timonium, Maryland (April 1972).
(33) Gregory, D.P., D.Y.C. Ng and G.M. Long, "The Hydrogen Economy," Chapter 8, *Electrochemistry of Cleaner Environments,* J.O'M. Bockris, (Plenum Publishing Corp., New York) (1972).
(34) Gregory, D.P., "Status of R&D Related to the Production, Transportation, and Utilization of Hydrogen as Fuel," Inst. of Gas Technology, Chicago, Illinois (testimony before the U.S. House of Representatives Committee on Science and Astronautics) (June 1972).
(35) Wiswall, R.H. Jr. and J.J. Reilly, "Metal Hydrides for Energy Storage," Proc. 7th Intersociety Energy Conversion Engineering Conference, San Diego, California, September 25-29, 1972, BNL-16889, Brookhaven National Laboratory, Upton, New York 11973.

GASEOUS HYDROGEN

The sources of material for this chapter were AD-A016 256, N74-11728-735, SAND-74-8219, AD-787 484 and ORNL-TM-4307. For a bibliography of these reports, see page 283.

COMPARISON OF HYDROGEN STORAGE METHODS

The three physical states of matter: gaseous, liquid, and solid, form a convenient means of classifying hydrogen storage systems. This chapter will be concerned with the storage and transfer of the gaseous form of hydrogen while the two subsequent chapters will deal with the handling of liquid and solid forms of hydrogen. A comparison of the likely short-and-intermediate-term hydrogen-storage alternatives is shown in Table 2.1 along with comparable data for gasoline, as a reference. Other alternatives such as slush and solid forms of pure hydrogen are not shown owing to the fact that energy densities and production costs for these forms are only modestly different (e.g., 15 to 25%) from comparable data for liquid hydrogen (3). Other more exotic possibilities such as atomic hydrogen are not shown because of their relatively undeveloped technologies.

Table 2.1 indicates approximate energy-storage densities (Btu/lbm and Btu/ft^3) for the fuel alone in the various physical states, as well as approximate energy-storage densities for fuel in existing storage vessels. Also shown in Table 2.1 are capital costs (\$/10^6 Btu) for selected storage capacities in the three physical states (1). Criteria other than those shown in Table 2.1 are relevant in particular cases; safety, reliability and maintenance, hardenability, ease of use, limits on storage duration, etc. can be important in specific cases, particularly military ones. Nonetheless, the data shown are useful in making preliminary comparisons of the relative feasibility of the various storage methods in particular applications.

All current and currently proposed methods for the production of hydrogen result in gaseous hydrogen rather than a liquid or a hydride. Thus, since hydrogen will be produced and ultimately used as a gas, it might seem advantageous to store and transmit hydrogen in the gaseous form.

TABLE 2.1: COMPARISON OF VARIOUS HYDROGEN-STORAGE MODES AND HYDROCARBON REFERENCE

Fuel	Selected Fuel Storage Capacity** 10^6 Btu	Selected Fuel Storage Capacity** ft^3	Capital Cost (1972 $/$10^6$ Btu)	Approximate Energy Storage Density: Btu/lbm, Fuel Only*	Approximate Energy Storage Density: Btu/lbm, Fuel & Vessel+	Approximate Energy Storage Density: Btu/ft^3, Fuel Only*	Approximate Energy Storage Density: Btu/ft^3, Fuel & Vessel+
Gaseous H_2	2	46	750	51,600	1,200	43,800	40,000
2,400 psig tank	50	1,140	720				
Liquid H_2	2	8.8	1,000				
	50	220	300	51,600	8,000	230,000	130,000
	36,000	158,000	26				
H_2 as metal hydride							
$FeTiH_2$++	0.72	4.3	12,000	980	580	200,000	140,000
Mg_2NiH_4	2	12	-	1,860 (incl hydride)	-	170,000 (incl. hydride)	-
H_2 as nonmetallic hydride							
Methylcyclohexane (C_6H_{11}-CH_3)	2	13	-	3,200 (incl. hydride)	-	150,000 (incl. hydride)	-
Hydrocarbon reference	2	2.4	15				
Gasoline (C_8H_{18})	50	60	2	19,100	15,000	835,000	830,000
	1,370,000	1,640,000	0.4				

*Based on:	Density (lbm/ft^3)	% H_2 (wt) Available
H_2	4.43 (liq.)	100
Gasoline	43.8	15.8
$FeTiH_2$	206	1.89
Mg_2NiH_4	90	3.6 (Not currently operational)
Methylcyclohexane	48	6.12 (Not currently operational)

**Capacities selected to coincide with available cost and energy density data.

+Typical values from existing units.

++From Brookhaven Nat. Lab. Unit supplied to PSE&G Co. (2) Cost calculated from formula. (3)

Source: AD-A016 256

Results from recent studies indicate that the storage of gaseous hydrogen in pressure vessels appears to be hopelessly noncompetitive with other alternatives over the full range of storage capacity (1). This is due to two factors: the low density of gaseous hydrogen even at high pressures, and the high cost of the pressure vessel. These factors cause compressed-gas storage of hydrogen to be bulky and/or heavy and capital cost per unit of stored chemical energy, for example, $\$/10^6$ Btu, to be high for gaseous hydrogen. Furthermore, economies of scale are very modest. There are specific, typically small-scale situations, for which compressed-gas storage of hydrogen might be warranted and is presently used. For example, hydrogen-gas tank trailers are currently in use, though the quantities handled are small. However, for the reasons cited, large-scale hydrogen use does not appear likely to involve storage in pressure vessels.

Bulk hydrogen fuel could be supplied by gas pipeline. Thus, the hydrogen would be distributed in the same form as it is produced and eventually used; and the pipeline could serve as an inventory device (line packing). Such an approach is used in the Ruhr-Valley hydrogen pipeline of 130 miles length in West Germany (4). For example, a 2-ft diameter, 300-mile long transmission line holding hydrogen at 1,000 psia would store 1.7×10^6 lb H_2 (9×10^{10} Btu). However, line-packing will probably be unimportant as a storage technique for hydrogen (5). Line-pack storage can only meet daily needs; seasonal peak-shaving requirements must be satisfied by large-scale storage containers (natural or fabricated).

It may be feasible to store large quantities of gaseous hydrogen underground in mined caverns, aquifers, depleted oil and gas wells, or in bladder devices. For example, an aquifer storage system at Baynes, France, with a capacity of 7×10^6 ft^3 at 570 psia, has been in operation for over ten years, initially with manufactured gas and currently with natural gas.

Economically, these approaches seem promising. Reference (1) cites a capital cost of \$3 to $\$6/10^6$ Btu for large underground storage in aquifers, and $\$350/10^6$ Btu for storage in an 866,000 ft^3 mined cavern at 900 psig (1972 dollars). However, more work must be done to establish the feasibility of these various approaches with hydrogen and their ultimate potential. Thus, if gaseous hydrogen is to be stored and transmitted on a large scale, the only practical means seems to be storage in large natural underground reservoirs and transmission by pipeline.

EXISTING HYDROGEN HANDLING SYSTEMS

Present hydrogen pipelines in refineries and chemical plants operate at pressures in excess of 1,000 psi (6.89×10^6 N/m^2) for short distances. Pipelines up to 320 miles (5.15×10^5 m) in length with pipe sizes from 6 to 12 inches (0.152 to 0.305 m) are presently operated in Germany and the United States. Smaller quantities of hydrogen gas are supplied in high pressure tanks. Manufactured gas containing 50 to 80% hydrogen is presently distributed in Europe. Some of these retail distribution pipeline systems are presently being converted to natural gas.

Similar conversion to natural gas has occurred in the United States, but some manufactured gas is still distributed to residential customers in parts of the United States. A direct analogy, however, between a hydrogen gas pipeline and a pipeline containing a hydrogen-rich gas may not be appropriate. A discussion of potential problems in a hydrogen gas pipeline is given later.

ECONOMICS OF TRANSMISSION

The transmission of energy may be accomplished economically by means of a hydrogen gas pipeline system. At the present time, the cost to transmit hydrogen in large quantities has been estimated to be at most 2 to 3 times the cost of natural gas transmission (5)(6). In comparison, the cost of overhead electrical transmission is 5 times the estimated cost of hydrogen transmission while the cost of underground electrical transmission is approximately 7 times the cost of overhead transmission (7). If the energy is required for heating only, the lower hydrogen transmission cost will offset the added production cost, if the electricity must be transmitted any appreciable distance. In fact, transmission of over 300 miles by overhead cable or 35 miles by underground cable will make the electrical energy cost more.

Even if the hydrogen is again converted to electricity at the load center, the hydrogen transmission system will reduce costs for long transmission distances. In this case, transmission by overhead cable of over 1,200 miles will make the cost of electrical energy higher. The cost of hydrogen derived from coal would be about one-half that of electrolytic hydrogen. Based on this assumption and an equal transmission distance the cost of hydrogen would be much less than the cost of electricity. Hydrogen could be competitive with electricity even when the hydrogen transmission distance is longer than that of electric transmission.

A HYDROGEN PIPELINE SYSTEM

The development of a hydrogen pipeline system can be accomplished in several ways but there are many indeterminate factors which would affect the development. Presumably some of the existing natural gas pipelines will be incorporated into the hydrogen network. The manner in which such a conversion might be accomplished warrants a detailed study. The location of the hydrogen plant with respect to the existing natural gas systems is important. Existing natural gas transmission lines could be used to transmit an equal amount of energy in the form of hydrogen gas, but such a system would require approximately four times the present compressor capacity and over five times the compressor horsepower (5). This would not be an optimum design for the compressors; however, if the existing pipe could be used, the overall cost of such a system would be favorable when compared to the cost of a new optimized system.

The existing transmission lines originate in the southwest and deliver gas to midwest, the east and west coasts but it is likely that hydrogen may not be generated in the southwest. It may be more economical to generate the hydrogen closer to the large markets in the coastal areas. If this were the case new transmission mains could be installed from the hydrogen plant to the existing distribution systems along the coast. Since the coastal areas are at the end of the present natural gas pipelines, it would be possible to convert a large coastal area to hydrogen while inland areas were allowed to continue using natural gas. This would provide a method of gradually converting all of the existing natural gas distribution systems to hydrogen.

If the implementation of a hydrogen pipeline system occurred in this manner, the inland areas would be at the end of the hydrogen pipeline. Since the energy demand in the inland areas is much less than in the coastal areas, it would now be possible to use the existing natural gas transmission mains to deliver hydrogen to the inland regions economically with a minimum of modification. Using the existing compressor capacity at approximately one-tenth the present compressor horsepower, the energy delivery rate of hydrogen would be approximately one-fourth the present natural gas energy delivery rate (5).

The possibility of converting existing natural gas distribution piping to hydrogen also exists. Since the distribution system is generally a low pressure system, it is possible that existing natural gas distribution piping would be adequate to deliver the required volume of hydrogen gas without danger of hydrogen environment embrittlement.

The manner in which a hydrogen pipeline system may evolve is not totally predictable. A hydrogen system may initially develop parallel to the existing natural gas system. More practical, however, may be a hydrogen pipeline system for only the newly developed areas. Also feasible is the use of hydrogen as an industrial and electric generation fuel in a new pipeline system.

In a hydrogen pipeline system, the large-scale storage of hydrogen will be necessary for peak-shaving. The feasibility of using existing storage tanks must take into consideration the fact that a natural gas storage tank will store at design pressure only one-third as much energy when filled with hydrogen. This may require that existing storage facilities be enlarged. The feasibility of using existing underground storage reservoirs will also be considered, since this storage method is much less costly than storage by any other means.

Underground storage of helium in the Cliffside field in Texas, a helium-bearing natural gas field, is an example of this type of storage facility (8). Helium has been successfully stored and withdrawn from the field since 1945. The storage field consists of a porous rock stratum 120 to 200 feet (36.5 to 61 m) thick, lying 3,500 ft (1.067×10^3 m) below the surface and occupying 50,000 acres (2.02×10^8 m^2). Other storage methods will have to be used if low cost underground storage in depleted gas fields or aquifers is not available. This will cause some slight increase in the cost of hydrogen.

When considering the use of existing natural gas pipelines for transmission of gaseous hydrogen (5)(6) there are two major factors that need to be resolved. One of these factors, discussed in greater detail later, is the change in compressor capacity needed to pump hydrogen. The other factor is the metallurgical problem that can occur if there is a switch from natural gas to hydrogen.

A discussion of the compatibility of hydrogen with materials is given in the next section. The particular metallurgical problem that is of most concern in using existing pipelines is known as hydrogen environment embrittlement. There is more concern than usual at this point in time because this particular embrittlement problem has been recognized only since the mid 1960's and extensive research has not been conducted in this area until the last several years (9)(10). The reviews of Jewett (10) and Gray (11) have been particularly useful in clarifying this metallurgical problem.

Hydrogen environment embrittlement may be defined as follows: When a metal susceptible to hydrogen environment embrittlement is plastically deformed in an environment containing hydrogen gas, cracking can occur at the surface with possible propagation of the crack into a catastrophic failure. This is an explanation describing the minimum requirements for the production of embrittlement. An important point is that this is a surface phenomenon and not a chemical or internal phenomenon as is the situation in most hydrogen embrittlement problems. The problem is usually observed in relatively pure hydrogen at high pressures and temperatures close to ambient. However, research (10) has shown that cracking can occur over a range of temperatures in the presence of some impurities at pressures less than one atmosphere. Gray (11) emphasizes the point, "Embrittlement might occur to a greater degree and over a broader range of temperatures and pressures than determined in laboratory tests to date." Presently known factors that affect the use of existing pipelines are: susceptibility of presently used alloys; plastic deformation due to high pressure; and transmission of high purity hydrogen.

There has been no known research performed on pipeline steels regarding susceptibility of metals to hydrogen environment embrittlement. There has, however, been work performed on steels that are in the same low carbon steel alloy class (10). These investigations indicate this class of steel undergoes severe embrittlement when exposed to high purity hydrogen at high pressure. This factor is considered an unfavorable aspect in utilizing existing pipelines for hydrogen transmission.

Plastic deformation of the metal in the presence of hydrogen is required before hydrogen environment embrittlement can occur. For an alloy in which there are no residual stresses involved, the permanent deformation is normally considered to take place at an applied stress equivalent to the yield strength of the material. Actually some permanent deformation takes place at a stress level slightly below the yield strength. Nevertheless engineering design, particularly pipeline design, is based on the yield strength of the material. Pipeline codes allow present natural gas pipelines to operate at 72% of yield strength or a 1.38 safety factor.

Pipeline companies generally operate as close to this figure as is economically possible. Combinations of residual stress and stress concentrations that occur in fabrication of the pipeline, or external stresses caused by pipeline movement, can add to the working stress value and actually cause yielding at localized points. In the presence of natural gas this localized yielding has not been serious, but in the presence of hydrogen the localized yielding might produce a crack ascribed to hydrogen environment embrittlement.

Proponents of utilizing existing natural gas pipelines cite examples of existing hydrogen lines such as those operated by Air Products and Chemicals, Inc. in Houston, Texas; and Chemishe Werke Huls AG in the Ruhr area of Germany (5). These pipelines have been operating for many years without an apparent degradation due to hydrogen; however, analysis of the pipeline operating conditions indicates they would not be susceptible to hydrogen embrittlement. In the case of the Houston pipeline the operating pressure is 200 to 250 psi (13.8 to 17.2×10^5 N/m^2), yielding a safety factor of 6 to 8.

The operating pressure of the German pipeline is only 150 psi (1.03×10^5 N/m^2). In these cases no localized yielding of the material would be expected since the working stress is low. Therefore, no hydrogen environment embrittlement would be probable. If existing natural gas pipelines were stressed at the 1.38 safety factor level, as is implied in the hydrogen economy proposals, localized deformation could possibly occur. This is one of the conditions needed for hydrogen environment embrittlement.

The purity of hydrogen appears to be a significant factor in the susceptibility of metals and alloys to hydrogen environment embrittlement. It has been determined that certain impurities can inhibit or eliminate the susceptibility of a metal to hydrogen environment embrittlement. One of these is an oxygen impurity content of 200 parts per million. Other impurities such as carbon dioxide, water vapor, ammonia and sulfur dioxide also inhibit embrittlement, whereas impurities such as nitrogen, helium and argon have little effect.

Most of the chemical and thermochemical processes proposed to produce hydrogen will not give gaseous hydrogen of ultrahigh purity. For example, 99.7% pure hydrogen, now being produced in one steam reforming plant would probably not cause embrittlement. But, electrolysis capable of producing hydrogen of 99.99% purity could cause embrittlement problems. It would probably be unreasonable to expect that hydrogen of purity greater than 99.98% would be piped in large quantities in a hydrogen economy. Taking into account these facts, it can be assumed that impurity would be a favorable factor in pumping hydrogen through existing pipelines.

Another case of note is hydrogen gas obtained from liquid hydrogen storage. In the liquefication process the hydrogen is purified to a high level. For example, oxygen content is reduced to less than one part per million to avoid the growth of oxygen crystals in the liquid hydrogen. When the liquid is later returned to the gas phase it is extremely pure. Perhaps impurities must be added at this point to protect downstream systems from hydrogen environment embrittlement.

HYDROGEN COMPATIBILITY PROBLEMS

Types of Hydrogen Embrittlement

The effects of hydrogen on, and its compatibility with metals and alloys have been investigated for over 50 years but extensive work has been performed only in the last two to three decades. The culmination of this work is thousands of publications concerning the interaction of hydrogen with metals. The work performed on the different aspects of hydrogen embrittlement has received several excellent reviews (10)(12)(13).

Until the mid 1960's, it was convenient to divide this work into two broad classifications: hydrogen reaction embrittlement and internal reversible hydrogen embrittlement. Recent work has identified a different phenomenon which is called hydrogen environment embrittlement (Table 2.2).

The first class is characterized by embrittlement caused by chemical reactions of hydrogen with the metal or alloying constituents. This form of hydrogen embrittlement manifests itself in the formation of compounds such as metal hydrides, water, and methane. The methane reaction in which carbon in steel combines with absorbed hydrogen is well known in the petroleum industry. Formation of the methane gas internally in the steel leads to fissures, blistering, and cracking.

Another example of reaction embrittlement is the chemical combining of hydrogen with oxygen in solid copper. Water in the form of steam, the reaction product, causes cracking of the copper metal. The anhydrous ammonia industry is well aware of the detrimental effects of hydrogen caused by the combinations of high pressure and temperature that are required to produce its product.

The second classification of metal-hydrogen interaction is the well researched and discussed time delayed hydrogen embrittlement. Even though the phenomena and the factors that influence it are well known, the exact mechanism is still not well defined. The phenomenon occurs by the absorption of hydrogen in the form of a monoatom or ion into the internal structure of the metal. After an interval of time, which can be as long as years, catastrophic failure of the metal results. The hydrogen in this case can come from many sources, particularly metal processing procedures such as electroplating, melting, or welding.

TABLE 2.2: CHARACTERISTICS OF TYPES OF HYDROGEN EMBRITTLEMENT

CHARACTERISTICS	TYPES OF EMBRITTLEMENT		
	HYDROGEN ENVIRONMENT EMBRITTLEMENT	INTERNAL REVERSIBLE HYDROGEN EMBRITTLEMENT	HYDROGEN REACTION EMBRITTLEMENT
USUAL SOURCE OF HYDROGEN	GASEOUS (H_2)	PROCESSING, ELECTROLYSIS, CORROSION (H)	GASEOUS OR ATOMIC HYDROGEN FROM ANY SOURCE
TYPICAL CONDITIONS	10^{-6} TO 10^8 N/m^2 GAS PRESSURE MOST SEVERE NEAR ROOM TEMP OBSERVED -100° TO 700° C GAS PURITY IS IMPORTANT STRAIN RATE IS IMPORTANT	0.1 TO 10 PPM AVERAGE H CONTENT MOST SEVERE NEAR ROOM TEMPERATURE OBSERVED -100° TO 100° C STRAIN RATE IS IMPORTANT	HEAT TREATMENT OR SERVICE IN HYDROGEN, USUALLY AT ELEVATED TEMPERATURES
CRACK INITIATION	(SURFACE OR INTERNAL INITIATION)[a]	INTERNAL CRACK INITIATION	USUALLY INTERNAL INITIATION FROM BUBBLES OR FLAKES
RATE CONTROLLING STEP	ADSORPTION = TRANSFER STEP (ABSORPTION OR LATTICE DIFFUSION)[a] = EMBRITTLING STEP	LATTICE DIFFUSION TO INTERNAL STRESS RAISERS	CHEMICAL REACTION TO FORM HYDRIDES OR GAS BUBBLES

[a]UNRESOLVED

Source: N74-11728-735

In the mid 1960's, another type of hydrogen embrittlement appeared unexpectedly in the aerospace industry (14)(15). Several storage tank failures, due obviously to hydrogen environment, were not explainable by either of the prior two classifications. The term hydrogen environmental embrittlement was given to this type of failure. This phenomenon is characterized by the formation and growth of surface cracks in a metal or alloy while being plastically deformed in a high purity hydrogen gas environment. Laboratory failures of many different engineering metals and alloys have been produced at hydrogen pressures of one atmosphere and higher. It has been only in the late 1960's and early 1970's that significant research studies have been carried out on this type of failure. The work of Jewett, et al (10) is an excellent comprehensive review.

A general analysis of how hydrogen embrittlement phenomena can affect the use of materials can be made in the following way. First, it must be understood that almost all hydrogen embrittlement phenomena are dependent at some point on the presence of monohydrogen atoms in the environment. It is now generally recognized that hydrogen cannot enter a metal or alloy in the form of molecular hydrogen, H_2; only monoatoms or ions can enter. However, it must be equally recognized that a molecular hydrogen gaseous environment is always in equilibrium with a certain amount of monohydrogen atoms as shown by the following reaction:

$$H_2 \rightleftharpoons H + H$$

At ambient conditions there is an extremely small number of monoatoms present in gaseous hydrogen and the general effect of hydrogen on metals is insignificant. However, the reaction above is moved to the right when temperatures are raised above ambient conditions. Therefore, at temperatures and pressures greater than ambient conditions more hydrogen embrittlement problems are encountered. This fact is well known in the petroleum, petrochemical, and anhydrous ammonia industries where severe combinations of pressure and temperature produce significant hydrogen embrittlement problems.

In a hydrogen economy many devices can be expected to operate at high temperatures, which cause molecular hydrogen to be dissociated into the monoatomic form. Therefore, a much greater range and amount of hydrogen embrittlement failures will be encountered. The hydrogen reaction type of embrittlement will become more acute, because of the increased rates of chemical reactions and diffusion through metals at these elevated temperatures. Embrittlement problems of this nature are normally reduced or alleviated by the use of alloyed steels. There will need to be a more general use of nickel, chrome, and stainless steels in hydrogen burning devices. The increased use of these more expensive metals and alloys will ultimately cause an increase in the price of many machines and processes that use hydrogen.

The problems resulting from hydrogen embrittlement of metals must be recognized, but should not be a cause for pessimism about the future uses of hydrogen. The petroleum and petrochemical industries have learned to provide for these problems and there is no reason to believe other industries cannot adequately deal with them. The hydrogen embrittlement problem is not one which will prevent the realization of a hydrogen economy, but it is a problem which will need constant awareness in order to build and operate viable machines and processes.

Embrittlement Mechanisms

Elevated Temperatures: When a steel is exposed to gaseous hydrogen at a temperature of several hundred degrees Celsius, the primary mechanism for hydrogen damage is blistering or void formation caused by a chemical reaction between hydrogen and carbon in the steel to form methane gas. While hydrogen can diffuse into the steel very rapidly, methane cannot diffuse out. Consequently, gas pockets containing high pressure methane may form over a period of days or weeks. At temperatures where the yield strength of the steel is lower and the creep rate is appreciable (T > 400°C), the structural integrity of the steel is adversely affected by blister formation.

Another mechanism of hydrogen attack may apply when a material is exposed to hydrogen at high temperatures, then cooled rapidly. Here the dissolved hydrogen does not have time to diffuse out of the material and precipitates as hydrogen bubbles at grain boundaries, inclusions, and other defects. Intergranular cracking may occur as a result. This mechanism is particularly important if a material is cycled frequently between low and high temperatures.

A third mechanism of damage involves the formation of high pressure hydrogen gas in existing voids in the material at a relatively low temperature, followed by enlargement of these voids to form blisters when the material is heated to a temperature of several hundred degrees. The voids may originate by the methane formation mechanism described above, or during initial casting, fabrication, or welding of the material. Welded material is particularly susceptible because of blowhole or shrinkage porosity.

An illustration of this mechanism of damage was obtained by subjecting several specimens of AISI 1020 steel to flyer plate impact treatments. By this technique, an array of voids were formed approximately 0.1 cm below the surface to simulate weld porosity. Hydrogen was introduced by equilibration with 24.1 MPa (3,500 psi) hydrogen gas at 200°C. The specimens were then heated to temperatures of several hundred degrees Celsius. The surface and a cross section of a specimen heated to 850°C showed blisters. Blisters similar to this also formed at 750°C; at lower temperatures no well defined blisters formed, although measurable swelling occurred.

Hydrogen damage at elevated temperatures has always been a concern in petroleum refining. Now there are also concerns in the development of coal liquefaction and gasification processes.

Temperatures Near Room Temperature: For many alloys which have stable microstructures, the most severe effect of hydrogen on mechanical properties occurs in the temperature range from -100° to +100°C. The mechanism responsible for embrittlement is a very controversial subject. A mechanism which a number of researchers favor is based on the belief that hydrogen present in the crystal lattice reduces the metallic binding energy or lattice cohesion to the point where a crack can propagate in the presence of applied or residual stresses near the crack tip.

When the hydrogen is transported to the crack tip from the gas phase, then dissociates on the surface, the behavior is termed environmental hydrogen embrittlement. When preexposure to hydrogen causes an appreciable amount of hydrogen to dissolve in the bulk of the alloy (including the crack tip area), the behavior is termed internal hydrogen embrittlement. If hydrogen is dissolved in the alloy, it can be transported to the crack tip either by diffusion or by the motion of dislocations which have hydrogen atoms bound to them. This mechanism is not as important at temperatures above approximately 100°C because the material is more plastic or forgiving at higher temperatures, which results in blunting or branching of the crack tip.

Cryogenic Temperatures: Although the mechanism is believed to be the same, hydrogen embrittlement at cryogenic temperatures is in general not as serious as at room temperature. The reason is that the rate of transport of hydrogen to the tip of a crack is slower. If hydrogen is dissolved in the material, the diffusion rate is slower. If the source of hydrogen is a gaseous or liquid environment, the dissociation rate of hydrogen molecules at the surface is slower.

The primary concern is the formation of an undesirable second phase in the material. In materials like 302 or 304 stainless steel, martensite can form over a long period of time whether or not hydrogen is present. In titanium alloys and A-286 stainless steel, a hydride phase may form in the presence of hydrogen. The formation of a second phase is not believed to be a problem in 310 stainless steel or in aluminum alloys. If the strength, ductility, and toughness of the latter materials are adequate at the temperature of interest, containment of hydrogen should not be a problem.

PRESSURE VESSELS AND PIPELINES FOR HIGH PRESSURE HYDROGEN

Pressure vessels and pipelines which contain high pressure hydrogen are the most critical applications of gaseous hydrogen storage and transfer technology. The quantities of material used in both applications are so large that there is a danger of having some areas with cracks or other metallurgical defects. There is a second danger in underground pipelines in that they can be damaged in service by construction equipment and by corrosion. In this section, two case studies will be presented, then some fracture mechanics calculations will be outlined which describe crack growth in pipeline steels.

Aerojet Pressure Vessels (15)

In the early 1960's, Aerojet General Corporation in Sacramento, California purchased four storage vessels for high pressure hydrogen. They were to be used as ground support equipment for the development of liquid hydrogen fueled rocket engines. These vessels had dimensions of 1.52 m (5 ft) by 21.34 m (70 ft) long and were designed to contain gas at 34.45 MPa (5,000 psi) pressure. The structure was laminated with a 1.27 cm (0.5 inch) thick inner shell surrounded by 22 layers 0.73 cm (0.29 inch) thick. The shells were made from type 1146a low alloy ferritic steel. Nozzles and end sections welded to the vessels were made from similar steels.

Three of the four vessels failed after less than a year in service because of large cracks which developed at welds. One of the four was repaired, but it failed again after approximately six months. An intensive investigation was held in 1965 to determine the cause of the failures. The results showed that the materials of construction had satisfactory chemical compositions and mechanical properties. A stress analysis, however, revealed that the applied and residual stresses around the nozzles were sufficient to cause failure. Since the vessels were pressure cycled, fatigue may have contributed to the failures.

It was clear that hydrogen embrittlement was an important factor in the failures because vessels of the same type had been used for high pressure nitrogen with no failures after operating for a period of years. Upon completion of the investigation, the vessels were derated to 24.12 MPa (3,500 psi) operating pressure.

When another failure occurred at the lower pressure, plans to use the vessels for hydrogen were abandoned.

At the time when the first failure occurred, a newly designed vessel was purchased. It was fabricated from A517-F (T-1) steel and consisted of four layers with leak paths through the outer three layers. The new vessel failed after only ten pressure cycles at a pressure of 26.87 MPa (3,900 psi). An inspection of the inner layer showed that two long cracks had formed, one 1.27 m (50 inches) long and the other 0.56 m (22 inches) long.

It is clear from the poor performance of the Aerojet pressure vessels that a detailed stress analysis, together with careful inspection for microcracks are necessities if low alloy steels are to be used in pressure vessels for high pressure hydrogen. The same would be true, but to a lesser degree, if the same vessels were made from stainless steels.

Los Alamos Hydrogen Pipeline (16)

In 1966, a 2.5 cm (1 inch) diameter pipeline was installed at Los Alamos Scientific Laboratory for carrying hydrogen at 13.78 MPa (2,000 psi) pressure over a distance of 4 miles from the Cryogenics Building to a test site. The original plans specified that stainless steel be used for the pipe material, but later it was decided to use a 5% Cr-Mo steel because of the high cost of stainless steel.

When the material was received, it was noticed that the plastic coating was damaged in many areas. After repairs were made, the 12.2 m (40 ft) sections of pipe were butt welded together and installed 1.22 m (4 ft) below the surface of the ground. Of the first 200 welds made, radiography showed that all but one were defective. The defective butt welds were then replaced with socket welds.

The first failure of the pipeline occurred 3½ years after installation. A large pressure drop was observed, but the leak could not be located until an odorant was added to the gas. The leak was caused by a hole in the line in an area away from welds. The hole did not result in the formation of a crack, and in the vicinity of the hole, several pits had formed. A year later two other leaks developed in welds. In one of these, a crack had propagated radially through the wall thickness. The weld failures were attributed to the following:

(a) The steel used was marginal in resistance to corrosion and hydrogen embrittlement.
(b) The welds contained blowholes.
(c) The weld penetration was inadequate.
(d) The lack of physical or electrochemical protection for the steel.

The pipeline has not been used since 1972 and there are no plans to use it in the future. This case study is an important one because the difficulties encountered could also arise if existing natural gas pipelines are used for hydrogen. Although there is a wide spectrum of materials and operating conditions used in the natural gas transmission system, in many of the lines there would be concerns about poor welds, corrosion in the ground, and how these could result in hydrogen stress cracking.

Fracture Mechanics Calculations

In this section will be described calculations of critical sizes for flaws of various types in a large pipe containing hydrogen at 13.78 MPa (2,000 psi) pressure. The calculations were performed for a 76 cm (30 inch) diameter pipe having a wall thickness of 1.8 cm (0.7 inch) and made from a low alloy steel with a yield strength of 413 MPa (60,000 psi). All loads due to expansion, traffic, and settlement were ignored.

Four types of flaws were considered: both axial and circumferential flaws through the wall, and axial and semielliptical surface flaws. Flaws through the wall could be formed during fabrication or could form by the growth of surface flaws. Surface flaws can be caused by careless handling during construction or by mechanical rabbits used for periodic cleaning of the interior of the pipeline.

A hard inclusion or a blowhole in a weld can act as a crack nucleation site. When one of these sites is near a free surface, the stress intensity in the small web near the surface is very high (17). A defect in this area can break through the surface and become a semielliptical surface flaw. Similar flaws can be created by incomplete fusion during welding.

For some of the calculations, analytical solutions were used, and in others computer techniques were used (18)-(22). The details of the calculations will not be reported, and only the results will be discussed here. In Figure 2.1, K_{ISCC} (threshold stress intensity) is plotted as a function of flaw size for the types of flaws considered. The fracture mechanics expression for a flaw in an infinite sheet (18) which can be used for an approximate calculation is $K = \sigma\sqrt{\pi a}$ where a is the crack length and σ is the tensile strength.

As one would expect, the critical flaw size increases with the threshold stress intensity for the material in a 13.78 MPa (2,000 psi) hydrogen environment. Also, the forgiveness of the material for various surface flaws as a group is less than for either axial or circumferential cracks passing completely through the wall because of higher stress levels near the surface flaws.

If values of K_{ISCC} for the appropriate steels and environment were available, the critical flaw sizes for crack propagation in hydrogen could be read directly from Figure 2.1. Since K_{ISCC} values have not been determined for pipeline steels, example calculations were performed for assumed K_{ISCC} values of 55 and 16.5 $MN/m^{3/2}$ (50 and 15 $ksi\sqrt{in}$). The value of 55 is an estimated K_{ISCC} for a typical pipeline steel with no metallurgical defects. The value of 16.5 might correspond to areas in pipelines or pressure vessels where there are hard spots, inclusions, or poor welds.

The results of these calculations are given in Tables 2.3 and 2.4. The results in Table 2.3 suggest that the chances of crack propagation in metallurgically sound material should not occur unless quality control during fabrication is very poor or the material is damaged severely in service. The critical flaw size for the most sensitive defect is 1.27 cm (0.5 inch). In Table 2.4 it can be seen that the critical flaw sizes are several times smaller when K_{ISCC} changes from 55 to 16.5 $MN/m^{3/2}$. Taking the most severe case of the line flaw, Table 2.4 shows that a groove, 0.07 cm (0.03 inch) in depth on the inside diameter, can result in crack propagation in a defective weld or otherwise imperfect material. This and other results in Table 2.4 indicate that extreme care must be taken to insure that the combination of metallurgical defects, together with corrosive attack or mechanical damage does not occur.

The U.S.A. Standard Code for Pressure Piping specifies that as much as a 2.5 cm (1 inch) length of incompletely fused material in welded joints per 30.5 cm (12 inches) of pipe length is acceptable. It is not known whether or not defects of this size are common in natural gas pipelines. Although the calculations performed were approximate and assumed values for K_{ISCC} were taken, it can be concluded that there are potential problems with pipelines and pressure vessels for hydrogen that deserve attention. The results obtained here help to explain the failures observed in the Aerojet pressure vessels and the Los Alamos pipeline discussed earlier.

FIGURE 2.1: PREDICTED RELATIONSHIP BETWEEN THRESHOLD STRESS INTENSITY FOR HYDROGEN STRESS CRACKING AND FLAW SIZE IN A PIPELINE STEEL

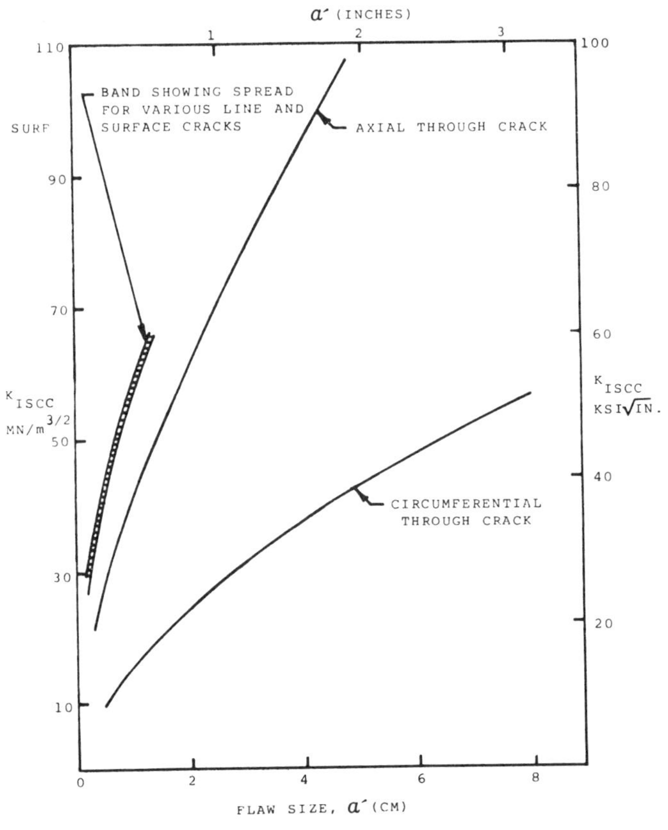

Source: SAND-74-8219

TABLE 2.3: CRITICAL FLAW SIZES FOR VARIOUS TYPES OF FLAWS IN A PIPELINE STEEL HAVING A K_{ISCC} OF 55 $MN/m^{3/2}$ (50 ksi$\sqrt{in}$)

Configuration		Critical a' (cm)	Geometry
Through Crack	Circumferential	8.23	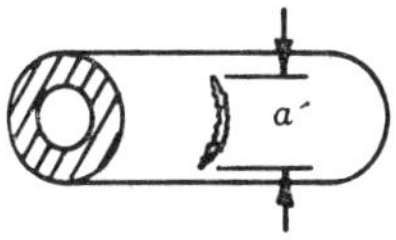
	Axial	1.63	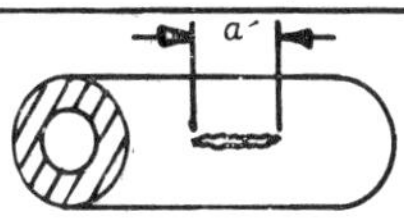
Surface Crack	Line flaw	1.27 inside 1.40 outside	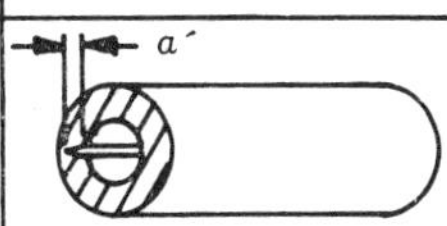
	Semielliptical	1.52 inside 1.65 outside	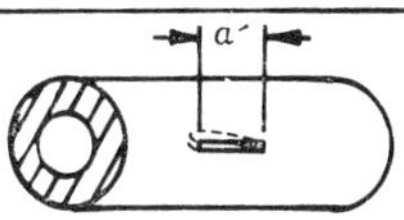

TABLE 2.4: CRITICAL FLAW SIZES FOR VARIOUS TYPES OF FLAWS IN A PIPELINE STEEL HAVING A K_{ISCC} OF 16.5 $MN/m^{3/2}$ (15 ksi$\sqrt{in}$)

Configuration		Critical a' (cm)	Geometry
Through Crack	Circumferential	1.23	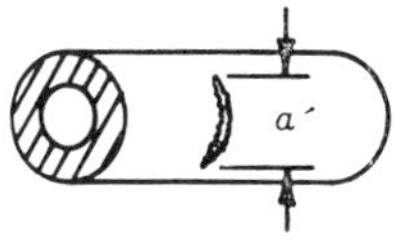
	Axial	0.25	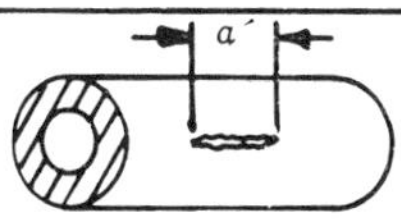
Surface Crack	Line flaw	0.07 inside 0.09 outside	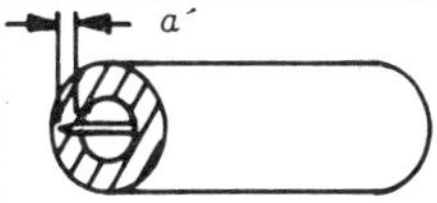
	Semielliptical	0.09 inside 0.09 outside	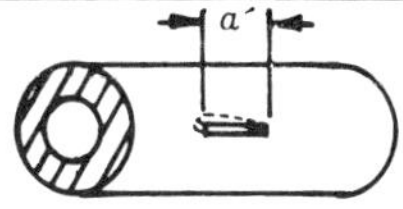

Source: SAND-74-8219

COMPRESSOR REQUIREMENTS FOR HYDROGEN PIPELINE TRANSMISSION

It has been suggested that the natural gas pipeline system be employed for the transmission of hydrogen over long distances in this country. Due to unique properties, problems in two broad categories are visualized if hydrogen were introduced. The first is the potential weakening of the pipeline itself due to hydrogen embrittlement, especially at those locations where existing stress concentrations create the condition of plastic (as opposed to elastic) deformation. This was described previously.

The second is the adverse effect on the existing pumping station machinery, both in performance insofar as ability to produce a sufficient pressure increase and flow rate in an efficient manner as well as operating lifetime. This is the issue considered in this section.

The existing pumping station compressors, reciprocal in the older ones and centrifugal in the newer, are examined from the viewpoint of their expected performance with hydrogen vis-a-vis natural gas, and potential problems are identified in detail. The potential of a third type of compressor, a regenerative compressor, is examined for hydrogen pipelines. It is concluded that it could be far superior to the other two assuming that its efficiency and capacity can be increased beyond that demonstrated to date.

Description of Existing Natural Gas Transmission System

The natural gas transmission system in the United States comprised in 1970 some 914,830 miles of pipeline carrying 63.8 x 10^9 ft^3/day of gas, representing an annual energy content of 23.34 x 10^{15} Btu. This energy distribution system served some 41,427,000 customers (23). A typical transmission pipeline is of the order of 36 inches in diameter and contains natural gas at 1,000 psi and 60°F, compressor stations are located approximately every 65 miles along the pipeline to boost the line pressure back up to 1,000 psi from the 750 psi level typical of the inlet pressure at the station. A compressor station may handle on the order of 500,000,000 ft^3/day of natural gas.

The pumping machinery used in compressor stations has been multicylinder reciprocating compressors driven by natural gas engines. This type of machinery has been supplanted in recent years by two-stage centrifugal compressors driven by gas turbine engines/expanders. The reasons for this will be discussed further in a later section. The composition of natural gas from various locations in the United States is shown in Table 2.5 (24). The principal component is seen to be methane. Of interest to note is the wide variation in composition of natural gas from different locations. Comparison of properties of methane with those of hydrogen appears in Table 2.6 in the next section.

TABLE 2.5: COMPOSITION OF NATURAL GAS AT WELLS

Location	Methane	Ethane	Higher HC	Nitrogen	CO_2	Other
Alaska	98.9	0.2	---	0.7	0.1	0.1
California	82.4	7.7	8.6	0.5	0.8	---
Colorado	68.0	7.2	5.9	18.2	0.1	0.6

(continued)

TABLE 2.5: (continued)

Location	Methane	Ethane	Higher HC	Nitrogen	CO_2	Other
Kansas	78.1	6.1	5.0	10.0	0.2	0.6
Montana	79.9	8.5	4.6	4.5	2.3	0.2
New Mexico	72.9	11.0	12.1	3.5	0.2	0.3
Ohio	86.8	5.9	3.3	3.6	0.2	0.2
Oklahoma	85.0	6.7	6.4	1.5	0.2	0.2
Pennsylvania	97.1	2.0	0.1	0.5	0.1	0.2
Texas	81.6	6.5	3.8	7.7	0.1	0.3
Texas	91.8	4.5	1.8	0.5	1.3	0.1
Utah	42.0	8.0	7.5	15.2	25.8	1.5
West Virginia	75.3	14.2	9.3	0.8	0.2	0.2
Wyoming	86.2	7.8	4.1	0.3	1.5	0.1
Australia	95.6	0.3	1.4	0.8	2.0	---
Canada	88.5	2.5	2.6	6.2	0.2	---
Venezuela	73.7	13.8	10.3	---	2.2	---
Transmission Line Linden, N.J.	94.5	3.3	1.2	0.3	0.7	---

Source: AD-787 484

Types of Pipeline Compressors

Before an appropriate compressor type can be selected for a particular application, certain basic information relating to its performance requirements should be at hand. This includes: pressure ratio, flow rate, efficiencies desired, and could include other special characteristics, as for instance avoidance of pulsations or gas contamination by lubricants. Long life and low maintenance requirements (especially at remote stations) are also usually desirable. One can then consider the type of machine desired from the range of types of compressors available, shown in Figure 2.2.

In this figure is shown the division of compressor types into the two broad categories of positive displacement machines which increase the pressure of a gas by confining it in a decreasing volume, and dynamic machines which operate by imparting kinetic energy to the gas which is then converted into pressure rise by diffusion. These latter machines utilize one or more rotating impellers and diffusor passages to achieve this. Not shown in this figure are devices which raise the pressure of a gas by mixing it with a previously pressurized medium (ejectors). Further differentiation of compressor types is also given in the figure.

One common way to categorize compressors as an aid to selection for a given application is in accordance with the specific speed, N_s, which combines the pressure ratio, capacity and a characteristic speed into a dimensional grouping as follows:

$$N_s = \frac{N\sqrt{Q}}{H^{0.75}}$$

where N = characteristic speed; Q = capacity; and H = pressure rise. On this basis the range of compressor types proceeds from positive displacement compressors, at low specific speed, through centrifugal and on to axial flow compressors at high specific speed. By examining the efficiency variation with specific speed for each type of compressor, a rational partial selection within overlapping ranges of specific speed may be made. It should be noted that this parameter (N_s) is most meaningful when applied to dynamic machines. Further compressor types can be found in References (25)(26)(27).

FIGURE 2.2: COMPRESSOR TYPES

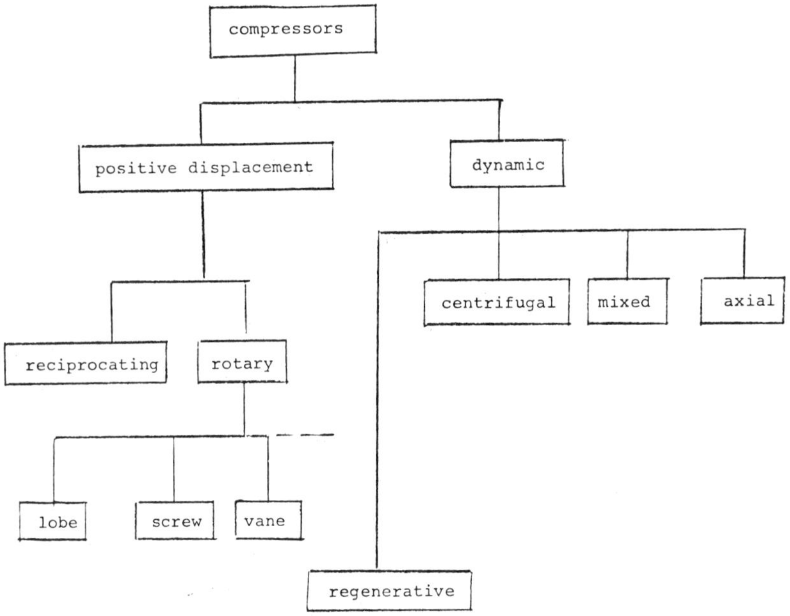

Source: AD-787 484

Reciprocating and Centrifugal Compressors

Before examining the question of hydrogen pipeline transmission and the related compressor requirements, it will be illuminating to examine the types of machines now used in natural gas transmission. At present in the natural gas industry two major types of compressors are used to boost the pressure along the pipe lines. These are reciprocating compressors (piston type) and centrifugal compressors. The latter machines have largely supplanted the former in pipeline service, due to the following reasons:

(a) An increase of capacity of pipeline systems and steadier operating conditions which led to:
(b) Lower pressure ratios and closer spaced compressor stations. Thus:
(c) Lower initial cost and maintenance expense were experienced for centrifugal compressors.

If the pipeline will be handling hydrogen instead of natural gas, several problems arise with both types of machines due to the difference in properties of hydrogen and natural gas, summarized in Table 2.6 below.

TABLE 2.6: PROPERTIES OF HYDROGEN AND NATURAL GAS

	H_2	CH_4
Specific gravity, 0°C, 700 mm Hg	0.08987	0.7169
Molecular weight	2.016	16.032
Ratio of specific heats	1.41	1.30

Source: AD-787 484

These problems will be discussed further in a later section but are briefly mentioned below:

(a) Both designs will suffer to one degree or another from hydrogen environment embrittlement depending on the materials chosen for the design.

(b) Reciprocating compressors will have a serious sealing problem due to the rapid diffusion of relatively small H_2 molecules and the attack by H_2 on sealing materials in nonlubricated designs.

(c) Centrifugal compressors will not produce the required pressure ratios without extensive multistaging; since pressure rise is the product of density and head rise, a lower density gas requires a higher head machine.

Furthermore, both machines will have to be built in expanded capacities to deliver the same energy content with hydrogen as they carry with natural gas, since only 26% of the present energy capacity at 750 psi with methane will be delivered when handling hydrogen (28), due to the lower density of hydrogen.

It is worth noting that in 1972, the installed compressor station horsepower exceeded ten million (29). The centrifugal machine has largely supplanted the reciprocating machine in natural gas pipeline service for the reasons previously discussed. However, in handling hydrogen, design difficulties arise in generating the required pressure rise (head) with centrifugal machinery and it becomes necessary to go to a multistage compressor in order to achieve the head required for a minimum of compressor stations along the pipeline, or to consider another type of compressor, to be discussed in the next section.

Before leaving this section, however, it is of interest to note that there has never been any hydrogen pipeline system developed to the point where compressor stations have been necessary. As mentioned earlier, the most developed hydrogen pipeline system is in Germany and comprises some 130 miles of steel pipe, twelve inches in diameter distributing hydrogen gas generated at 600 psi. The gas simply blows down the pipeline and is delivered to users at 225 psi. Details of this pipeline system can be found in reference (4). There has been some hydrogen compressor design and development experience to date, mostly related to liquification and process work. No such design or development experience appears to exist on the scale required for pipeline application.

Regenerative Compressors

A possible solution to the design difficulty discussed in the last section is to utilize the regenerative compressor for hydrogen gas pipelines. Regenerative compressors are capable of making the head in a single stage that would normally require seven to eight centrifugal stages. This is because the head coefficient obtainable from this type of machinc is of an order of magnitude higher than that obtainable from a centrifugal machine. However, regenerative machines have been efficiency limited until very recently and have also been generally applied in the low end of the centrifugal compressor specific speed range.

Head coefficient is the ratio of the pressure rise produced by an impeller to the square of the tip speed. When taken at the design point (maximum efficiency) it is usually referred to as the design head coefficient. The pressure rise produced by a dynamic compressor is a function of the head rise and the density of the gas.

A fairly conventional regenerative turbomachine has a multiblade impeller with open side channel on one side. A barrier in the side channel between the inlet and outlet ports causes the establishment and maintenance of the regenerative flow pattern. As the impeller rotates the fluid circulates in and out of the blade buckets several times during one revolution gaining energy with each reentry. This type of flow pattern allows the generation of high pressure at low rpm.

Assuming a conservative design point head coefficient for a regenerative machine of 3.5 (this design point head coefficient is of the order of five times higher than a comparable design point heat coefficient for a high speed single-stage centrifugal machine), Figure 2.3 shows the head which can be made with a regenerative machine at various speeds of rotation for various diameter impellers.

The efficiency limitation of regenerative turbomachinery is shown in data presented in Figure 2.4 giving the comparative efficiency of low specific speed pumps of centrifugal, Barski, regenerative and advanced regenerative designs (Garret) as a function of specific speed. It can be seen that the regenerative machine has tended to peak at about 50% efficiency in the low specific speed range, while the centrifugal picks up from this point. Also it can be noted that in very low specific speed ranges the regenerative is more efficient than the centrifugal.

Recent developments carried out at Garret Company based on earlier work at Sunstrand Corporation carried out by Dubey under government sponsorship (31) have shown that it is possible to substantially improve the efficiency of regenerative turbomachinery by the improvement of the ordering of the flow pattern in the side channel, reduction in the extent of impeller blading and better aerodynamic design of the bladed elements. The improvement in efficiency obtainable is shown in Figure 2.4. Figure 2.5 (30)(31) shows a diagram of this improved type of regenerative turbomachine. It now becomes possible to contemplate the use of this design on hydrogen pipeline service, where high horsepower designs are needed.

Historical Development: The regenerative turbomachine appears to have originated in either Germany or Holland shortly after the First World War. It was the subject of a number of doctoral theses at German technical universities during the period between the wars. Among these are the thesis of Carl Ritter at the Technische Hochschule of Dresden (1930), the thesis of Walter Schmeidchen also at the Technische Hochschule of Dresden (1931), the thesis of Carl Schmidt at the Technische Hochschule of Hannover (1931) and the thesis of Heinrich Engels at the Technische Hochschule of Hannover (1940).

These earlier investigators were concerned with exploring over a wide range, the effects of varying geometry on the design of regenerative machines. The experiments involved work varying the proportions of blading to coverage of the side channel over the blading, shape of side channel and blading and the like. As a result of these investigations, a great deal of information was derived allowing one to reach the optimum proportions for conventional regenerative turbomachinery design. These efforts have been summarized in a book by Pfleiderer (30).

FIGURE 2.3: REGENERATIVE TURBOMACHINE HEAD-SPEED-SIZE PERFORMANCE

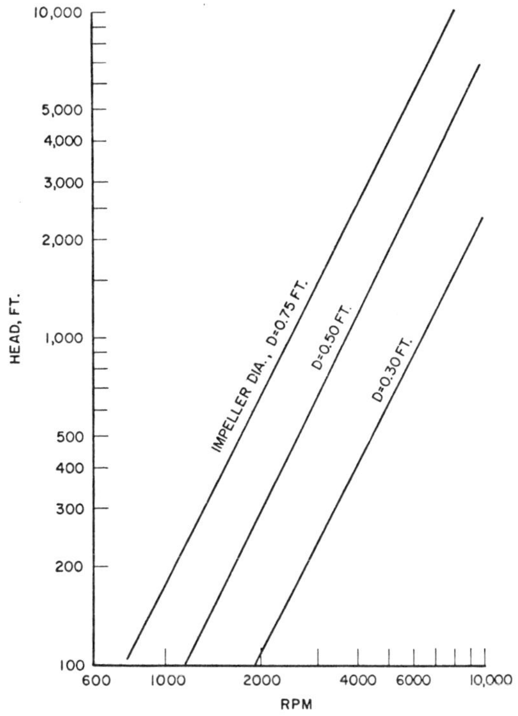

Source: AD-787 484

FIGURE 2.4: COMPARATIVE EFFICIENCY OF LOW SPECIFIC SPEED PUMPS

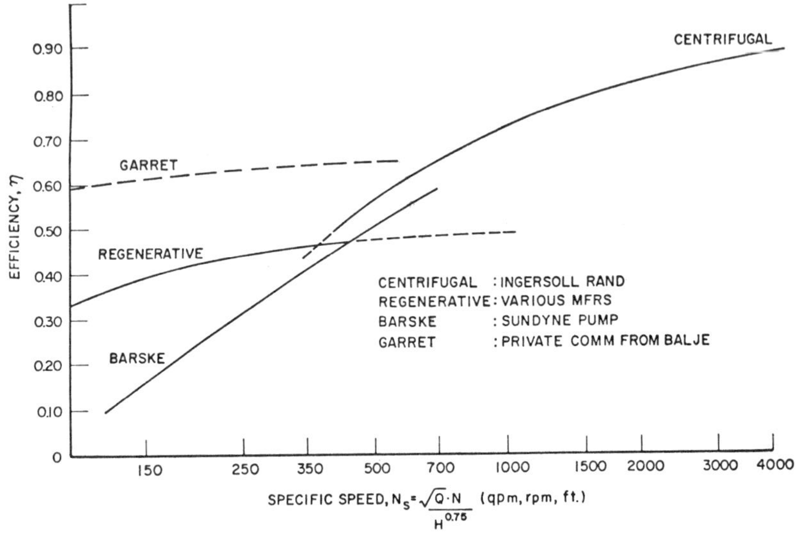

Source: AD-787 484

Of special interest is the work of Engels dating from 1940 in which he not only concentrated on the circular radial section for impeller and side channel but also introduced the concept of a guiding ring [which turned out to be the concept used by Silvern (31) in his research at Sunstrand] in order to better guide the flow and reduce the chaotic mixing process, thereby raising the efficiency level. However, in Engel's work the ring was attached to a full-bladed impeller and rotated with it, which limited its effectiveness. This point will be further discussed later.

Other investigators, following the Second World War, picked up this early work and attempted to develop a rational theory for the type of flow developed within the regenerative machine. One school of thought, the leading exponents of which were Crewdsen (32) and Iversen (33) postulated the essential flow mechanism as being one of drag induced by the impeller on the stationary fluid in the side channel. Certain explanations of observed performance were obtainable by means of this hypothesis which however, proved to be limited in explaining many of the other features of performance which are observed.

Work was also carried out at Massachusetts Institute of Technology (34), National Aeronautics and Space Administration (35), and Oak Ridge National Laboratory (36)-(39). The interest at the latter location was one, significantly enough, of utilizing the regenerative machine as a light gas compressor and circulator in the gas-cooled reactor designs which were being developed there. Mention may also be made of the work done by Shimasaka (40)(41) and Senoo (42)-(45) where the emphasis was on the development of the regenerative turbomachine as a pump where its low NPSH performance characteristics are attractive. NPSH, or net positive suction head is a term usually taken to signify the pressure available at the pump inlet above the cavitation pressure at that point.

Recent investigations at universities have included work by Burton (46) and Grabow (47) where attempts were made to define and develop further the theory upon which a rational design approach to this type of turbomachine could be made. In addition, there were investigations directed toward utilization of this type of turbomachine as a turbine. Emphasis in this area was given in research conducted by Balje (48).

In the meantime, commercial development of this type of turbomachine continued, principally as a pump in Germany and elsewhere in Europe and in the United States as well, where its high head, favorable low NPSH characteristics allowed it to enjoy a special application advantage and where its efficiency limitations were not especially important.

This latter factor however, tended to limit the sizes in which this type of machine was built and sold to less than 50 horsepower. This is two orders of magnitude lower than the horsepower requirements of a typical compressor station on a natural gas pipeline.

In the last fifteen years, the machine has begun to be developed as a gas handling device—first, as a high pressure blower (49)(50) and more recently moving into the range where compressors would normally be applied. Again, this work was pioneered in Germany but has been picked up in the United States and is actively under development at this time in several locations. Among these may be listed Garrett Corp. in California; Rotron Manufacturing Co. in Woodstock, N.Y.; and others.

This brief review has served to point the direction in which development of this type of machine is occurring, and to lend support to the concept that it may well be a suitable machine to apply for the purpose of compression of hydrogen gas in pipelines.

Technical Problems: This section and the next will be devoted to a discussion of the technical problems which will have to be faced in applying the regenerative turbomachine for the compression of hydrogen. These fall into groups as follows:

(a) the need to raise the efficiency levels of the machine to those levels which are now enjoyed by centrifugal compressors;

(b) the need to extend the range of specific speed to higher values than those where the machine has been applied before;

(c) the need to eliminate the noise which the machine will generate at the blade passing frequency at high sound pressure levels, due to the close running clearances between impeller and casing port barrier;

(d) the need to provide adequately for the somewhat peculiar loads that the impeller will throw onto the shaft and bearings as a result of the regenerative turbomachine geometry;

(e) the need to choose proper alloys or other means to avoid hydrogen embrittlement. This difficulty is shared with other compressor designs.

FIGURE 2.5: ADVANCE DESIGN REGENERATIVE TURBOMACHINE

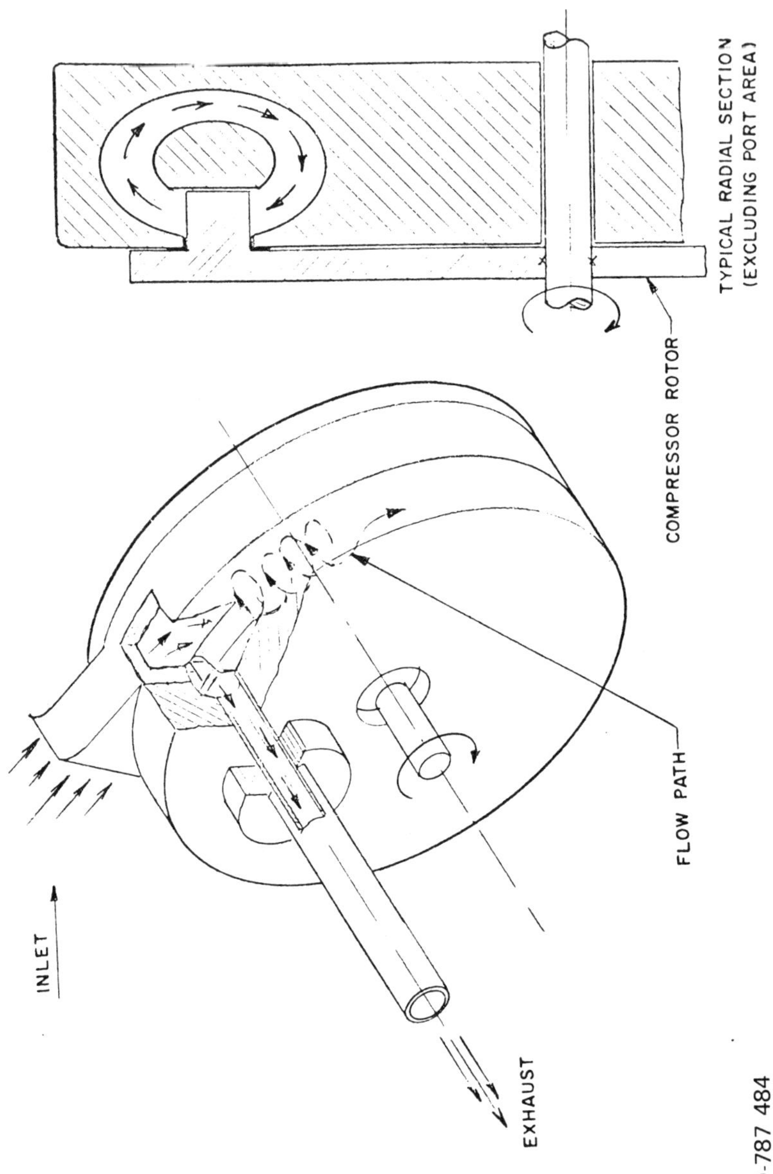

Source: AD-787 484

Possible Solutions, Design Approaches, Recommendations: The technical problems which were raised previously may be approached in a number of ways which will now be discussed corresponding to the order in which they were raised in the previous section.

(a) It may be possible to raise the efficiency of regenerative machinery beyond the levels which have been obtained thus far by means of improving flow guidance in the side channel through proper proportioning of the turning ring and channel geometry in relation to the blading so that the diffusion process occurs in a more guided manner.

Another approach would be to utilize the turning ring itself which can be made hollow, as a heat exchanger so that the compression may approach the isothermal condition; that is, circulate a coolant through the hollow turning ring and reject the heat of compression outside the machine via an auxiliary heat exchanger.

(b) The range of specific speed may be extended; that is, it may be possible to apply the design to a higher flow rate through improvement in the inlet and outlet porting creating a better match to the blading geometries. As a side benefit, this may also contribute to the raising of efficiency if the inlets and outlets can also be designed to provide some turning for the flow. The extent to which this may be achieved is unknown at this time.

(c) Noise problems may be approached through two basic means: first, tuned silencers may be provided at the inlet and outlet of the machine designed so that maximum attenuation occurs at the blade passing frequency. These silencers can also be designed so as to provide some attenuation for broad band or aerodynamic noise as well.

In addition, it may be possible to reduce the noise level of the machine through staging, that is dividing the total work among two or more impellers although this may increase the complexity of the machine to a certain degree. It may also provide a benefit with respect to accommodating the loads developed on the impeller and transferred to the shaft and bearings.

(d) The bending and twisting loads that a regenerative impeller develops tend to cause the impeller to rotate in a plane not perpendicular to the shaft. This creates problems in maintaining a uniform clearance between the impeller and the casing port barrier, and also contributes to unbalance and reduction of bearing life.

One way in which it becomes possible to solve this problem is by staging where the impellers are arranged so that the net eccentricity developed is reduced to zero. It also is possible to design impellers with double-sided blading and to arrange the inlets and outlets in stages in a multistage design alternately around the periphery of the casing. One should also try to provide as large a bearing as possible, particularly in overhung designs at the impeller end of the shaft.

(e) The problem of embrittlement in a regenerative compressor will be in many respects similar to that experienced in other types of compressors with the exception of the following: the impeller itself because of its geometry can be fabricated from castings or forgings essentially in one piece; therefore, stress raisers that may exist in fabricated centrifugal impellers due to welded construction are eliminated.

COST OF COMPRESSION EQUIPMENT FOR HYDROGEN STORAGE

Figure 2.6 presents the investment of hydrogen compression equipment for evaluating the cost in 1972 dollars of compressing gas into high-pressure cylinders or into caverns for storage. Because of the low molecular weight of hydrogen, compression is normally carried on in reciprocating compressors which operate at a relatively low speed. These reciprocating machines require a relatively high investment on a unit energy basis as compared to modern centrifugal and axial compression equipment.

Since energy systems considering the use of hydrogen as the transport medium would most likely involve much larger volume requirements than are normally processed in today's market, new design concepts may be developed to achieve lower unit investment costs on the basis of employing rotating machinery for gas compression.

FIGURE 2.6: COST OF COMPRESSION OF HYDROGEN GAS

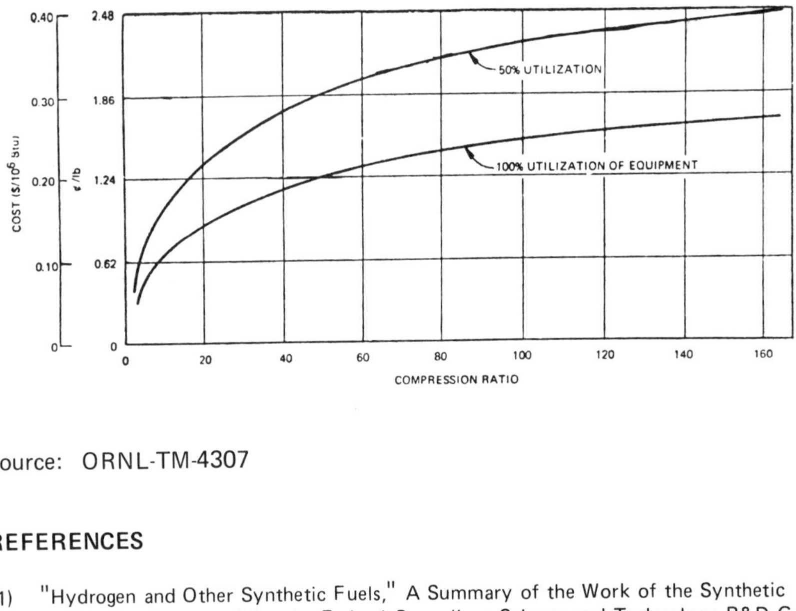

Source: ORNL-TM-4307

REFERENCES

(1) "Hydrogen and Other Synthetic Fuels," A Summary of the Work of the Synthetic Fuel Panel prepared for the Federal Council on Science and Technology R&D Goals Study, September 1972, TID-26136.

(2) Strickland, G., Reilly, J.J. and Wiswall, R.H., Jr. "An Engineering Scale Energy Storage Reservoir of Iron Titanium Hydride," *Proceedings of The Hydrogen Economy Energy (THEME) Conference*, 18-20 March 1974, Miami Beach, Fla., p. S4-10.

(3) Hord, J. (ed.), "Selected Topics on Hydrogen Fuel," NBS Special Publication 419, May 1975, p. 2.29.

(4) Weil, K.H., "Trip Report on Hydrogen Technology in Germany (1972)," A report to the National Academy of Sciences, Comm. on Motor Vehicle Emissions, Alternative Power Systems.

(5) Gregory, D.P. "A Hydrogen-Energy System," Report prepared for American Gas Association by Institute of Gas Technology, August 1972.

(6) Reynolds, R.A. and Slager, W.L., "Transportation and Storage of Hydrogen for Eco-Energy," GE72TMP-54, General Electric Company, December, 1972.
(7) Hausz, W., Leeth, G., Lueck, D., and Meyer, C., "Hydrogen Systems for Electric Energy," 72TMP-15, General Electric Company, 1972.
(8) Seibel, C.W., *Helium, Child of the Sun,* The University Press of Kansas, 1968.
(9) Walter, R.J. and Chandler, W.T., "Effects of High Pressure Hydrogen on Metals at Ambient Temperature," Report No. N70-18637, Rocketdyne, Canoga Park, California, February 28, 1969.
(10) Jewett, R.P. et al, "Hydrogen Environment Embrittlement of Metals," NASA CR-2163, March, 1973.
(11) Gray, H.R., "Hydrogen Environment Embrittlement," NASA TM X-68088, Lewis Research Center, Cleveland, Ohio, 1972.
(12) Triona, A.R., *Transactions of American Society of Metals,* V. 52, p. 54, 1960.
(13) Flsea, A.R., and Fletcher, E.E., "Hydrogen-Induced, Delayed, Brittle Failures of High-Strength Steels," Defense Metals Information Center, Report 196, January 20, 1964.
(14) McPherson, W.B., and Cataldo, C.E., "Recent Experience in High Pressure Gaseous Hydrogen Equipment at Room Temperature," Technical Report No. 8-14.1, American Society for Metals, Metals Park, Ohio, October, 1968.
(15) Laws, J.S., Frick, V., and McConnell, J., "Hydrogen Gas Pressure Vessel Problems in the M-1 Facilities," NASA CR-1305, Washington, D.C., March, 1969.
(16) Edescuty, F.J., et al, Los Alamos Scientific Laboratory, Los Alamos, New Mexico, 1974.
(17) Shaw, R.C. and Kobayashi, A.S., *The Surface Crack,* Amer. Inst. of Mechanical Engineers, New York, 1972, p. 79.
(18) Paris, P.C. and Sih, G.C., ASTM Special Technical Publication No. 381, Philadelphia, Pennsylvania, 1970.
(19) Bowie, O. and Freese, C.B., *Eng. Fracture Mech.,* 1972, Vol. 4, p. 315.
(20) Srawley, J.E. and Gross, B., *Eng. Fracture Mech.,* 1972, Vol. 4, p. 587.
(21) Erodogan, F., and Kibler, J.J., *Int. J. Fracture Mech.,* 1969, Vol. 5, p. 229.
(22) Folias, E.S., *Eng. Fracture Mech.,* 1970, Vol. 2, p. 151.
(23) Savage, R.L., et al, editors, "A Hydrogen Energy Carrier," Vol. II, Systems Analysis, 1973 NASA-ASEE, NASA Grant NGT 44-005-114.
(24) Gibbs, C.W., ed., "Compressed Air and Gas Data," Ingersoll-Rand Co., New York, 1969.
(25) Chlumsky, V., "Reciprocating and Rotary Compressors," E.&F.N. Spon Ltd., London, 1965.
(26) Wislecenus, G.F., "Fluid Dynamics of Turbomachinery," Dover, 1965.
(27) Vavra, M.H., "Aerothermodynamics and Flow in Turbomachines," John Wiley, New York, 1960.
(28) Wurm, J. and Pastris, R.F., "The Transmission of Gaseous Hydrogen," SPE 4526, 1973.
(29) "Gas Industry Research Plan, 1974-2000," AGA, January 1974, M-20272.
(30) Pfleiderer, C., "Die Kreiselpumpen fur Flussigkeiten und Gase. Wasserpumpen, Ventilatoren, Turbogeblase, Trubokompressoren," 5th Edition, 1961, Springer-Verlag, Berlin, pp. 604-619.
(31) Dubey, M., "Study of Turbine and Turbopump Design Parameters," Final Report, Vol. III, "Low Specific Speed Turbines Based on Tangential Flow Theory," S/TD No. 1735, Sunstrand Turbo Contract NONR 2292(00) TO NR 094-343, 30 January 1960.
(32) Crewdson, E., "Water-Ring Self-Priming Pumps," Proceedings of Institution of Mechanical Engineers, Volume 170 n 13, 1956, pp. 407-417.
(33) Iverson, H.W., "Performance of the Periphery Pump," ASME Paper 53-A-102, *Transactions of the ASME,* January 1955, pp. 19-28.
(34) Wilson, W.A, Santalo, M.A., Oelrich, J.A., "Theory of the Fluid-Dynamic Mechanism of Regenerative Pumps," *Transactions of American Society of Mechanical Engineers,* November 1955, Volume 77, pp. 1303-1316.
(35) Weinig, Friederich S., "Analysis of Traction Pumps," Wright Air Development Center, WADC Technical Report 54-554 (AD 67 339), June 1955.
(36) Burton, D.W., "Review of Regenerative Compressor Theory," AEC TID 7631, *Rotating Machinery for Gas-Cooled Reactor Application,* 1963, pp. 228-242.
(37) Gates, P.S., "Performance Characteristics of a Peripheral Compressor," AEC TID 7690,

Rotating Machinery for Gas-Cooled Reactor Application, Meeting Nov. 4-6, 1963; pp. 76-101.

(38) Gates, P.G., "Peripheral-Compressor Performance on Gases with Molecular Weights of 4 to 400," American Society of Mechanical Engineers paper 64 WA/FE-25, for meeting Nov. 29-Dec. 4, 1964.

(39) Namba, I.K., "Development of Regenerative Compressor for Helium Circulation," Oak Ridge National Laboratory ORNL-TM-218, July 20, 1962.

(40) Shimosaka, Minoru, "Research on the Characteristics of Regenerative Pump," *Bulletin of the Japan Society of Mechanical Engineers,* v 3 n 10, 1960, pp. 191-199.

(41) Shimosaka, Minoru and Yamazaki, Shinzo, "Research on the Characteristics of Regenerative Pump," *Bulletin of the Japan Society of Mechanical Engineers,* v 3 n 10, 1960, pp. 185-190.

(42) Senoo, Yasutoshi, "Theoretical Research on Friction Pump," Reports of Research Institute for Fluid Engineering, Kyushu University, Volume 5, n 1, 1948, pp. 23-38.

(43) Senoo, Yasutoshi, "Researches on the Peripheral Pump," Report of Research Institute for Applied Mechanics, Kyushu University, Volume III n 10, July 1954, pp. 53-113.

(44) Senoo, Y., "Influence of the Suction Nozzle on the Characteristics of a Peripheral Pump and an Effective Method of Their Removal," Report of Research Institute for Applied Mechanics, Kyushu University, Volume III n 11, 1954, pp. 129-142.

(45) Senoo, Y., "A Comparison of Regenerative-Pump Theories Supported by New Performance Data," *Trans ASME,* Vol 78, pp. 1091-1102, July 1956 (Paper 55--SA-44).

(46) Burton, J.D., "The Prediction and Improvement of Regenerative Turbo-Machine Performance," Thesis, Southampton University, 1966.

(47) Grabow, G., "Investigation on Peripheral Pumps," Second Conference on Flow Machines, Budapest, Oct. 1966, pp. 147-166.

(48) Balje, O.E., "Drag Turbine Performance," ASME paper No. 56-AV-6, March 1956.

(49) Hollenberg, J.W., "Vortex Blower Development," The Singer Company, Corporate Research Report No. CR-146, July 1967 (Proprietary).

(50) Hollenberg, J.W., "Vortex Flow Study," The Singer Company, Corporate Research Report No. CR-97, 1966 (Proprietary).

LIQUID HYDROGEN

The sources of material that were used for this chapter were the following reports: AD-A016 256, N74-11728-735, SAND-74-8219, CONF-730807-3 and ORNL-TM-4307. A complete bibliography starts on page 283.

Hydrogen as a liquid fuel has certain obvious advantages: relatively high density, ability to absorb large quantities of heat as its temperature is raised, and low-pressure (low-weight) containers. Some of these properties of liquid hydrogen are shown in Table 3.1 below.

TABLE 3.1: PROPERTIES OF LIQUID HYDROGEN*

Normal boiling temperature	36.7°R
Critical temperature	59.74°R
Critical pressure	12.98 atm
Density	4.43 lbm/ft^3
	(0.59 lbm/gal)
Heat of vaporization	191.9 Btu/lbm
	(113.7 Btu/gal)
Sensible heat	
Saturation temperature to 540°R	1710 Btu/lbm
Saturation temperature to 1460°R	5500 Btu/lbm

*For composition in equilibrium at 70°F (25% para, 75% ortho)

Source: AD-A016 256

However, liquid hydrogen is not without its disadvantages as well: theoretical and practical thermodynamic inefficiencies, transfer losses, chilldown losses, steady-state boil-off losses, and the potential for a large fire from a small liquid-hydrogen spill.

TECHNOLOGY OF LIQUEFACTION AND STORAGE

Although liquid hydrogen was first produced as a mist of fine droplets in 1884 and a true liquid in 1898, the first large-scale liquefaction plant for hydrogen was not constructed in the U.S. until 1952. This first plant, constructed at the National Bureau of Standards Cryogenic Engineering Laboratory in Boulder, Colorado, could produce about 0.5 ton/day (450 kg/day) of liquid para-hydrogen.

The output from this plant was used primarily in a program to develop materials, equipment and procedures for handling and storing liquid hydrogen safely. From this modest beginning the technology has expanded to the point where there are now several liquefaction plants in operation, the largest having a capacity of 60 tons/day (55,000 kg/day). Liquid hydrogen is being shipped by tank truck, rail tanker and barge almost routinely and stored in insulated tanks with capacities up to 900,000 gallons (3,400 m^3).

The liquefaction process for hydrogen is somewhat more complicated than for most other common gases, because the Joule-Thompson coefficient for hydrogen is negative at temperatures above 364°R (202°K), which is well below ambient temperature. Thus high-pressure hydrogen gas will get warmer if allowed to expand while at room temperature. For this reason, liquid nitrogen precoolers are usually used and the liquefaction of nitrogen is a part of the hydrogen liquefier.

Added to this complication is the "ortho-para problem" of hydrogen. Ortho-hydrogen is different than para-hydrogen in that the electrons have an opposite spin direction. At room temperature gaseous hydrogen is 75% ortho and 25% para, while the equilibrium liquid is nearly pure para.

The latent heat of conversion from ortho to para is greater than the latent heat of vaporization so that the conversion must be accomplished in the liquefier where this heat may be removed if the liquid is to be stored. Catalysts are incorporated in the liquefier plant to accelerate conversion to the para state (1). The ortho-para problem of hydrogen will be discussed in greater detail later.

The theoretical work of liquefaction of hydrogen is 5,002 Btu/lb (1.16×10^7 J/kg), but it appears that there are rather severe practical limitations to approaching this value. While the figure of merit for nitrogen liquefaction is approaching 9/10, that is, the actual work required is only 10/9 times the theoretical work, it is felt that achieving a figure of merit greater than about one-third will not be feasible for hydrogen in a commercial plant. The work required for liquefaction is therefore taken as 15,000 Btu/lb (34.9×10^6 J/kg). This includes conversion to the para state.

Transport and storage of liquid hydrogen has undergone considerable improvement in recent years, mainly due to the large quantities of liquid hydrogen used as fuel in the space program and the liquid hydrogen stored at the Nuclear Rocket Development Station in Nevada.

The quantities needed in these two programs were so large that there was a strong impetus in developing improved pumps, pipes, tankers and storage containers.

Improvements in the area of insulators are indicative of what can be accomplished when it is apparent that there is economic justification. The standard of insulation in 1952 when the first sizable liquefier was built was high vacuum with silvered walls. Addition of one or more radiation baffles could reduce heat transfer but was expensive.

Development of powder insulation such as perlite, which serves as a multiple radiation shield reduced heat transfer by about an order of magnitude from a simple vacuum. A vacuum in the powder-filled space was still required but the pressure did not need to be as low.

Further improvements were made by using alternate layers of glass fibers and metalized plastic film, also in vacuum. One such insulator called Quilted Super Insulation (QSI) by Linde Corporation further reduces the heat leak by another order of magnitude.

The type of storage or transport container used and the type of insulation employed will depend upon the particular application. For example, 3 feet (91.44 cm) of evacuated perlite may be appropriate for a stationary storage tank. Since the fabrication cost of QSI is high, two inches (5.08 cm) would obviously be better than evacuated perlite for a rail transport tank where size rather than weight is the limiting factor (2).

The rate of development in this field makes it very difficult to project into the future. Simply looking back at the last ten years' work in Liquefied Natural Gas (LNG) indicates what magnitude of change might be envisioned. Developments in LNG storage for peak-shaving capability have been substantial. Not all of these developments appear to be directly applicable to liquid hydrogen storage, but some may be with minor technological changes. In attempting to envision how hydrogen could replace natural gas, the work recently performed in LNG may be used as an indicator of what can be done, not of how it will be done.

For example, the cryogenic in-ground (CIG) storage system which utilizes frozen, moisture-saturated soil as the actual tank wall and insulation does not appear to be applicable to hydrogen at this time. The CIG concept makes possible large storage containers for LNG at low unit cost, but it appears that such containers would have to be lined and possibly insulated for hydrogen storage.

Thus, simplicity, the major advantage of CIG, is largely lost. Technological development may change this situation, and hopefully such changes will make the use of liquid hydrogen more economical.

LIQUEFACTION

Hydrogen is typically produced as a gas. The production of liquid hydrogen requires additional capital investment, energy expenditure, and plant operating costs. In this section, an estimate of these costs is presented for liquefaction plants of various sizes. The costs are given in 1973 dollars.

As a context in which to view these costs, current hydrogen-gas production costs can be cited. Most industrial hydrogen today derives from natural gas or petroleum-refinery gases. From natural gas, hydrogen costs about 0.045 to 0.07 \$/lb (0.9 to 1.4 $\$/10^6$ Btu).

As energy sources for hydrogen production from water, coal and electricity are higher-cost alternatives, increasing hydrogen production costs to 5 to 6 $\$/10^6$ Btu in the case of electricity at 12 mills/kw-hr.

Energy Requirements

The minimum work requirement (ideal work of liquefaction) to liquefy gaseous hydrogen from 70°F and 1 atm to saturated liquid at 1 atm is 5,002 Btu/lbm (1). Strobridge (3) has published a survey and correlation of current refrigerator/liquefier capital costs and energy requirements.

Parrish and Voth (4) have used these data to determine the efficiencies for various-sized units. The efficiencies shown in Table 3.2 were based on existing plant operations (the largest existing plant is approximately 7,700 lb/hr) with a reasonable extrapolation to a 40% efficiency for a 220,000 lb/hr plant. Also shown in Table 3.2 are the liquefaction energy requirements for the different sized plants.

TABLE 3.2: PRACTICAL ENERGY REQUIREMENTS FOR PRODUCING LIQUID HYDROGEN FROM GASEOUS HYDROGEN

Plant size, lb/hr	Liquefaction efficiency, %	Practical energy requirements, BTU/lb
220,000	40	12,500
15,000	35	14,300
7,700	34	14,700
3,300	33	15,200
1,100	30	16,700

Source: AD-A016 256

Thus, it requires 24 to 33% of hydrogen's lower heating value (51,600 Btu/lbm) to convert it into liquid.

Capital Costs

Strobridge's survey of current refrigerator/liquefier capital costs yielded the following correlation: $C = 6{,}000P^{0.7}$ where C is the capital investment cost in 1973 dollars and P is the installed compressor power in kilowatts. Parrish and Voth used the above equation to calculate the capital investment costs shown in Table 3.3. These costs reflect one-of-a-kind construction, and large-scale construction of liquefaction plants could lead to modular design and lower costs.

TABLE 3.3: CAPITAL COSTS FOR PRODUCING LIQUID HYDROGEN FROM GASEOUS HYDROGEN (1973 DOLLARS)

Plant Size, lb/hr	Cost, $
220,000	94.6×10^6
15,000	16.2×10^6
7,700	10.1×10^6
3,300	5.7×10^6
1,100	2.8×10^6

Source: AD-A016 256

Liquefaction Costs

In order to determine the cost of producing liquid hydrogen, the cost of the required power, the operating and maintenance costs, and the fixed charges on the capital investment must be added. Using a power cost of 12 mills/kw-hr, a 90% plant-use factor, a 15% interest rate and a 25-year plant life, Parrish and Voth have calculated the liquefaction costs shown in Table 3.4.

TABLE 3.4: LIQUEFACTION COSTS (1973 DOLLARS)

Plant Size, lb/hr	Cost, $/lb	Cost, $/10^{6}BTU
220,000	0.068	1.32
15,000	0.098	1.90
7,700	0.110	2.13
3,300	0.127	2.46
1,100	0.166	3.22

Source: AD-A016 256

The costs shown in Table 3.4 are based on current technology and extrapolations of trends in efficiency.

Recovery of Liquefaction Energy

As indicated above the energy and cost requirements to liquefy hydrogen are significant and are a deterrent to the use of liquid hydrogen as an energy carrier. However, when gaseous hydrogen at ambient temperature and pressure is desired and liquid hydrogen is available, it is possible to recover some of the liquefaction energy.

The maximum amount of energy that can be recovered in obtaining gaseous hydrogen (STP) from liquid at 1 atm is the ideal work of liquefaction, 5,002 Btu/lb, or 10% of hydrogen's lower heating value. A practical process of reconversion to mechanical work might recover one third of this energy, resulting in an overall recovery of about 12% of the actual work of liquefaction or only 3% of hydrogen's lower heating value. While small as a percentage, this could yield a significant amount of recoverable energy in large systems. Still, in light of the low efficiency, it appears that benefit must be gained from the low temperature of cryogenic hydrogen beyond that deriving simply from its availability as a heat sink for heat-engine operation.

Para-Hydrogen

Hydrogen can exist in two different molecular forms: ortho-hydrogen and para-hydrogen. The difference between these two forms of hydrogen is the relative directions of spin of the nuclear particles which make up the hydrogen molecule. The relative concentrations of these two forms at equilibrium is strongly temperature dependent.

The liquefier efficiencies presented in the previous section were based on the production of para-hydrogen. Hydrogen in equilibrium at 70°F and 1 atm is 25% para-hydrogen and 75% ortho-hydrogen. On liquefaction in the absence of a catalyst, the liquefied hydrogen retains this composition. However, the mixture undergoes a slow conversion to an equilibrium composition at 36.7°R of virtually 100% para-hydrogen. This involves an exothermic reaction of second order with respect to the ortho-hydrogen concentration. This conversion is not instantaneous but takes place over a period of time because the change involves energy exchanges by molecular magnetic interactions.

If the unconverted normal hydrogen is placed in a storage vessel, the heat of conversion (304.8 Btu/lbm) will be released in the vessel. Since the latent heat of vaporization is 191.9 Btu/lbm, significant boil-off of the liquid occurs due to ortho-para conversion. The extent of this storage loss, for a perfectly insulated vessel, is shown in Figure 3.1 (5).

FIGURE 3.1: LIQUID HYDROGEN STORAGE LOSSES FOR VARIOUS INITIAL PARA-HYDROGEN CONCENTRATIONS

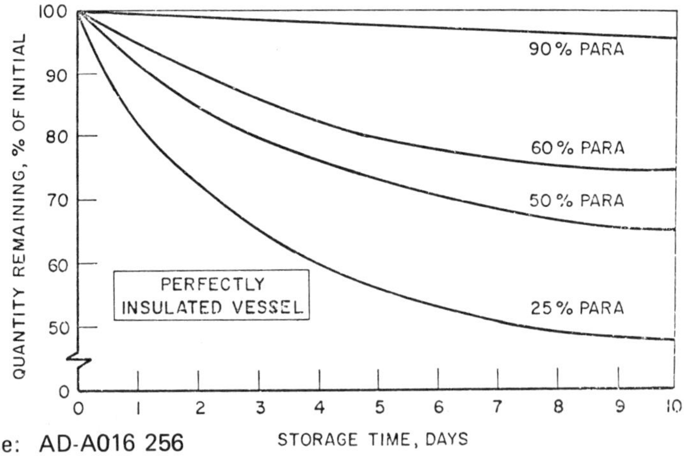

Source: AD-A016 256

The high initial boil-off rate of 25% para(normal)-hydrogen implies a need for either venting, vapor storage or reliquefaction if the initial boil-off vapor is not usable at a rate of about 20% per day.

Therefore, when the vessel to be filled is large (necessitating a long production period), or when there may be a delay between production and use, a catalyst is employed to speed up the conversion so that the energy is removed during the liquefaction process rather than while the liquid is stored.

However, conversion to para-hydrogen reduces the liquefier output—about 35% if conversion of normal hydrogen to para-hydrogen is done at 36.7°R. (Similarly, if vapor phase conversion to 60% para-hydrogen at liquid nitrogen temperature is used prior to conversion at 36.7°R to 100% para-hydrogen, the liquefier output is only reduced by about 23%.) Thus, the question arises as to whether this conversion is justified at the expense of reducing the liquefier output.

For most practical applications, relatively long storage is unavoidable for user and supplier. Under this condition, production of 100% para-hydrogen is advantageous unless most of the boil-off caused by ortho-para conversion can be used as vapor. However, some situations require high output for immediate use, and then it is advantageous to increase the liquefier output at the expense of the para-hydrogen concentration of the liquefied product.

For example, since the production of 100% para-hydrogen from normal hydrogen (25% para-hydrogen) typically results in about a 23% decrease of liquefier output, Figure 3.1 indicates that when the liquefied product is stored in perfectly insulated vessels, conversion to 100% para-hydrogen is justified only when storage beyond one and one half days is anticipated. Thus a knowledge of liquefier output versus per cent of para-hydrogen in the liquefied product allows optimization of a liquid hydrogen production and storage system for a specific mission.

STORAGE AND TRANSFER

Once cryogenic hydrogen has been liquefied and then purified to the required level, a transfer and storage system is required. The primary objectives of any cryogenic storage and transfer system are to minimize cryogenic fluid losses and maintain the desired transfer rates, with a system that is economical, reliable and safe.

The design of cryogenic-hydrogen storage vessels and transfer lines progressed rapidly during the space program. An attempt will be made to point out some of the operating difficulties of such systems and to present available data from actual operating experience. Differences between the transient (or start-up) behavior of these systems and their steady-state characteristics are major considerations.

Representative Storage and Transfer Losses

Losses during handling and transfer are significant, and it is not difficult for large quantities of liquid to be lost when improper handling techniques are used. Some general experience data support this point.

At a liquid hydrogen plant operated by Air Products, Inc. in Painesville, Ohio, 85% of the liquid produced is shipped and 15% is lost during handling at the plant; while at another plant in West Palm Beach, Florida, losses amount to less than 10% of full capacity (7). NASA has gathered data on liquid-hydrogen usage and losses in the space program (8). 460,000 gallons of liquid hydrogen are supplied to fill the Saturn V vehicle for launch. Since the second and third stages of the Saturn V vehicle have capacities of about 270,000 gal and 73,000 gal respectively, transfer losses of 117,000 gal or 25.5% of the total are apparent.

Finally, experience at the Los Alamos Scientific Laboratory obtained while handling liquid hydrogen in 5,500-liter (1,450 gal) batches, indicated that losses during transfer from the transport Dewar to a similar stationary Dewar amount to 10 to 20%, depending critically on the temperature of the receiving Dewar (9). This handling and transfer experience with liquid hydrogen is summarized in Table 3.5.

TABLE 3.5: LH_2 HANDLING AND TRANSFER EXPERIENCE

Operation	Scale	Loss
Liquefaction plant [Ref (7)]	¾ ton/day (Ohio)	15%
	30 tons/day (Fla)	10%
Saturn V launch [Ref (8)]	460,000 gallons	25.5%
Los Alamos batch transfer [Ref (9)]	1,450 gallons	10-20%

Source: AD-A016 256

The data shown in Table 3.5 indicate losses from only one portion of the particular liquid hydrogen system referenced. Of greater interest would be the total losses at a typical facility due to the handling, transfer and storage of liquid hydrogen, but such information was not available. However, for comparison, it is useful to review early experience with liquid oxygen (LOX) losses at Edwards Air Force Base (10).

A detailed review of LOX records indicated the enormity of the liquid oxygen boil-off losses. As much as 6.8 gallons of LOX was purchased for every gallon of LOX used in a rocket firing, i.e., a loss of 85.3%. A detailed analysis indicated the origins of these losses.

LOX was obtained at a nearby railway depot. It was unloaded from a railway tank car, transferred by gas pressurization through a flex hose, filter, valves, flanges and piping, transported 30 miles to the test stand site, and rapidly unloaded by gas pressurization and a pump through a complex network of metal piping that required chilldown. During this transfer, 13% LOX boiled away.

This loss is attributed to the gas pressurization, chilldown of transfer lines, heat leak through surfaces, liquid trapped in the transfer line (and eventually drained at the end of each operation), and leakage. The remaining lossess occurred at the test stand site. A summary of the losses is shown in Table 3.6 (10).

TABLE 3.6: LOX LOSSES AT EDWARDS AIR FORCE BASE

Operation	LOX Loss
Transfer from railroad car to base storage vessel	13%
Transfer from storage to missile (childown)	8%
Missile-tank childown	13.7%
Liquid trapped in transfer system	14.5%
External heat-leak during storage and hold	36.1%
Total	85.3%

Source: AD-A016 256

About 50% of the liquid oxygen is lost aside from missile-tank heat leak. These figures are for an early liquid oxygen experience but, similarity of the transfer losses (13%) with those of Table 3.5 for hydrogen suggests similarity of LOX and LH_2 losses.

Storage Dewars

Liquid-hydrogen storage vessels have been built that range in size from 1 liter laboratory flasks up to 850,000-gallon Dewars used to store hydrogen for space applications. The performance of these storage vessels varies and depends primarily on the type of insulation employed, vessel size and shape, and vessel structural-support considerations. In this section, some of the parameters which affect hydrogen storage-vessel design are considered. Also of interest is the performance of existing Dewars (e.g., boil-off losses, chilldown losses, energy-storage density, etc.).

Insulation: The past two decades have seen marked improvement in thermal insulations for cryogenic-hydrogen applications. In fact, material development and insulation concepts have advanced beyond our ability to optimize their use in engineering design, since heat loss through supports, ports and piping is significant. The proceedings of past Cryogenic Engineering Conferences (6) document insulation-technology development.

Insulations frequently used for the storage of liquid hydrogen are: vacuum with a liquid-nitrogen shield; evacuated foams; evacuated perlite; and evacuated multilayer "super-insulations." The choice of insulation for a particular application is usually a compromise in which such factors as cost, ruggedness, weight, volume and reliability are considered along with the insulating properties of the material.

Insulating properties are typically reported as an apparent thermal conductivity, even though the heat flux per unit area is a result of the combined conduction, convection and radiation heat flux. A summary of these results is shown in Figure 3.2.

For a given design heat loss, the required insulation thickness can vary greatly depending on the type of insulation chosen. For example, the 26,000-gallon storage tank frequently employed for rail shipments, contains a 15-inch thickness of perlite insulation, and the hydrogen loss rate is about 0.5% per day. Since hydrogen transport vessels are typically volume limited rather than weight limited, Perkins and Frainier have calculated that by retaining the vessel's outside dimensions and decreasing the insulation thickness to maintain the same performance with Linde SI-4 super-insulation, the storage tank capacity may be increased from 26,000 to 35,000 gallons (11).

In addition to thermal effectiveness and volume considerations, weight of the required insulation may be a consideration. The densities of the various insulation materials are shown in Table 3.7.

TABLE 3.7: DENSITY OF CRYOGENIC INSULATIONS

Material	Density, lb/ft^3
Polystyrene foam	2-3
Fiberglas	0.5-10
Perlite powder	5-6
Superinsulation (Linde SI-4)	4.7

Source: AD-A016 256

Insulation weight is important not only to total container weight, but also to cooldown losses; the larger the mass of insulation, the greater the storage vessel cooldown loss. As an example, Ishaghoff and Canty (12) have calculated that it would require 109,000 pounds of perlite to reduce the steady-state boil-off loss from a 105,000-gallon (62,200-pound) liquid hydrogen tank to 0.3%.

The cooldown loss for this 109,000 lb of perlite would amount to 24,900 lb of liquid hydrogen or 40% of that stored. On the other hand, 1,050 lb of quilted superinsulation would yield the same thermal performance and the cooldown loss would only be about 240 lb or 0.4% of the stored hydrogen. Consideration of cooldown loss may control the design in instances when the storage Dewar is frequently emptied and later refilled. In such a case, vacuum insulation alone might be the best choice since it minimizes the cooldown loss.

The material cost per unit volume for expanded foams, powders and fibrous materials is low. The cost per unit volume for the superinsulations, on the other hand, is high. However, since less material is required for equal performance, the cost of the insulation for a given container makes the choice of superinsulation attractive in some applications (generally smaller Dewars with large surface-to-volume ratios).

FIGURE 3.2: APPARENT THERMAL CONDUCTIVITY OF CRYOGENIC INSULATIONS

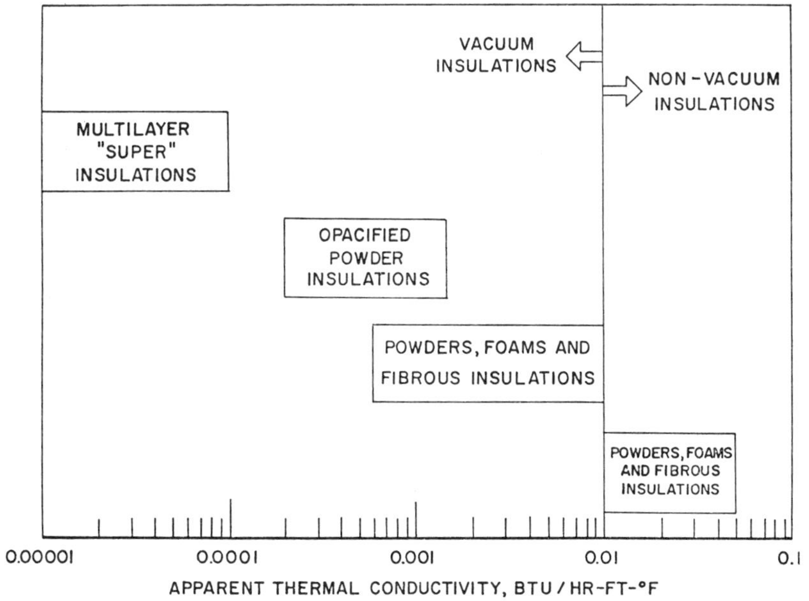

Source: AD-A016 256

Stoy (13) has shown that while the installed material cost of superinsulations is nearly 90 times that of perlite on a unit-volume basis, the total insulation cost for the same steady-state thermal performance is only 4 times higher, owing to the smaller volume of superinsulation required.

The practical performance of these different insulations varies. Expanded foams have the disadvantage that their large thermal-expansion coefficients can cause cracking in the foam during cooldown, while diffusion of H_2 gas into the voids can increase the thermal conductivity by a factor of 3 or 4.

Evacuated powders may pack under vibrating loads and thermal cycling, increasing the heat flux. In evacuated powders the residual gas pressure must remain below 10^{-2} torr to achieve low apparent thermal conductivity. The multilayer superinsulations suffer for two reasons. They must be evacuated to pressures below 10^{-4} torr to be effective. They are also difficult to apply to complicated shapes. These practical performance factors must be considered when evaluating insulations for a specific application.

Steady-State Performance: Design of a stationary liquid hydrogen storage tank frequently resolves itself into balancing initial cost against boil-off rate. For transport Dewars, weight, volume, and ruggedness also become a design consideration. It is therefore useful to be able to make a rough estimate of the insulation requirements imposed by a given evaporation rate. However, heat loss from liquid hydrogen storage tanks is not restricted to heat flux solely through the insulation. Unfortunately in present-day storage vessels, supports, ports, and piping can account for 20 to 50% of the total heat leak (13)(14)(15) depending on the container size, g-loading requirements, etc.

A means for estimating insulation requirements was developed by Perkins and Frainier (11) and is shown in Figure 3.3. Figure 3.3 shows thermal resistance (x/k) versus tank capacity for a boil-off loss of 1% per day of liquid hydrogen. This curve allows one to estimate quickly the thickness of a given insulation (x) required to satisfy a given hydrogen boil-off rate with an insulation of a given thermal conductivity. The curve accounts for typical heat leaks through supports and piping, and therefore, unusual tankage incorporating such factors as high "g" loadings, large manhole, or high L/D ratios would require adjustment of the values shown.

FIGURE 3.3: INSULATION REQUIREMENTS FOR TYPICAL LIQUID HYDROGEN TANKAGE

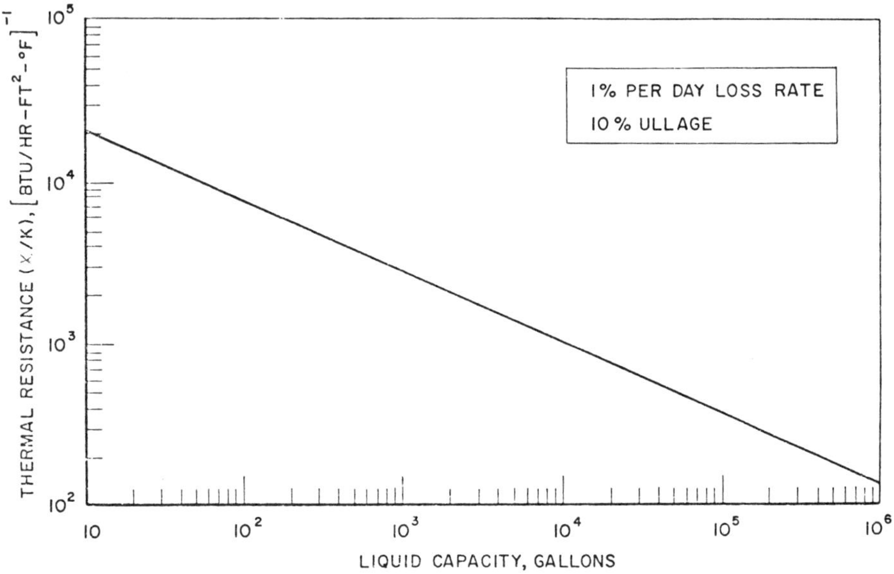

Source: AD-A016 256

For example, a 10,000-gallon tank might demand a loss of 1% per day. From the figure, the thermal resistance required is 1,000. If an evacuated perlite insulation (k = 0.007 Btu/hr-ft-°F) is to be used, the required thickness is 0.7 foot.

For losses other than 1%, the thickness required is obtained by dividing the thickness obtained from Figure 3.3, by the desired loss rate in per cent per day. For

TABLE 3.8: LH_2 STORAGE DEWAR PERFORMANCE

Dewar Identification	Capacity Gal	Capacity 10^6 Btu*	Insulation	Boil-Off %/day	Dewar wt, lb	Dewar vol, ft^3	Btu/lb*	Btu/ft^3*	Cost (1975)
Linde-LSH 150 [Ref (11)]	39.7	1.21	Linde SI-4 (1")	2.5–3.0	166	10.7	7,300	113,000	$50/gal [Ref (15)]
Balloon Flight [Ref (42)]	120	3.67	Evacuated Santocel	-	350	37.9	10,500	97,100	-
Linde LSH 1000 [Ref (11)]	264	8.07	Linde SI-4 (1¼")	1.0	2,400	158	3,370	51,000	-
Beech Transport Trailer [Ref (11)]	1,585	48.4	Powder	1.9	-	-	-	-	$12/gal
Semi-Trailer [Ref (11)]	7,800	238	Linde SI-4	0.5	32,200	1,780	7,400	134,000	-
Conventional Trailer [Ref (8)]	13,000	397	Evacuated Super-insul.	0.8 0.25	40,700	-	9.700	-	-
- [Ref (41)]	14,000	428	Evacuated Super-insul.	0.22	-	-	-	-	-
Rectangular Trailer [Ref (8)]	16,000	489	-	0.4	40,300	-	12,100	-	-
Rail Tank [Ref (11)]	26,000	795	Evacuated Perlite	0.5	-	6,540	-	122,000	-
Nevada [Ref (18)]	50,000	1,530	Evacuated Perlite	0.18	-	-	-	-	-
- [Ref (41)]	200,000	6,110	Evacuated Super-insul.	0.07	-	-	-	-	-
Nevada [Ref (16)]	500,000	15,300	Evacuated Perlite	0.05	-	-	-	-	$2/gal [Ref 16)]
KSC [Ref (8)(11)]	850,000	26,000	Evacuated Perlite	0.026 0.03	-	180,000	-	144,000	$1-2/gal [Ref 16)(43)]

*Based on lower heating value of 51,600 Btu/lbm.

Source: AD-A016 256

FIGURE 3.4: BOIL-OFF LOSSES AND CAPITAL COSTS FOR LIQUID HYDROGEN STORAGE DEWARS FROM TABLE 3.8

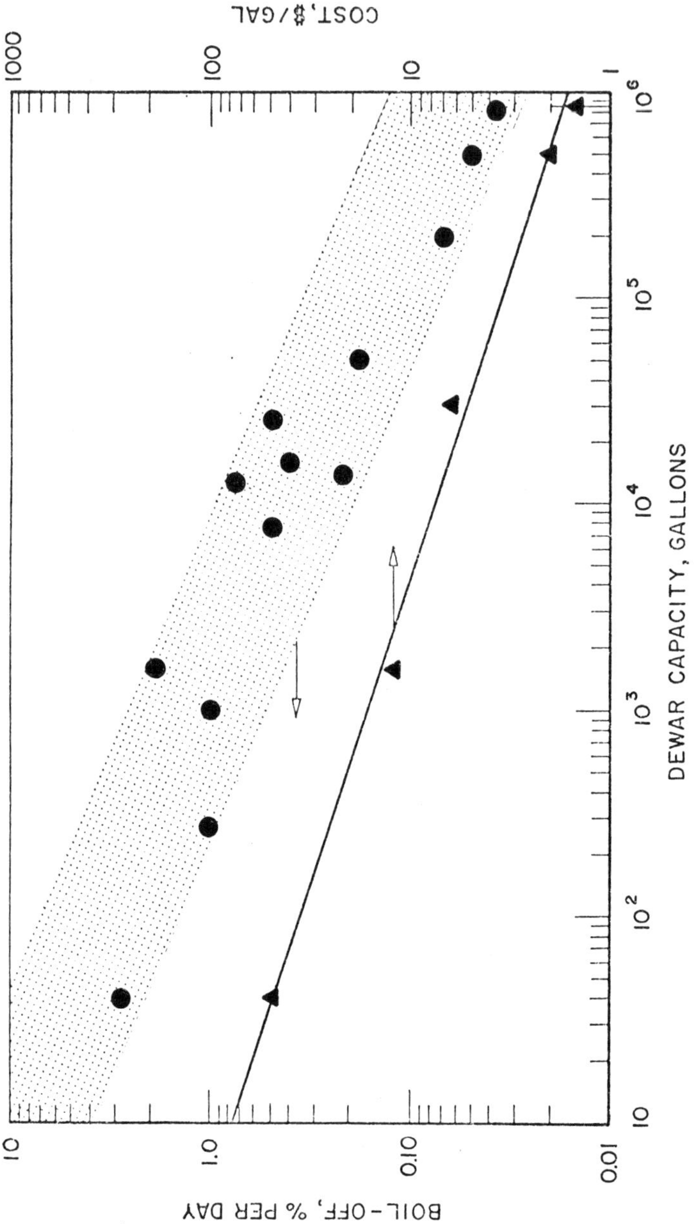

Source: AD-A016 256

example, with a desired loss rate of 0.1% per day, the thickness required would be 7 feet.

Current LH_2-Dewar Performance: Design data and operating experience have been gathered for a variety of liquid hydrogen Dewars which range in size from 40 to 850,000 gallons. These data are shown in Table 3.8. Boil-off loss data and some cost data are plotted in Figure 3.4 versus tank capacity.

These data indicate a rough trend for boil-off losses versus capacity; however, it is not surprising that the correlation is poor, since the vessel shape and type of insulation used varied widely. Detailed specifications for the various insulations used are not readily available so no attempt has been made to correlate subsets of the total body of data for specific types of insulation.

The cost data seem to correlate surprisingly well and indicate that costs may be approximated (in 1975 dollars) from $C = 160\ G^{-1/3}$ where C = cost (dollars per gallon) and G = tank capacity (gallons). It should be noted, however, that the correlation reflects a small number of data and that the costs reflect small-volume production and are subject to substantial decreases with large-volume production.

A 50-gallon liquid hydrogen storage Dewar for an auto would currently cost about $2,500. However, manufacturers estimate prices of from $200 to $750 based on large-volume production (16).

Also of interest are liquid-hydrogen storage densities. Table 3.8 indicates these typically range from 97,000-144,000 Btu/ft^3 and 7,300-12,000 Btu/lbm in most existing storage Dewars. Results from a hydrogen-tank weight optimization study for lunar storage (17) indicated a possible storage density of 9,100-15,200 Btu/lbm for storage of 75-150 lbm of liquid hydrogen.

Cooldown Losses: When a liquid hydrogen storage Dewar at ambient temperature is filled with cryogenic hydrogen, considerable time and quantities of liquid are required to chill the storage Dewar. In one study, cooldown of a 50,000-gallon Dewar required vaporization of 8.1% to 16.9% of the total LH_2 supplied to fill the Dewar (18).

A slow chill was much more efficient, boiling off only about one half the liquid hydrogen of the faster chill. Small Dewars, with larger surface-to-volume ratios may require even higher percentages for cooldown. Unfortunately, studies of Dewar cooldown and filling techniques are very limited (18)(19)(20).

The optimum design of storage Dewars must involve the cooldown period. In fact, in many practical instances, the heat transfer process in the Dewar may never reach steady-state conditions. And while the steady-state losses from a storage Dewar may always be decreased by the addition of more insulation, this added mass to be chilled down increases the cooldown loss.

Transfer Systems

The transport of liquefied hydrogen is accomplished by either of two methods. LH_2 can be transported either by batch lots in Dewars or by continuous or intermittent flow through special piping. Some aspects of these have already been discussed above. The proper design of piping transfer systems is critical, especially if transport efficiency is important and/or transfer distance long. A number of problems arise in long lines: pressure and flow surges, large cooldown losses, possible delays in response due to long cooldown times, and thermal-contraction bowing of the line under partial-fill conditions.

A major consideration for liquid-hydrogen transfer lines is the expected steady-state heat leak through the line into the fluid. Besides wasting liquid hydrogen through evaporation, the resulting two-phase flow reduces the carrying capacity of the line due both to fluid-density decrease and to flow-resistance increase.

For example, Jacobs et al (21) have calculated if 1% of the liquid hydrogen evaporates in a pressurized-transfer system, the capacity of the line falls to 6% of the design maximum flow, while if 10% of the liquid evaporates the capacity falls to 2%. Thus, it is advantageous to avoid two-phase flow. However, as will be discussed later, two-phase flow during the cooldown of the pipeline is unavoidable.

Liquid hydrogen-transfer engineering problems are discussed in this section, and those factors which limit the performance of transfer systems will be emphasized. Liquid hydrogen pumps are discussed in a later section.

Types of Transfer Lines: A cryogenic-fluid transfer line is generally of one of three types: (1) uninsulated, (2) porous-insulated and (3) vacuum-insulated. The last may be insulated by vacuum alone or in combination with multilayer or evacuated-powder-insulation.

Uninsulated Lines – Liquid air, oxygen, and nitrogen are frequently transferred for relatively short distances through uninsulated lines. Scott (22) states "it is difficult to imagine a situation in which an uninsulated line for liquid hydrogen would have any merit." During transfer of liquid hydrogen, solid air forms on the outer surface of the line for free-convection conditions. This condensation causes an increase in the heat-transfer rate since the latent heat of air is transferred to the line in addition to the convection heat flux.

Despite the shortcoming of uninsulated piping, Richards et al (23) performed a study to determine the heat-leak and pressure-drop characteristics for uninsulated liquid hydrogen transfer lines. Their results are incorporated into single-phase flow design curves in such a way that design variables—flow rate, pressure drop, pressure, pipe diameter, pipe length—may be related and optimum performance determined.

It was concluded that the heat flux is sufficiently low to make practical the use of uninsulated liquid hydrogen transfer lines, provided either that mass flow rates are high, the installations are temporary or experimental, or the period of use is brief and infrequent.

It was also determined that the pressure drop caused by the flow of boiling liquid hydrogen could be calculated with reasonable accuracy using the correlation of Martinelli and Nelson (24) and that the pressure levels necessary to maintain single-phase flow were reasonable.

Porous-Insulated Lines – Lines transferring liquid oxygen or nitrogen are sometimes insulated by applying glass wool, polystyrene foam, plyurethane foam, etc. to bare pipes. Since water vapor degrades insulation performance, a vapor barrier is applied to the outside insulation surface. When these insulations are employed on pipes carrying liquid hydrogen, an additional problem arises.

Air condenses within the insulation, and the oxygen in the condensed air presents a fire and explosion hazard if the adjacent insulation is combustible. Consequently, porous insulations are generally not employed for liquid-hydrogen transfer.

Vacuum-Insulated Lines – The most effective means of transferring liquid hydrogen seems to be vacuum-insulated lines. These lines consist of an inner line in which the liquid flows and an outer concentric vacuum jacket. Thousands of feet of vacuum-insulated pipe have been used in the space program. Sizes have ranged from one quarter to 20 inches diameter, pressures from a few psi to 3,000 psi, and lengths from a few feet to more than 8,000 feet (25).

The annular space may be solely vacuum, or filled with evacuated powder or multilayer insulation. Vacuum integrity is essential to maintaining thermal performance. Multilayer (radiation shield) insulation offers one-fifth to one-thirtieth of the heat-transfer rate of vacuum only and also is much less sensitive to changes in vacuum level.

Evacuated powder has the advantage that the vacuum required for effective insulation is two to three orders-of-magnitude less than that required without powder (22). However, the heat capacity of powder insulation increases the cooldown losses of a transfer line. It is also difficult to avoid the formation of voids in the powder in the annular region.

Long sections of vacuum-insulated transfer line contain spacers to prevent contact between the inner and outer lines and to support the weight of the hydrogen in the line. These spacers must be designed to minimize heat leak by conduction; otherwise, the vacuum insulation becomes ineffective. Techniques for spacer design are well developed (1).

Design Considerations: The design of vacuum-insulated piping systems requires the consideration of many factors—cost, thermal contraction effects, thermal performance, vacuum integrity, field handling and installation, and maintenance. Generally, however, the factor which requires the most consideration—once the pipe size and system layout have been determined—is adequate provision for the thermal contraction of the system. Thermal contraction problems can be minimized in various ways:

1. Inherently flexible systems
2. Expansion joints
3. Expansion loops
4. Low thermal-expansion material (e.g., Invar)

The first three methods are frequently employed in conventional piping systems (e.g., steam). However, because of the need for minimizing heat transfer to the liquid hydrogen during steady-state operation and the necessity for minimizing line cooldown losses, the design of vacuum-jacketed lines for liquid hydrogen requires additional considerations. The use of Invar in cryogenic piping systems is attractive since it has a coefficient of contraction of less than one sixth that of stainless steel at cryogenic temperatures. When fully restrained and cooled to liquid hydrogen temperature, the stress induced in Invar is approximately 10,000 psi, well below the allowable stress and an order-of-magnitude below the corresponding stress with steel.

There are no general rules to follow when choosing among these alternatives for a given system. Examples of all these approaches and combinations of these approaches are given in Reference (25). Factors generally considered in specific cases are: relative cost; preexistence of trenches, supports and other equipment; maintenance; and reliability.

Bowing – In some instances, thin-walled transfer piping is subject to severe temperature gradients and the phenomenon of pipeline bowing occurs. Bowing is the tendency of the centerline of the pipe to be deformed into an arc. This condition occurs when the piping is only partially full of liquid, a situation common during a normal operating cycle when liquid trapped in the line between valves is allowed to boil-off (e.g., during transfer of liquid hydrogen to a receiver Dewar, start-up of a gas turbine, etc.).

Studies with liquid nitrogen have indicated that when the pipe is half-filled with liquid, the top fibers of uninsulated stainless-steel, thin-walled pipes did not cool substantially below ambient temperature at steady-state, while the bottom fibers of the pipe were at the liquid nitrogen temperature (26). This extreme temperature gradient causes the pipe to bow. This situation is magnified with LH_2.

The severe temperature gradient obtained in this situation can cause large additional stresses in pipes that are restrained in any manner, particularly at the support point and anchors. Therefore, unless liquid hydrogen pipelines are properly designed for the condition of bowing, severe pipe stresses, intolerable pipe displacements, and pipe-support failures will occur.

Couplings – For long liquid hydrogen transfer lines it is difficult to construct the line in a single section. Hence, a means of coupling sections together is necessary. Three general types of couplings are employed. A simple connection, consisting of a bolted-flange mechanical connection surrounded with fiber glass, foam or powder insulation meets the need for simplicity and ease of maintenance; however, this type of connection imposes a large cooldown penalty and liquid hydrogen leakage may result from improper tightening of the bolts and thermal cycling.

For applications which do not require frequent dismantling of the line a welded joint connection is employed. This joint has the advantage of high thermal performance and low cost, at the expense of troublesome, costly disassembling methods.

A high-performance joint frequently employed is the reentrant bayonet system. It has several advantages: [1] it introduces no additional pressure drop in the system, [2] the heat leak through the joint is small, [3] the joint can be easily assembled and disassembled without special tools, and [4] there are no leakage problems with the joint. However, the bayonet joint rates low in maintenance since installation requires significant axial clearance and since care must be taken not to damage the male portion during assembly and disassembly. Finally the bayonet joint is a high-cost component [$900 for a 3 x 5-inch joint, 1966 price, (1)].

Little development work seems to be proceeding on improving the above coupling methods. However, Randall and Sullivan have developed a novel coupling concept that provides a simple, rugged, relatively low-heat-loss joint for vacuum-insulated lines (27). This coupling concept is applicable to systems needing low heat-loss joints that require easy separation for modification and/or maintenance.

Pressurized Transfer: The transfer of liquid hydrogen from storage Dewars to points of use has been often been effected by pressurization of Dewars. Because of direct contact between the stored liquid and gas, pressurizing gas is limited to helium and hydrogen. The high cost of helium restricts its use in large systems.

The quantity of gaseous hydrogen required to pressurize and transfer a given amount of liquid is an important parameter in the selection and design of the type of transfer system to be used. The quantity of gaseous hydrogen required is dependent on the heat and mass transfer that occurs at the agitated, liquid-gas interface in the ullage of the tank. These processes are complicated, and determination of gas requirements rests largely on correlations of experimental results.

Unfortunately, there is little experimental data available on which to make such a correlation. In one study (28), eight transfers of liquid hydrogen were effected with pressures up to 100 psig and at rates between 31 and 560 gpm while initial and final ullages were also varied. The results showed a pressurized-gas requirement of about 1% of the mass of liquid transferred at 50 psig and 2% at 100 psig.

A preliminary analysis of the process also indicated that not all of the liquid is transferred, but some vaporizes and remains in the tank along with the pressurized gas. Much more experimental and theoretical work is required if hydrogen losses during pressurized transfer are to be accurately predicted.

Transient Transfer Phenomena: A severe limitation imposed on the rapid transfer of cryogenic hydrogen is the transient period that exists during pipeline cooldown. When liquid hydrogen is admitted to a transfer line which is initially at ambient temperature, liquid vaporizes while the line is cooling.

Frequently, the initial surge of liquid into the line causes vaporization of sufficient liquid to exceed the venting capacity of the line. The resulting increase in pressure causes backflow into the storage Dewar and is followed by a period of surging flow and pressure during the cooldown. The magnitudes of these surge peaks are sufficient to concern designers and users of transfer lines. Obviously, systems requiring transfer of a given quantity of hydrogen in a given period of time must include the lower and fluctuating rates of flow during the cooldown period.

In addition to knowledge of flow and pressure surges during cooldown, cooldown times and quantities of liquid hydrogen consumed during cooldown are important parameters. Therefore, these various aspects of the cooldown of cryogenic liquid-transfer systems have been investigated intensively (29)-(36).

Many of these studies have dealt with liquid nitrogen. However, some data are available for liquid hydrogen equipment cooldown requirements.

Transfer Line Surge — Figure 3.5 is an example of the pressure and inlet flow rate recorded during the first few seconds of cooldown (33). The fluid was liquid nitrogen; the time span shown in the figure was 15 seconds; and the total cooldown time was 93 seconds. The plotted pressure was at a station 80 feet from the inlet end, which registered the highest pressure of the four stations in this particular test run.

FIGURE 3.5: TYPICAL VOLUME FLOW RATE AND TRANSFER LINE PRESSURE IN THE FIRST 15 SECONDS OF COOLDOWN

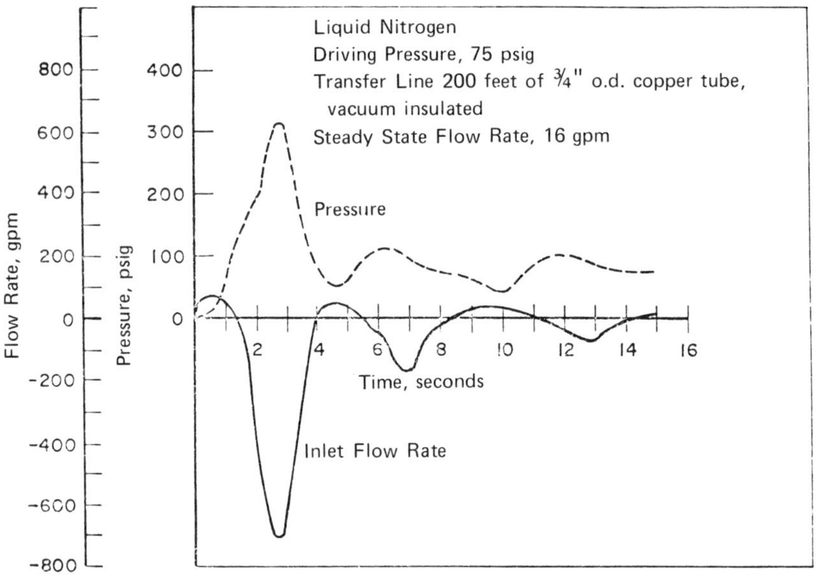

Source: AD-A016 256

It has been speculated that the greatest peak pressure was probably slightly higher and occurred somewhere between the stations. The peak pressure shown on Figure 3.5 is 312 psig, over four times the supply pressure of 75 psig. As the line pressure exceeded the supply pressure, liquid and gas were driven back into the supply vessel at an indicated backflow rate of 710 gpm, whereas the steady-state forward flow rate after cooldown was 16 gpm.

Liquid hydrogen produces smaller surges over much shorter periods than liquid nitrogen. For example, peak surge pressure for a 20-ft LH_2 transfer line was roughly twice the line driving pressure (35).

Steward et al developed a model which fairly accurately predicts LH_2 trends (35). Both model and experiment indicate that peak surge pressure increases with increasing inlet pressure, but decreases when the inlet liquid is saturated at the driving pressure. Also both the model and experiments show that a small amount of line precooling (as might occur if the inlet valve leaked) will cause larger peaks. This is the result of a larger amount of liquid entering the transfer line before contacting a warm surface.

The results of investigations of transfer-line surge indicate that important parameters influencing surging are: driving pressure, inlet liquid temperature, liquid type, pipe diameter and length, and pipe precooling. Long transfer lines aggravate surging, and peak surge pressure can exceed the inlet pressure approximately in proportion to transfer line length.

Steward concluded that one of the most effective ways to reduce surging cooldown would be to warm the liquid supply nearly to the saturation point of the driving pressures, thus minimizing the amount of liquid initially entering the line before vaporization starts.

The high pressures and reverse flow rates which result from surging must concern designers of liquid hydrogen cryogenic systems, since they can lead to severe control problems and/or seriously affect the mechanical integrity of the transfer system.

Cooldown Times and Cooldown Liquid Requirements – Frequently it is necessary to estimate the time and amounts of liquid hydrogen required to cool cryogenic equipment to operating condition. Time and quantity are of significance in relation to fast starts of hydrogen-fueled power plants and the accompanying loss of available liquid.

Several significant factors which influence line cooldown include transfer pressure, quantity of system mass to be cooled (which includes pipe wall thickness and pipe length, valves, pumps, etc.), external heat leak, and line flow impedance. It is obvious that the liquid required to cool a system to its operating condition is a function of the fraction of the heat sink available in the coolant employed. This fraction depends on both the cooldown procedures and the system characteristics.

Jacobs (32) has presented curves which allow computation of cooldown requirements for liquid hydrogen with commonly used materials—stainless steel, copper and aluminum. Table 3.9 indicates these requirements for three initial equipment temperatures.

The minimum liquid requirement is based on the condition that all of the refrigerating effect available from the liquid is utilized, while the maximum liquid requirement assumes that only the latent heat of vaporization of the liquid is utilized. A knowledge of these requirements defines a probable range in which the actual cooldown requirement will be.

Jacobs compared his calculated results with the experimental data of Edeskuty and found good agreement using the minimum requirements. On the other hand, data on LH_2 cooldown of 50 feet of 1.38-inch-ID vacuum-jacketed transfer line containing a flow meter, flow regulator, and shutoff valve indicated more than the minimum requirements; specifically the refrigeration utilized accounted for all the latent heat of vaporization of the liquid plus 45% of the refrigeration from the gas (31).

Thus, in order to estimate the cooldown requirement accurately the engineer must evaluate both the specific cooldown procedures and the system design.

TABLE 3.9: LH_2 REQUIREMENTS FOR THE COOLDOWN OF EACH POUND OF CRYOGENIC EQUIPMENT

Initial Equipment	---- 18-8 Stainless Steel ----		------- Aluminum -------	
Temperature	Maximum, lb	Minimum, lb	Maximum, lb	Minimum, lb
300°K	0.200	0.036	0.380	0.073
240°K	0.135	0.030	0.260	0.060
180°K	0.080	0.022	0.155	0.043

Source: AD-A016 256

Cooldown times can be appreciable. Bronson, et al (31) report a 5-minute cooldown time on their 50-foot vacuum-jacketed transfer line. In addition, complete cooldown of concentrated masses, such as pumps, valves, and flanges, lagged behind the filling of the transfer line with liquid by as much as twenty minutes. Steward, et al (35) measured cooldown times of from 35 to 160 seconds in a 200-foot LH_2 transfer line; the higher the driving pressure, the shorter the cooldown time.

Studies by Cowley (37) and others (38)(39)(40) have shown that the application of a thin layer of a poor thermal conductor between the fluid and metal wall decreases the body cooldown time by shortening the film boiling time and establishing nucleate boiling by which heat is removed at a more rapid rate.

Storage and Transfer Problems

The above section on storage and transfer covers some of the major design considerations to be addressed when selecting or evaluating a liquid hydrogen storage and transfer system. Of major concern are the transient and steady-state response of two major components of such a system—the transfer line and the storage Dewar.

It has been shown that large losses in liquid hydrogen may occur during the handling, transfer and filling phases of liquid hydrogen transport. Liquid hydrogen boil-off occurs both in the transient and steady-state transfer operations. Unless the system is designed to use this boil-off, gaseous hydrogen must be vented and thus lost. A summary of these various losses is indicated in Figure 3.6. The values shown are representative of actual LH_2 handling, transfer, and storage experience.

FIGURE 3.6: LIQUID HYDROGEN LOSSES DURING TRANSPORT, TRANSFER, AND STORAGE

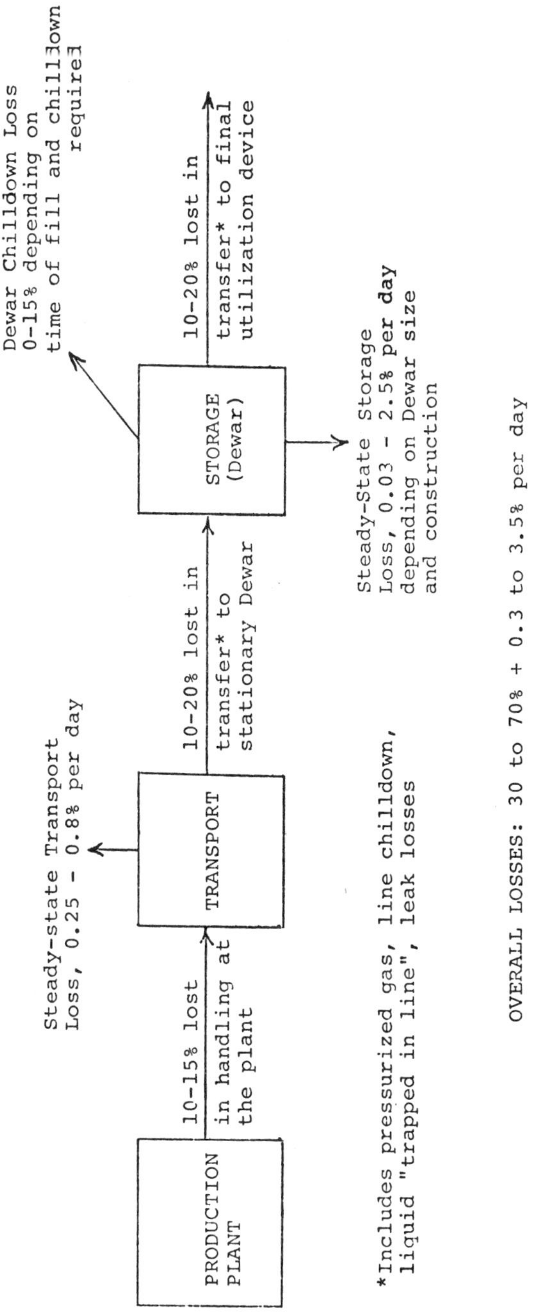

Source: AD-A016 256

Material development and insulation concepts have advanced markedly; and liquid hydrogen storage vessels are easily designed to yield high steady-state thermal performance ($<$3% per day boil-off for small 40-gallon Dewars, $<$0.03% per day for 850,000-gallon vessels).

The development of vacuum-insulated lines and high-performance couplings has resulted in high steady-state thermal performance also in transfer lines. Thus liquid hydrogen boil-off during steady-state transfer and storage does not generally appear to present a serious problem.

However, a severe limitation imposed on the rapid transfer of liquid hydrogen is the transient period that exists while the pipeline, initially at ambient temperature, cools down. This cooldown process is characterized by a period of surging flow and pressure, which can lead to severe control problems and/or seriously affect the mechanical integrity of the transfer line. In addition, the chilldown of transfer lines and storage vessels results in large liquid hydrogen boil-off losses (10 to 20%). Much more information is required if LH_2 transfer and storage systems are to be designed, and procedures developed, to minimize the effects of liquid hydrogen transient transfer phenomena.

SAFETY

The growing applications for liquid hydrogen in low-temperature research and propulsion systems have made the control of liquid hydrogen hazards a matter of major concern. This control problem exists in small laboratories transferring a few liters of liquid hydrogen as well as large facilities handling thousands of gallons per day. A liquid hydrogen accident can jeopardize the lives of personnel and destroy scientific equipment.

When liquid hydrogen is released to the environment at standard conditions, it rapidly evaporates expanding its volume approximately 850 times (44). This hydrogen gas, when mixed with sufficient air to yield the 4% lower flammability limit forms a combustible hydrogen-air mixture with a volume 21,000 times the volume of the original liquid. This indicates the potentially destructive capacity of a small liquid hydrogen spill.

Experience at both Union Carbide-Linde Division (45) and NASA (46) has shown that LH_2 accident potential is low when trained competent personnel are employed and adequate safeguards are provided. For the most part, mishap occurred when the safety guidelines and prescribed safety procedures were neglected.

Linde's experience is that rail accidents have occurred at the rate of 1 in one billion cubic feet of product transported across the country. Truck accidents that have resulted in property damage or loss of life have occurred at the rate of 1 in 1.5 to 2 billion cubic feet, which corresponds to 1 in 1.3 to 2 million miles of highway travel (45).

A review of NASA's records of accidents and incidents which have occurred in the development and operation of NASA's propulsion and power systems provides excellent information on the overall safety of hydrogen (46).

Fire and Explosion Hazards

Fire and explosion constitute the greatest hazard associated with the use of liquid hydrogen. The wide flammability limits (4.1% to 74.8%), low ignition energy (0.02 millijoules), and short quenching distance (0.06 cm) of hydrogen-air mixtures combine to increase the fire and explosion hazard following a liquid hydrogen spill.

Ignition: One of the most common accidents involving liquid hydrogen is a release of liquid or gaseous hydrogen. Of the total of 96 mishaps experienced by NASA, 80 (83%) of the mishaps involved a release of liquid or gaseous hydrogen. Of these 80 mishaps, 66 involved a release of hydrogen to the atmosphere, 20 involved release to enclosures, and in 6 mishaps hydrogen was considered as being released to both locations (46).

When hydrogen was released to the atmosphere, ignition of the mixture occurred 62% of the time, while for release to enclosures, ignition occurred every time (100%). This emphasizes the danger when hydrogen is released inside an enclosure.

An investigation was carried out by Cassutt, et al at A.D. Little, Inc., to determine if an ignited hydrogen-air mixture would detonate in free space (47). Five-foot and eight-foot-diameter balloons were filled with near-stoichiometric mixtures of hydrogen plus air and ignited with explosive and flame sources, hot wires, and sparks, at the center of the balloon.

Detonation only occurred with a 2-gram pentolite initiator at certain mixture ratios. It was concluded that since the probability of these idealized conditions occurring in practice is extremely remote, the chance of detonating a large mass of hydrogen gas released at an unconfined site as a result of an accident is low.

In the same investigation, a number of outdoor spills of liquid hydrogen in quantities from 1.25 to 500 gallons were made. Ignition of the vaporized gases was by spark or flame sources. The ignition time was varied from prior to release to 8 seconds after release, and the depth of the pools varied from 2 to 12 inches. In all cases, no detonation or tendency towards detonation resulted.

For the 1.25-gallon spill, partial confinement by the walls of a test bay was responsible for a pressure pulse roughly equivalent to that produced at the same distance from a completely unconfined 500-gallon spill. These results indicate that barricading at storage vessels can produce a pressure build-up which may be damaging.

A number of spill tests were also made in which the vapor clouds were ignited by means of explosive igniters such as 2-gram and 4-gram pentolite charges. In no case was there any tendency toward detonation or significant increase in combustion pressures. It was concluded that the nonideal mixing occurring in actual spills considerably inhibits detonation.

The dramatic pressure build-up caused by partial confinement of the 1.25-gallon spill previously cited points up the potentially serious consequences when ignition occurs in a completely confined hydrogen-air mixture. This situation yields conditions necessary to produce a detonation along with its destructive effects. In another series of experiments with liquid hydrogen (48), pressures above 200 psi were obtained from ignition of the hydrogen-air mixtures formed by rapidly spilling 40 liters of the liquid in a 3,800 ft^3 enclosure fitted with a weak wall. The damage to equipment located within the enclosure indicated that a detonation had occurred.

Even the smaller overpressures developed by a deflagration can be disastrous in an enclosed area, since very few structures are designed to withstand the pressures developed at even the lower limits of flammability of hydrogen-air mixtures. In many of the NASA mishaps which resulted in an ignition following a hydrogen release in an enclosure, an "explosion" resulted which seriously damaged equipment and buildings. While some of these ignitions seemed to result in a detonation, others caused only a 10-psi overpressure. Nevertheless, serious damage occurred in many cases.

Hydrogen Flames: Ordinary flames such as those from hydrocarbons have an emissivity of about one and consequently radiate a large amount of heat. Hydrogen flames, on the other hand have low emissivities [about 0.09, (47)] and hence low radiant heat fluxes. This can be a hazard since one may actually walk into the flame before realizing it is there. However, the lower radiant energy is advantageous to fire fighters, allowing them to work closer to the fire.

Extent of Hazards From LH_2 Spills: Two important concerns associated with liquid hydrogen spills are the duration of the hazardous period following such a spill and the extent to which the vapor cloud in the vicinity of the spill forms a combustible mixture downwind of the spill.

In one study (47), the spill tests indicated that vapor-cloud ignition produces a hot fireball which ignited combustible material within the confines of the fireball. The major emphasis of the study was to provide data for the prediction of evaporation rates from the ground and to determine the distance downwind a hazardous condition will exist. The results indicated:

1. That, initially, all heat supplied to evaporate the liquid comes from the ground. In later stages of evaporation (i.e., after about 3 minutes) some heat contribution is made by condensation of air into the hydrogen pool. The evaporation rate has an initial value of the order of 5 to 7 inches per minute. It was also found that ignition of the vapor did not significantly affect the rate of evaporation, but that use of a pebble bed of crushed rock greatly increased the evaporation rate.
2. That in spill tests vapor clouds were formed extending up to 200 feet downwind. Upon ignition at the pool the flame traveled downwind for over 100 feet.
3. That in discharges of liquid hydrogen from a pipeline at rates varying from 30 to 300 gpm, a vapor cloud is formed which persists near ground level for 500 to 700 feet downwind. Ignition of the cloud was only accomplished within 100 feet of the vent (the data were too preliminary to conclude whether or not under certain conditions, the vapor could not be ignited at greater distances).

In another study (49), initial evaporation rates of liquid hydrogen from paraffin were of the order of 3 inches per minute, while steady-state evaporation rates were about 1 inch per minute. These evaporations rates are almost identical to those above and allow one to estimate the time for a hydrogen spill to evaporate, once the spill area is known.

A frequently mentioned figure is that a spill of 500 gallons of liquid hydrogen in an unconfined area will diffuse to a nonexplosive mixture in about one minute. However, this would seem to depend strongly on the rate of spill and the surface on which the spill occurred. Thus, one cannot generalize as to the duration of hazard from a given spill. The dimensions of the flames above 0.5 to 7.4 liter spills (48) could be represented approximately by the equation

$$H_{max} = W_{max} = 7\sqrt{V} = 17.8\sqrt{M}$$

where H_{max} and W_{max} are maximum flame height and width respectively in feet, V is liters of liquid hydrogen spilled, and M is the mass of liquid hydrogen in pounds.

Experience at Air Reduction Co. led to the conclusion that the most credible accident at a consumer site is a spill caused by careless connecting and disconnnecting of the liquid hydrogen trailer filling hose and that this volume at no time should exceed 50 gallons (50). Interviews with some experienced distributors of liquid hydrogen indicated this spill is more often on the order of 5 to 15 gallons.

Flaring: To eliminate uncontrolled ignition after hydrogen enters the outside atmosphere, the gas is frequently flared at the vent. However, there seems to be little agreement as to the discharge level at which the gas should be flared. Release rates of 15 lb/min, 30 lb/min, and 60 lb/min have been cited as the rate above which the gas should be burned (51)(52)(53). On the other hand, the Bureau of Mines concluded that flaring itself can create hazards and recommends that: "In general, . . ., flare stocks should not be used to dispose of excess hydrogen." (48).

Until more definitive information is available each facility must base its decision on whether to flare or not to flare by considering the location and type of nearby structures, general wind direction and velocity, etc. NASA experienced a number of ignitions due to high venting rates of hydrogen through venting systems. Hence, the disposal of hydrogen by venting or flaring can cause a fire hazard.

Operation

"Hydrogen operations must be carried out in such a way that the life and health of personnel are not jeopardized and that the risk of damage to property is minimized. Safety of hydrogen handling is therefore built around three basic axioms: adequate ventilation, leak prevention, and elimination of ignition sources. It is necessary to observe all three for safe handling" (45).

General Safety Precautions: [1] Liquid hydrogen should not be exposed to air. The low temperature of liquid hydrogen can solidify air. Vents from liquid hydrogen storage Dewars may be plugged by accumulations of frozen moisture condensed from the air. The resulting pressure buildup can be sufficient to rupture the container, releasing hydrogen to the air. Thus openings in liquid hydrogen containers should be examined periodically to make sure that they do not become plugged with frozen moisture. Also, if air or oxygen is allowed to condense and solidify in liquid hydrogen, a potential explosion hazard can result.

(2) Liquid hydrogen should always be stored and handled in well-ventilated areas. Large quantities of hydrogen should be stored outdoors. Indoors, liquid hydrogen must be stored and handled in well-ventilated rooms and buildings. Hydrogen gas is nontoxic, but can cause asphyxiation by diluting air enough to exclude oxygen. This condition will be reached well after the lower flammability limit of hydrogen in air is achieved.

(3) Exclude all potential ignition sources from areas where liquid hydrogen is handled, stored, or used. This prohibits smoking, restricts open flames and necessitates the use of properly approved electrical equipment.

(4) Avoid contact with hydrogen liquid or gas, or with uninsulated pipes or vessels containing liquid hydrogen. Due to its extremely low temperature, liquid hydrogen can produce an effect on the skin similar to a burn; these "burns" can also be produced by the very cold gas formed by evaporation of the liquid. Particularly hazardous is contact of these cold gases with delicate tissues, e.g., the eyes. Splashing of liquid hydrogen is a common hazard. It is good practice to wear protective clothing (particularly over eyes, face and hands) when handling liquid hydrogen.

Operational Experience: NASA has compiled extensive experience with the use of liquid hydrogen (46). Development of a comprehensive liquid hydrogen safety program, coupled with the employment of trained personnel, has resulted in an operation in which relatively few accidents have occurred. The factors which contributed to the 96 mishaps recorded are categorized below in Table 3.10 (46).

TABLE 3.10: CAUSES OF NASA'S HYDROGEN ACCIDENTS

Procedural deficiencies	25%
Planning deficiencies	14%
Materials failures	$<$3%
Design failures	22%
Operation and work area deficiencies	26%
Malfunction	8%
Materials incompatibility	$<$3%
Contamination	$<$1%

Source: AD-A016 256

Detailed analysis of these mishaps revealed that the number would have been reduced considerably if established rules and regulations had been carefully followed.

Industrial experience to date (45) has shown that: [1] liquid hydrogen can be handled safely for commercial applications, [2] existing specifications, regulations and standards are adequate for use as a base in expanding the application of hydrogen, and [3] safe operation of hydrogen facilities requires trained, competent personnel.

It is not clear that the excellent safety records achieved in industry and at NASA could be duplicated if hydrogen were employed on a much wider scale. The safe handling, storage, transfer and use of liquid hydrogen requires trained personnel, with knowledge of comprehensive safety guidelines, strictly adhering to prescribed safety procedures.

LIQUID HYDROGEN PUMPING

Pump Types

A wide variety of types of pumping equipment is generally available for the solution of a particular pumping problem, such as the pumping of liquid hydrogen. The choice is narrowed somewhat as various system constraints are introduced, but one is usually still left with a number of different possible choices.

The spectrum of types of pumping equipment ranges from positive displacement machines (reciprocating or rotary) through centrifugal (single or multistage) and out to axial-flow (again single or multistage) as the head requirements lessen and the capacity requirements increase. A parameter which aids in the rational selection of pump types is the specific speed defined by

$$N_s = \frac{N\sqrt{Q}}{H^{0.75}}$$

where N_s is the specific speed, N is rpm, Q is the flow, in gpm, and H is the head, in feet of fluid. The various types of pumps available for a given application tend to exhibit their maximum efficiencies in different specific speed ranges. This is shown in Figure 3.7, for centrifugal and axial flow pumps (the definition of N_s loses precision for positive displacement pumps).

An additional type of pump, not shown in this figure, but well-adapted for low-specific-speed operation is the regenerative turbine pump. This pump and its characteristics have been described in the preceding chapter in which this pump type was examined for application as a hydrogen pipeline compressor.

Although some work has been directed toward the adaptation of positive-displacement pumps, particularly of the swash-plate axial-piston type, for airborne liquid hydrogen pumping, the current trend appears to favor turbomachinery. This is because turbomachines are usually lighter and more reliable, even though capacity control is more difficult to obtain with them than on a positive-displacement machine where varying the stroke or speed gives linear capacity control at relatively constant head.

Accordingly, this section will be devoted to discussion of turbomachinery, specifically centrifugal and regenerative pumps in their application to liquid hydrogen pumping.

FIGURE 3.7: PUMP TYPES AND EFFICIENCY

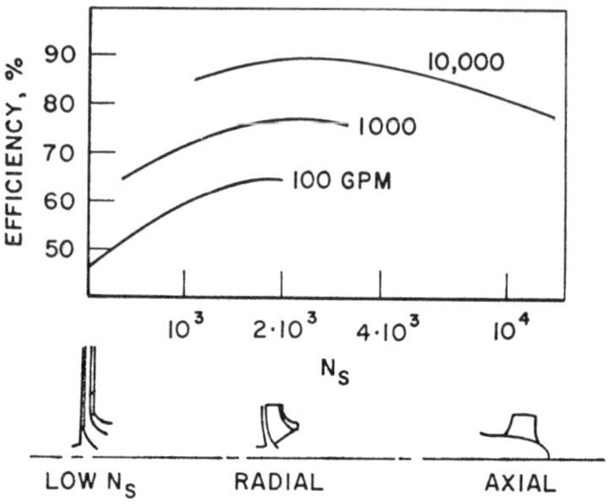

Source: AD-A016 256

The ability of a turbomachine to deliver a given quantity of fluid against an operating system pressure is expressed in terms of its head-capacity diagram and the system resistance characteristic. This is displayed schematically in Figure 3.8.

In this figure are shown the head capacity curve ($\psi - \varphi$), the system resistance curve (R) and the efficiency capacity curve (η). The head and capacity are usually nondimensionalized by means of a head coefficient defined by:

$$\psi = \frac{gH}{u^2}$$

where ψ is the head coefficient, g is the acceleration of gravity in ft/sec^2, H is the head in ft, and u is the impeller tip speed in ft/sec. The capacity coefficient is usually defined as

$$\varphi = \frac{Cm}{u}$$

where φ is the capacity coefficient, Cm is the average "flow through" velocity and u is the impeller tip speed.

Sometimes, the flow coefficient will be nondimensionalized as

$$\varphi = \frac{Q}{ND^3}$$

where Q is the flowrate, in ft^3/min, N is impeller rpm and D is the impeller diameter, in ft.

In any case, the intersection of the pump characteristic with the system resistance line (R) determines the operating point of the pump. This is usually arranged so that for the pump chosen, the efficiency is a maximum at this point as shown in the figure. This point is usually referred to as the "design point."

FIGURE 3.8: SCHEMATIC TURBOMACHINE-SYSTEM CHARACTERISTIC

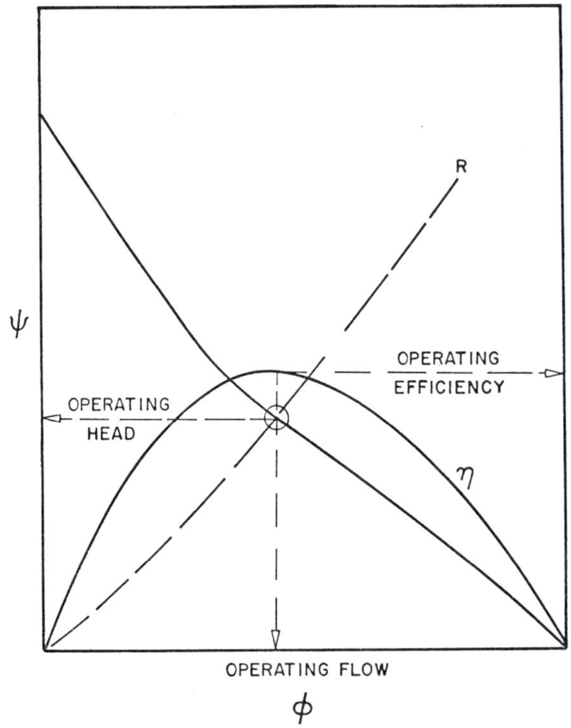

Source: AD-A016 256

Operation of the pump at other points on its char--' ristic due to varying system characteristics is referred to as "off design" operation and is necessary in many applications, as for example, in the case where capacity control is required.

An additional operating factor of crucial importance in pumping of liquid hydrogen is the NPSH performance of the pump. NPSH or net positive suction head is that head available to suppress cavitation (boiling) of the fluid being pumped at the pumping temperature. (Although specific definitions of NPSH vary, in general it is a measure of the local pressure excess over the vapor pressure of the liquid, at the local temperature.)

Cavitation usually occurs where the velocities in the pump are high and the pressures low, or near the inlet. It is a performance-limiting factor and can cause destruction of the pump if allowed to continue over a prolonged period of operation. Figure 3.9 shows the effect of low NPSH and consequent cavitation on centrifugal pumps of low specific speed design.

FIGURE 3.9: CAVITATION CHARACTERISTICS OF LOW N_S CENTRIFUGAL PUMPS

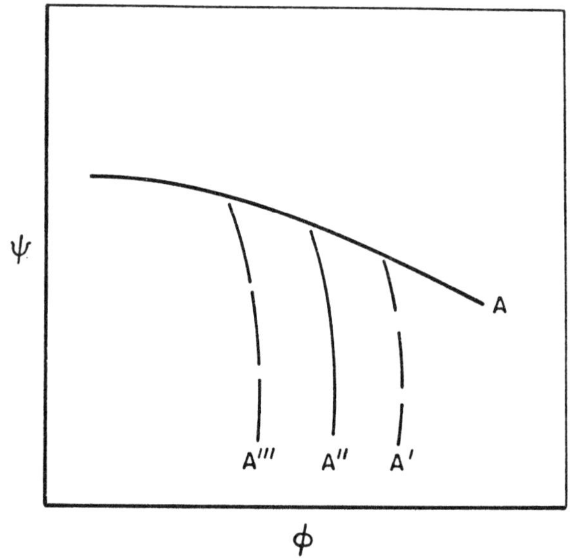

Source: AD-A016 256

As the NPSH is reduced, the normal characteristic, labelled **A** in the Figure, will be replaced by the curves labelled **A', A'', A'''** in turn. This flow reduction will be accompanied by loud noises and eventual destruction of the pump, by the well-known cavitation erosion phenomenon. It becomes necessary therefore to examine the cavitation characteristics of pumps selected for LH_2 applications.

A parameter which describes the limitation of performance of centrifugal pumps due to cavitation is the suction specific speed, given by

$$N_{ss} = \frac{Q\sqrt{N}}{(NPSH)^{0.75}}$$

where N_{ss} is the suction specific speed, Q is the flow rate, in gpm, N is impeller rpm and NPSH is the net positive suction head, in ft. Most low N_s centrifugal pumps are limited to a value of N_{ss} = 12,000 although this can be extended through the use of inducers to N_{ss} = 20,000 allowing for a correspondingly lower value of NPSH.

Figure 3.10 shows the minimum NPSH required as a function of head for values of N_{ss} between 12,000 and 20,000 respectively. Additionally shown in the figure is the effect of specific speed variation (and consequent efficiency) for the range N_s between 400 and 1,200. These ranges are of interest in the pumping of liquid hydrogen, especially for airborne applications.

FIGURE 3.10: NPSH REQUIRED BY LOW N_s CENTRIFUGAL PUMPS

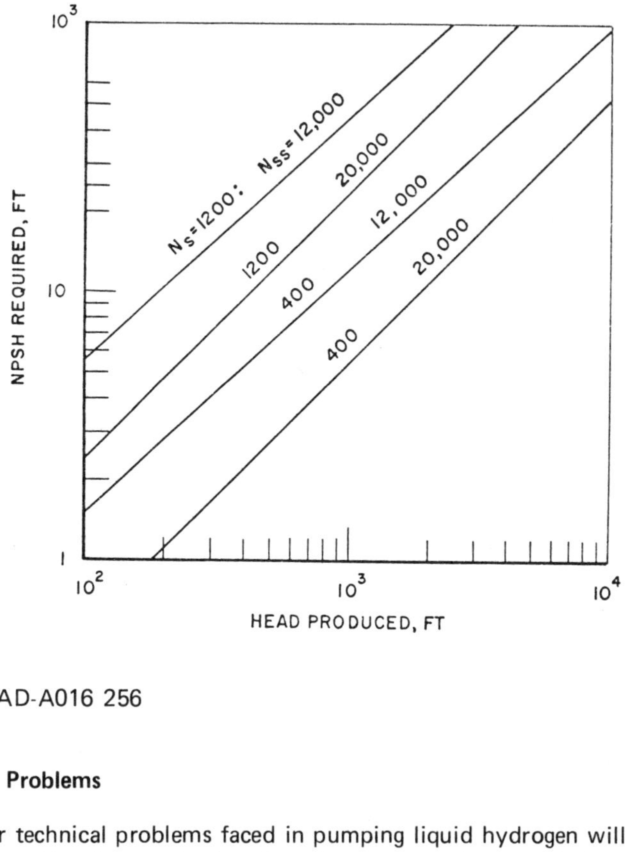

Source: AD-A016 256

Technical Problems

The major technical problems faced in pumping liquid hydrogen will be discussed next, both with reference to the discussion of pump types and characteristics given previously and also with respect to the actual experience (evaluation and experimental) in the field. It should be stated at the outset that although virtually every major type of pump has been considered for pumping liquid hydrogen (regenerative pumps have received insufficient consideration, most likely due to lack of general knowledge about them) the emphasis has been on axial piston and multistage high-speed centrifugal pumps.

NPSH: Since liquid hydrogen in a pumping system is usually close to its boiling point, a slight reduction in suction pressue due to inlet velocities can bring about the inception of cavitation in the pump. This will affect performance of all pumps and ultimately lead to their failure if cavitation operation is prolonged.

The cavitation characteristics of centrifugal pumps have been described previously. The cavitation effect on reciprocating pumps is to decrease their volumetric efficiency (defined as the ratio of delivered volume flow to swept volume of the piston). The relation between VE and NPSH for a reciprocating pump in the cavitation region is usually of the form

$$VE = \frac{K\sqrt{NPSH}}{N}$$

where VE is the volumetric efficiency, N is the pump speed, K is a constant (depending on valving) and NPSH is the net positive suction head.

For a given NPSH available at the inlet of a pump to be selected for a liquid hydrogen pumping system, one should select a pump that will meet the system head and flow requirements with a minimum of velocity at the inlet and one that has the capability of pumping the vapor phase together with the liquid.

Although the cavitation characteristics of regenerative pumps are not well developed, one might expect in view of this type of pump's ability to make the head (and flows) of low specific-speed centrifugals with much lower impeller tip speeds at comparable efficiency that it would be a good candidate to consider for systems where centrifugals are currently favored.

In addition, it handles the vapor phase quite well and has self-priming characteristics lacked by centrifugals. The promise of this approach warrants further investigation.

Capacity Control: A fuel pumping system for an engine or turbine has the requirement of delivering a widely varying flow rate in accordance with the engine or turbine load and speed. This range of capacity can extend as much as 40 to 50 times from idle to full power and speed.

Even with provision for fuel storage at pressure in an accumulator, the extent of capacity control required of the pump is very large. This is why most fuel pumps tend to be of the positive displacement type where capacity can be varied (at relatively constant delivery pressure) by speed control or as in the case of the axial piston/swash plate driven pump by means of varying the stroke (through variation of wash plate angle). The difficulty with these types lies in their reliability and weight penalty in aerospace applications.

Physical Problem: In this category one can combine the problems of weight (relating to pump type), reliability (relating to seals, bearings and rubbing or sliding pumping elements if any) and efficiency, both as to power consumption and cryogen loss during chilldowns and boiloff (relating to pump type and volume of pumping chamber). This is where trade-offs in design are required right away, i.e., between high efficiency as in positive displacement pumps and reliability at some sacrifice in efficiency as in turbomachinery.

To illustrate this point and by way of summary of this section, Table 3.11 was abstracted from reference (54). Shown in this table are some factors of comparison for liquid hydrogen pump systems proposed for the space-shuttle. Noted on this table is the specific speed of 835 for the centrifugal pump system.

TABLE 3.11: PUMP SYSTEM COMPARISON FACTORS

Pump System	Efficiency, %			Weight, lb (kg)			Risk (LH_2 Data Availability)		
	Pump	Drive System	Net	Pump	Sub System	Net	Durability	Performance	System Component Required
Variable Displacement Vane	65	90	58.5	53 (24.04)		53 (24.04)	None	None	Boost Pump
Fixed Displacement Vane	65	70	45.5	40 (18.14)	10 (4.54)	50 (22.68)	Minimal	Minimal	Boost Pump Variable Speed Air Turbine Drive
Fixed Displacement Gear	60	75	45	65 (29.48)	115 (52.16)	180 (81.65)	None	None	Boost Pump Variable Speed Drive
Axial Piston	75	70	52.5	90 (41.82)	100 (45.36)	190 (86.18)	Minimal	Minimal	Variable Speed Drive
Turbopump (Staged Centrifugal) $N_s = 835$	49	70	34	20.1 (9.12)	10.2 (4.63)	30.3 (13.74)	50 hr (requires extension to 500 hr)	RL10 Data	Air Turbine Drive Cavitating Venturi Pump Controller

Source: AD-A016 256

A regenerative pump operating at that specific speed would be of comparable efficiency to the centrifugal and would run at a lower impeller tip speed to make the same head. Consequently it should require less NPSH. Further it should be smaller in size and hence reduce chilldown and boiloff losses. The merits of this approach do not appear to have been sufficiently investigated.

THE ROLE OF LIQUID HYDROGEN IN AN ENERGY SYSTEM

It is envisioned that in a hydrogen energy system there will be several areas where liquid hydrogen may be used to improve the system. These improvements may be in the overall economics, in the reliability, or even in safety, depending on the particular system considered.

In any energy supply system the system must furnish energy on demand, and the demand is not generally constant. The fluctuating demand for energy makes it important to have a flexible supply system. If hydrogen gas were being used only for domestic heating the fluctuations in demand would be somewhat predictable since there would be a higher demand on cold days.

If, however, hydrogen takes over a large share of the uses of energy the pattern becomes more complex. An example of the use of liquid hydrogen for load leveling or peak-shaving will be described later. The assumptions may not be valid for many years but they should indicate the economics of using liquid hydrogen in this type of situation.

Another area where liquid hydrogen could possibly be useful is in supplying energy to a small town where the total use rate and distance from source might make a pipeline for hydrogen gas uneconomical. A set of relatively basic assumptions are used to develop a model to investigate this situation in a later section.

One further suggested use of liquid hydrogen is in cooling a cryogenic electrical transmission cable. In this concept electrical energy and hydrogen energy flow in the same pipe. The economies are a result of the reduction of electrical energy lost through line resistance.

THE ENERGY PIPE

Liquid Hydrogen-Cooled Electrical Transmission Cable

The suggestion was made several years ago that although underground electrical transmission cables are very expensive, some economy might be achieved by cooling the electrical conductors to reduce resistive losses of electricity. If the cooling were achieved by pumping liquefied natural gas through the cable, a double economy could result since both electrical energy and natural gas fuel could be delivered (55).

However, at LNG temperatures the electrical resistance of copper is reduced by only a factor of four. The reduction factor is ten at liquid nitrogen temperatures and two hundred to five hundred for liquid hydrogen temperatures, depending on the purity of the copper conductor.

Belanger and Jefferies (56) examined the potential of an underground electrical cable cooled by a cryogenic fluid. They chose liquid nitrogen rather than liquid hydrogen as the coolant for their studies. The advantages for liquid nitrogen are many. They include lower energy requirement for liquefaction, higher heat of vaporization, better electrical insulator, less stringent thermal insulation requirements and fewer safety problems in handling.

On the other hand, hydrogen further reduces the resistivity of the conductors. If the resistance is lowered by a factor of ten using nitrogen, however, 90% of the electrical losses are saved, and saving a portion of the remaining 10% cannot be justified economically by using liquid hydrogen.

The Edison Electric Institute has been funding a major study of cryogenic underground cables (57)(58) which should provide the technological developments necessary to prove the system. While these studies use nitrogen, many of the construction problems will be similar for other cryogenic systems.

In cryogenic systems, it appears that the use of liquid hydrogen may compete with liquid nitrogen provided that the three following requirements are met:

[1] The electrical transmission line must be placed underground anyway. (For reasons such as right-of-way problems, aesthetics, environmental problems, etc., this is so).

[2] The transmission line must be long enough and the capacity high enough so that significant losses in energy occur if the line is not cooled. (A 1,000-MVA line may have losses of 20% in 250 miles. This is equivalent to $1,000/hr or more in lost energy costs.)

[3] Gaseous hydrogen is available at one end of the line and liquid hydrogen is required at the other end. The requirement may be for peak-shaving or some other use but the use must bear the cost of liquefaction.

It must be emphasized that all three requirements must be present. If, for example, gaseous hydrogen is available at one end of the line and is to be used in the gaseous state at the other end, it is probably more economical at the present time to put a hydrogen gas pipe and a nitrogen-cooled electrical cable side by side in the same right of way. The electrical cable cannot bear the cost of liquefaction of the hydrogen. As the concept is developed then large amounts of energy can be transferred in the energy pipe.

One recent development which improves the thermodynamics of liquid hydrogen transmission involves the use of "slush hydrogen" which is a mixture of up to 60% solid hydrogen in liquid. The latent heat of fusion as well as the heat capacity in raising the liquid temperature to the boiling point significantly increase the apparent heat capacity of the system.

Use of the slush hydrogen has been investigated by Minnic and Fox (59) and the properties of the mixture have been established. These properties are given in an article by Sindt (60). Efforts to develop this system for future use are recommended.

Superconducting Transmission Cable

Related to the energy pipe concept is the superconducting cable which at the present time, uses liquid helium as the coolant. However, there are indications that superconducting alloys can be used at slush hydrogen temperatures with sufficient conduction capacity to be worthwhile. Personnel at Los Alamos Scientific Laboratory have been engaged in development work in this area and the design information which follows was obtained from them (61).

The principle behind the development of superconducting transmission lines is that resistive electrical losses can virtually be eliminated by taking advantage of the electrical properties of certain materials at cryogenic temperatures. The superconducting material which has probably been studied more extensively than any other is Nb_3Sn.

Its superconducting transition temperature (T_C) is about 18K. Newer materials such as alloys of Nb, Al, and Ge, have T_C values higher than 20K. To maintain superconducting properties in these materials, coolants such as liquid helium (4K), slush hydrogen (14K), or a combination of the two must be used. There are economic advantages to using slush hydrogen, but the possible containment problem must again be examined.

A drawing of one of several prototype designs is shown in Figure 3.11 (64). It is proposed that this design be used in a DC transmission line one kilometer long, with a power rating of five gigawatts. The superconducting material is plated onto the outer surfaces of two concentric aluminum tubes. Helium or slush hydrogen refrigerant will flow through the system at a pressure near one atmosphere. Surrounding the conductors and coolant are a stainless steel containment tube, a vacuum jacket, radiation shield, insulation and a structural steel case.

An estimate of component costs made in 1972 for the transmission line showed that the cost of the stainless steel would be approximately 10% of the overall component cost of $10,500/km. The adequacy of stainless steel and the possibility that cheaper materials could be used for containing slush hydrogen merits some discussion.

Reasonable specifications for the containment material are a Charpy impact strength of at least 20.3 Nm (15 ft-lb) and a tensile ductility of at least 15% elongation (or 15% reduction in area), all at 14K. Furthermore, there should be no significant degradation in properties in the presence of hydrogen.

In reviewing 1970 handbook data on the mechanical properties of steels at cryogenic temperatures (62), it is found that low-carbon steels, as typified by AISI 1020, undergo a ductile to brittle transition just above liquid nitrogen temperature (77K). The impact strength at 14K is 4.1 Nm (3 ft-lb), and the corresponding percent alongation is 4%, neither of which is adequate for the intended application.

If the same properties are examined for 304 stainless steel (62), it is found that the impact strength is 80.0 Nm (59 ft-lb) and the elongation is 30% at 14K. A recently published report (63) on the ductility of 304L and 310 stainless steels gives reduction of area values at 4K and 53 and 59%, respectively.

FIGURE 3.11: COAXIAL SUPERCONDUCTING TRANSMISSION LINE (CONCEPTUAL)

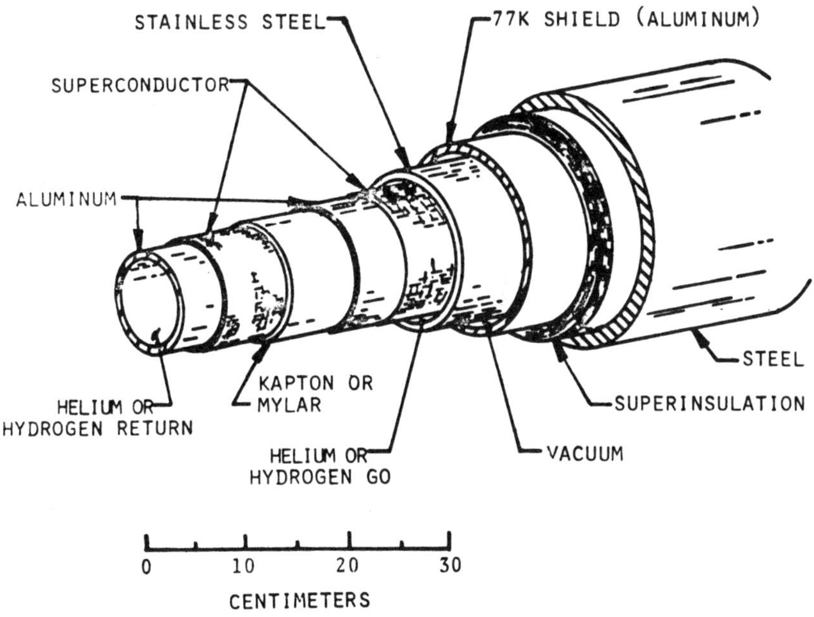

Source: SAND-74-8219

All of these properties are well above the specified limits, but they were measured with no hydrogen present. Work on the degradation of mechanical properties of steels in the presence of hydrogen at cryogenic temperatures has been reviewed by Keeton (64). None of the austenitic stainless steels studies (304, 310, 321, 347, and A-286) was seriously degraded unless cathodic charging was used to drive large amounts of hydrogen into the material. Very little work has been done, however, at temperatures as low as 14K.

One uncertainty remains in using an austenitic stainless steel for long-term service at low temperatures. There is a tendency for martensite to form to a greater or lesser degree depending on steel composition. Some martensite will form immediately on cooling in type 304L, but not in 310 or 316. Isothermal martensite formation (65) over long periods of time has never been studied in these steels as far as is known. Rate of formation data and the tolerance level for martensite should be determined before the best choice of materials can be made.

Alloy steels other than stainlesses should also be considered as containment materials for slush hydrogen. Many of them are lower in cost and could have acceptable properties. For example, a 9% nickel steel has a Charpy impact strength of 28.5 Nm (21 ft-lb) and a ductility of 22% elongation at 14K (62). Normally it is heat-treated to form tempered martensite. This structure is thermodynamically stable, so in contrast to austenitic stainless steels there would be no tendency for a phase change to occur at cryogenic temperatures over a long period of time.

The properties of 9% nickel steel in hydrogen at low temperatures have not been determined, but it is being used currently in liquified natural gas hardware. A new alloy containing 12% Ni and 0.25 to 0.50% Ti has much higher toughness at low temperatures, but it is not commercially available (66). Pressure vessel steels such as 4340 are possible candidates only if they are not heat-treated to maximum strength properties (67).

In general, the above discussion for containing hydrogen in superconducting transmission lines applies also to superconducting magnetic energy storage. The principal application for steels in prototype magnetic energy storage Dewars is a one-meter (40-inch) diameter containment vessel for the coolant (68). A commercial unit would be many times larger. The use of a ferritic steel instead of stainless should not introduce technical problems because of the steel being ferromagnetic.

In reviewing prototype designs for both cryogenic applications above, it is clear that steels could be used for components in addition to those specified. For example, in the transmission lines expensive material with a high electrical conductivity (99.9% Al or OFHC Cu) is now specified as the substrate material for the superconductor.

High electrical conductivity is needed to prevent hot-spot formation due to a temporary or localized high resistance in the superconductor. With the development of better superconducting materials and better fabrication processes, it may be possible to replace the Al or Cu with a steel.

PEAK SHAVING FACILITY

As a basis for calculation the following assumptions can be made:

> The annual energy demand in the area, which may be supplied in the form of hydrogen gas, is about 10^{15} Btu (10^{18} J).
>
> The hydrogen supply is by gas pipeline from a remote production site. The supply pipeline will deliver 3×10^{12} Btu/day (9×10^{11} J/s) when operating at full capacity, 9.5% above average demand.
>
> A typical weekly demand curve is similar to that shown in Figure 3.12. This weekly demand curve is superposed on the "average" annual demand curve shown in Figure 3.13. This latter curve is assumed to be relatively flat if most of the energy is for commercial and industrial use and only a small part for space heating.

Assuming that the price of energy in the form of hydrogen gas is $\$1/10^6$ Btu ($1 per million Btu) which is equivalent to $\$1/10^9$ J and that no suitable underground gas storage such as aquifers or depleted gas fields is available nearby, it is apparent that two types of peak-shaving must be provided: one, to take care of the rather wide variation in daily loads and the other to handle the seasonal swing. It will be noted that the oversizing of the pipeline can almost handle the seasonal variation.

FIGURE 3.12: HOURLY HYDROGEN ENERGY DEMAND FOR A TYPICAL FALL OR SPRING WEEK

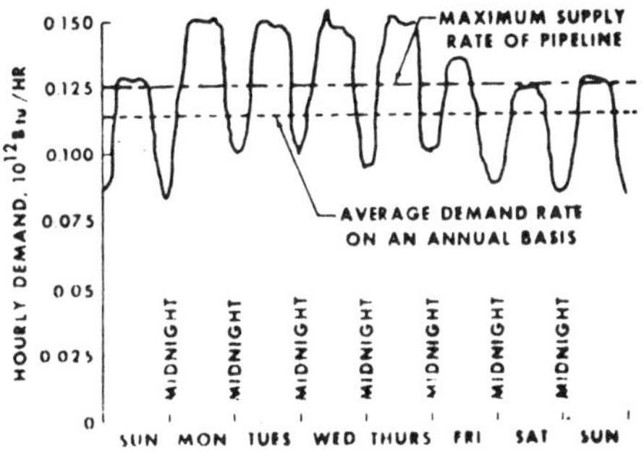

FIGURE 3.13: AVERAGE DAILY HYDROGEN ENERGY DEMAND FOR A TYPICAL YEAR

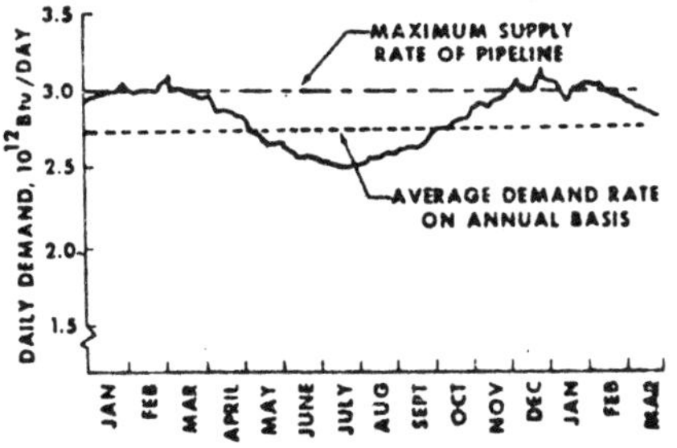

Source: N74-11728-735

This is not the ideal situation and would not be used if an inexpensive means of storing gas for long periods were available. As will be seen in the calculations, however, liquefaction of hydrogen and subsequent storage of the liquid is not inexpensive. A third consideration should probably be included: the possibility of pipeline failure. Repairs in such a siuation could probably be accomplished in a few days. If a three or four-day supply of hydrogen were stored on a more or less permanent basis it would provide an emergency reserve for the entire system.

It should be possible to combine the emergency reserve with the peaks on the annual curve in Figure 3.13. If these peaks are not too far above the maximum rate of supply and do not last for too long, they actually do not compromise the reserve capacity of the system. The total area above the maximum supply line in a typical winter is less than one total daily supply. Thus a further assumption can be made. A nominal four-day supply of hydrogen will be kept in liquid storage for safety and handling the peaks on the annual curve due to extreme weather. This supply will be replenished during low energy demand periods.

It is also apparent that daily demands which are above the maximum supply rate cannot be supplied unless additional storage facilities are provided. The actual quantity which must be stored is not large, but could be equivalent to a three or four-day supply during the winter.

However, the liquefaction capacity must be greatly increased to account for the normal, short-term, cycle. From Figure 3.12 it is apparent that during the fall or spring when the average demand is near the maximum supply, the peak demands are significantly greater. Thus, a rather sizable liquefaction capacity must be constructed but it need be operated only on an intermittent basis.

This will add significantly to the total peak-shaving cost; unfortunately there appears to be no logical solution to this dilemma. Fortunately, very little added storage capacity is required since it is filled and emptied on nearly a daily basis. One final assumption will consequently be made: a one-day supply of liquid hydrogen must be stored for daily variations but sufficient liquefaction capacity must be maintained to reliquefy this amount during any seven-day period when gas is available.

Reviewing the assumptions, it appears that the storage requirements are about eight average days, which is equivalent to 2.2×10^{13} Btu (2.3×10^{16} J). The liquefication capacity is rather large and is mainly controlled by the last assumption. If this amount of capacity is installed then by proper scheduling of operation it will be able to fill all storage in an eight-week period during the summer. This means that the capacity of the liquefier should be about 3.9×10^{11} Btu/day (1.14×10^{11} J/s), which is nearly 15% of the average daily demand.

Admittedly, an eight-day supply in addition to the 9.5% over-supply capacity is rather large for the ideal assumptions made, but the attempt here has been to be conservative. If the situation should improve to where reliability of transmission is assured, then storage capacity could be reduced. An estimate of the costs for this size liquefaction and storage capacity may be obtained from a study made by N.C. Hallet (69).

He indicates that storage in large, vacuum perlite insulated tanks can probably be made to hold ten to twelve million gallons each. In this size the losses due to boiloff will be about 0.015% per day and will be reliquefied. Thus, characteristics of the storage system are:

Total liquid hydrogen storage capacity = 427×10^{13} Btu (2.3×10^{16} J)
Mass of liquid hydrogen stored = 427×10^{6} lb (939×10^{6} kg)
Capital cost of storage tanks = \$$427 \times 10^{6}$
Boiloff = 64,000 lb/day (140,800 kg/day)

The liquefier capacity must be 7.6×10^{6} lb/day (1.67×10^{7} kg/day) to satisfy the previous assumptions. Fuel for this plant capacity will cost over \$114,000 per day of operation if hydrogen at \$1/10^6 Btu (\$1/10^9 J) is used. However, assuming 20 weeks of operation, mostly during the summer, the annual fuel cost will be about \$16,000,000. (Costs are quoted in 1968 dollars.)

Again from Hallet, the plant capacity could be built for a capital investment of \$580,000,000 and operating costs in addition to the fuel costs would be about \$25,000,000 (69). Assuming that operating costs of the tank farm are negligible the final totals are as follows:

Capital cost, liquefier	\$ 580,000,000
Capital cost, storage	427,000,000
Total Capital	\$1,007,000,000
Liquefier fuel cost	\$ 16,000,000
Plant operating costs	25,000,000
Annual variable costs	\$ 41,000,000

Assuming 12% investment charge and 20-year life, the equivalent annual cost is \$176,000,000 of which \$135,000,000 is investment charge.

While this seems at first glance like an excessive cost, it must be compared to the value of the fuel handled to bring it into perspective. Remember that it was assumed that the system handled 10^{15} Btu/year (10^{18} J/year). At a wholesale cost of \$1/10^6 Btu (\$1/10^9 J) this fuel has a value of one billion dollars. Thus the cost has been increased by only 18% in order to improve system reliability and handle a variable load demand.

This size facility would require 60 of the largest proposed tanks, (12,000,000 gallons [45,420 m^3] each), and the liquefaction capacity would probably be split among two or more plants.

SMALL TOWN USAGE OF LIQUID HYDROGEN

The small town which is not located along the route of a gas line represents an area where liquid hydrogen might play an important role in supplying local energy needs. The economics of this situation are entirely different from those of the peak shaving plant described in the previous section.

A hypothetical example will be evaluated which may illustrate the basic considerations. Future energy predictions indicate that energy use may increase to about 900,000 Btu (10^9 J) per capita per day. A small town of about 5,000 population, some regional commerce and a little light industry might consume 4×10^9 Btu/day (4.2×10^{12} J/day). It will be assumed as a calculation base that about half of this energy can be supplied by hydrogen. In order to make representative calculations the following arbitrary assumptions will be made.

The community is located 250 miles (4×10^5 m) from an industrial center which has liquefaction plants and liquid storage for peak shaving.

Seasonal variations may be ignored since the rate of supply from the industrial center may be adjusted.

A one-week supply will be kept on hand to offset equipment failure or other difficulties in delivery.

The hydrogen will be delivered as a liquid, stored as a liquid and gasified for use.

Storage capacity at the town would be about 14 x 10^9 Btu (14 x 10^{12} J). This is about 470,000 gallons (1,780 m^3) or 272,000 lb (124,000 kg) of hydrogen. A storage tank of this capacity is well within present capability. Lower cost insulation can be used since continual gasification of the supply liquid will be required.

In fact, a boiloff rate of several percent per day would be acceptable. With such a high boiloff rate a relatively inexpensive tank using internal foam plastic insulation with a hydrogen barrier liner would probably be suitable. The cost of large tanks of this type would probably be less than half that of tanks discussed in the previous section.

In the size needed here, however, the cost of the simpler tank will still probably be about $1/lb ($0.45/kg) of hydrogen capacity. Extrapolating from calculations made by Hallet (69) the following numbers are representative.

Storage capacity:	272,000 lb (124,000 kg)
Capital cost:	$300,000
Annual operating cost:	$10,000/year

The boiloff of 5% per day equals 7 x 10^8 Btu/day (7 x 10^{11} J/day). In addition, to supply the average daily demand twice this heat transfer will be needed to vaporize the liquid hydrogen.

The cost of transporting the liquid hydrogen from the liquefaction plant to the town is more difficult to determine. Both truck and rail tankers will be considered in an effort to cover the possibilities. Some improvement in present systems will be assumed but these will be relatively minor.

When transporting liquid hydrogen by tanker, size rather than weight, is the limiting factor. This is true because of the low density of liquid hydrogen. Regulations controlling tank truck size set the maximum net cargo at about 13,000 gallons (49 m^3). Some improvement may be possible by redesign of the tankers but unless site limits are changed it appears that 15,000 gal (57 m^3) is a reasonable upper limit. With this size tanker an average of about 31 deliveries per week are needed. If the shape of the seasonal demand curve is similar to that shown in Figure 3.13, the peak delivery rate would be about 35 deliveries per week, and the minimum about 28.

For a round trip of 500 miles (8 x 10^5 m) the actual time on the road will probably be 14 to 16 hours. If multiple crews are used and loading and unloading are handled by other than the drivers, it is conceivable that each tanker could complete one trip every 24 hours. However, maintenance, trip scheduling and questions of reliability dictate that the number of tankers should be greater than five. If several such systems are served by one center and if the tankers are interchangeable then 20% over capacity may provide an adequate safety margin.

This means that the system being considered would be charged with six tankers, each of 15,000-gallon (57 m^3) capacity. A reasonable estimate of the cost of these tankers is $100,000 per unit. Operating costs of 50¢ per mile (0.03¢/m) would probably be typical and a crew of four men would be the minimum which could keep a truck operating. The system costs for truck transportation would therefore be the following:

Capital investment	$600,000
Operating cost	$450,000/year
Crew for 6 trucks	$600,000/year

If the trucks are amortized on a ten-year basis at 12% the equivalent annual cost for the transportation system is $1,156,000. This is a relatively large cost and adds $1.58/10^6 Btu ($1.50/10^9 J) to the cost of hydrogen fuel.

An alternative method of supply is by rail tanker rather than highway tanker. The capital costs increase since more tank car capacity would be required in order to properly schedule shipment and return, but operating costs should be reduced. If rail tank cars of 25,000 gallons (95 m^3) capacity can be built, and this seems reasonable, then only 19 cars per week, on the average, will be required and 21 per week during peak demand.

If two shipments per week can be made then 12 tank cars should be adequate allowing as before for an emergency reserve. It is estimated that these cars will have a unit cost of $150,000 for a total capital cost of $1,800,000. Operating costs will be lower and in general will be a part of the freight rates.

In trying to arrive at a reasonable hauling cost for future shipments of liquid hydrogen the estimates have been based on the assumption that such shipments will be routine and will be charged at relatively low bulk rates. One cent per ton mile (0.08¢/kg-m) is probably in the proper range provided that the gross car weight is used for the round trip.

A tank car holding 25,000 gallons (95 m^3) of liquid hydrogen will probably weigh 80 to 100 tons (3.52×10^5 to 4.4×10^5 kg) empty while the hydrogen weighs only 725 tons (3.19×10^4 kg). Thus a round trip of 500 miles (8×10^5 m) at one cent per ton mile (0.08¢/kg-m) would cost about $500 per tank car load.

One thousand tanks per year will thus result in an annual transportation cost of $500,000. If the capital costs are amortized on a ten-year basis at 12%, the equivalent annual transportation cost for rail shipment will be about $820,000. Although this is much less than truck transport it requires that the town be located on a rail line.

In order to complete the economic analysis, the liquefaction costs as well as the cost of the gas liquefied must be included. In the previous section a plant used for peak shaving was considered. If the capacity is increased slightly, such a plant used for peak shaving could supply liquid for a small town at the incremental cost.

The equivalent annual cost for only the liquefier in the previous section was about $119,000,000 for an assumed 20 weeks of operation during which hydrogen with a heating value of 5.5×10^{13} Btu (5.8×10^{16} J) was liquefied. The cost of liquefying can then be taken as $\$2.16/10^6$ Btu ($\$2.04/10^9$ J) and the cost of the gas liquefied is assumed to be $\$1/10^6$ Btu ($\$1/10^9$ J). If the liquefaction plant charges $\$3.50/10^6$ Btu ($\$3.32/10^9$ J) for liquid hydrogen it should recover the cost of the gas and the liquefaction process, and receive a fair handling charge in return.

The cost of obtaining energy in the small town in the form of hydrogen is thus comparatively high. The cost breakdown is as follows:

Cost of liquid hydrogen	$\$3.50/10^6$ Btu ($\$3.32/10^9$ J)
Cost of rail transport	$\$1.12/10^6$ Btu ($\$1.06/10^9$ J)
Cost of terminal	$\$0.19/10^6$ Btu ($\$0.18/10^9$ J)
Total	$\$4.81/10^6$ Btu ($\$4.56/10^9$ J)

Cost of liquid hydrogen	$\$3.50/10^6$ Btu ($\$3.32/10^9$ J)
Cost of highway transport	$\$1.58/10^6$ Btu ($\$1.50/10^9$ J)
Cost of terminal	$\$0.19/10^6$ Btu ($\$0.18/10^9$ J)
Total	$\$5.27/10^6$ Btu ($\$5.00/10^9$ J)

To evaluate these numbers they must be compared with the costs of other forms of energy or other means of hydrogen delivery.

While it is difficult to find any data on the cost of building small pipelines it is not difficult to see that the cost would have to be very high to equal the system cost of the liquid hydrogen transport case. For example, assuming construction costs of $25,000 per mile ($15.5/m) plus one pumping station at $100,000, a capital investment of $6,350,000 would be required (70).

For a twenty-year amortization and 12% interest, the equivalent annual cost is $851,000 which adds $\$1.15/10^6$ Btu ($\$1.09/10^9$ J) for delivery of gas. While no system operating costs have been added it is apparent that the cost of energy with the gas system will never exceed the costs using the liquid delivery.

It must be emphasized that the cost of liquefaction makes the system economical, not the cost of transporting the liquid. Thus, the use of low-cost transportation schemes—barges, airships, supertanker planes, or dirigibles—will not make the system economically competitive with the gas pipeline; the cost of liquefaction must be drastically reduced for the transportation costs to have much influence.

COMPARISON OF ENERGY TRANSMISSION COSTS FOR A NEAR-URBAN ENVIRONMENT

The costs of transporting various forms of energy in a near-urban environment are summarized in Table 3.12. An average cross-country cost for natural gas transmission today is reported to be 1.5 to $1.7¢/10^6$ Btu per 100 miles (71).

Hydrogen transmission cost would be 2.6 times this cost since compressor size and horsepower must be considerably increased and a more expensive fuel (hydrogen) used to power the compressors. By increasing the pipeline operating pressure from the usual 750 psia to 2,000 psia, this factor could be reduced to about 1.5.

TABLE 3.12: ENERGY TRANSMISSION—RELATIVE TRANSPORTATION COSTS (NEAR-URBAN ENVIRONMENT)

	Volume (10^6 Btu/hr)	Size	Investment per 100 Miles (10^6 \$)	Total Cost, Operating Plus Investment[a] per 100 Miles (¢/10^6 Btu)
H_2 gas	21,000	36-inch pipe	68[b,c]	5.7[c]
NH_3	21,000	36-inch pipe	62[b,c]	5.0[c]
CH_3OH	21,000	30-inch pipe	52[b,c]	4.2[c]
Natural gas	21,000	36-inch pipe	62[b,c]	5.1[c]
H_2 liquid	550	3-inch pipe	79[d]	250.0
H_2 liquid	1,800	1,100,000-gal barge	2.4	6.9
H_2 liquid	62	13,000-gal trailer	0.15	34.0
Electric (superconducting line)	13,600	4,000-MVA line	140	19.3
Electric (above ground)	8,500	2,500 MW, 750 kv	30	8.3
Electric (below ground)	8,500	2,500 MW, 345 kv	~300	~100

[a]Unit cost assumption: 100% utilization of facility, 15% = fixed charge rate (depreciation, interest, return on investment), operating power 6 mills/kwh.

[b]Reported pipeline costs can vary by as much as a factor of 4, depending on terrain, congestion, etc.

[c]For cross-country straight lines, 50% reduction is likely.

[d]No reliquefaction included.

Source: CONF-730807-3

The relative cost comparisons approximate the largest energy transmission systems considered practicable for each fuel system in a near-urban environment. Since synthetic fuels can be generated relatively near the use point, a greater degree of congestion and resultant higher unit cost per distance increment will be encountered over that which is generally experienced in contemporary transcontinental systems.

However, since average energy delivery distances will be shortened with synthetic fuels systems, overall delivery costs of energy will be reduced. The results of this type of analysis, although not directly comparable, show that many methods of energy transfer may be competitive, subject to the above considerations. Therefore, as a result of any detailed systems analysis, it is unlikely that the transportation of energy alone will be decisive in the selection of an optimum energy distribution system or medium.

A major advantage of pipelining fuels is the ability to scale up this type of system to very large capacities, with the resultant effect of lowering the unit-energy transfer cost as an economy of scale. The use of pipeline systems for energy transport will therefore be superior for transmission over long distances and/or transmission of very large blocks of energy.

The relative flow characteristics of the various synthetic fuels yield designs that are very similar. Although the volumetric flow of a fuel such as hydrogen is quite high because of its low specific heat content, it is compensated for by the low molecular weight of the gas. As a result, hydrogen system performance is similar to that of the other hydrogen-bearing fuels.

The only penalty in transmitting hydrogen, as compared with methane, is the increased cost of recompression which amounts to a minor fraction of the transmission system cost in a near-urban environment. This penalty should be more than offset by the potential efficiency gain of hydrogen energy-conversion devices and the resultant flow reduction in the hydrogen energy transmission system.

Pipeline transport of liquid hydrogen is by far too costly to be a competitive method of energy transport except over very short distances, or for transfer of energy blocks very much larger than are common in existing practice. However, where it might be practicable to transmit electricity via cryogenic or superconductive cable, simultaneous transmission of liquid hydrogen through the cryogenic power conduit could alter this conclusion.

Although some generalizations are apparent from the above considerations, it is not possible to clearly identify a superior mode of energy transport. This can only be determined by a detailed systems and engineering analysis, giving due consideration to the synergistic benefits which may be realized by combining several forms of energy transport. All transportation schemes have been practiced on scales large enough to constitute prototype demonstration of feasibility, with the exception of large-volume pipelining of high-purity hydrogen gas.

REFERENCES

(1) Barron, R., *Cryogenic Systems,* McGraw-Hill Book Co., New York, N.Y., 1966.

(2) Commander, J.C. and Ratter, L., "An Economic Analysis of Perlite Versus Superinsulation in Liquid Hydrogen Storage Vessels", NASA CR-54720, October, 1965.

(3) Strobridge, T.R., "Cryogenic Refrigerators—An Updated Survey", Nat. Bur. of Stand. (U.S.) Tech. Note 655, June 1974.

(4) Hord, J. (ed.), "Selected Topics on Hydrogen Fuel", Nat. Bureau of Standards, (U.S.), NBSIR 75-803, January 1975.

(5) Bainbridge, R. and Horton, T.R., "The Production of Liquid Hydrogen at the Rocket Propulsion Establishment", Technical Report No. 71/17, Rocket Propulsion Establishment, Wescott, U.K., Dec. 1971.

(6) Timmerhaus, K.D., (ed.), *Advances in Cryogenic Engineering*, Vols. 1-18, Plenum Press, New York, 1960-1973.

(7) Vander Arend, P.C., "Large-Scale Production, Handling, and Storage of Liquid Hydrogen", Ref. (6) above, Vol. 5 (1960), p. 49.

(8) Bain, A.L. and Boggs, W.H., "NASA Space Program Experience in Hydrogen Transportation and Handling", Paper presented at the Hydrogen Energy Fundamentals Symposium, University of Miami, Coral Gables, Fla., March 3-5, 1975.

(9) Laquer, H.L., "Handling Liquid Hydrogen on a Pilot-Plant Scale", Ref. (6) above, Vol. 5 (1960), p. 85.

(10) Liebermann, P., "E.R.E.T.S. LOX Losses and Preventative Measures", Ref. (6) above, Vol. 2 (1960), p. 225.

(11) Perkins, W.E. and Frainier, R.J., "Practical Storage and Distribution of Liquid Hydrogen and Helium", Ref. (6) above, Vol. 5 (1960), p. 69.

(12) Ishaghoff, I. and J.M. Canty, "Quilted Insulation", Ref. (6) above, Vol. 9 (1964), p. 45.

(13) Stoy, S.T., "Cryogenic Insulation Development", Ref. (6) above, Vol. 5, (1960), p. 216.

(14) Arnett, R.W., Mullen, L.O. and Warren, K.A. "Some Methods for Reducing Heat Leak Through Support Members in Liquefied Gas Storage Vessels", Ref. (6) above, Vol. 4 (1960), p. 410.

(15) Kropschot, R.H., et al., "Multilayer Insulation", Ref. (6) above, Vol. 5 (1960), p. 189.

(16) Bartlit, J.R., "Liquid Hydrogen Handling, Transport, and Storage", *Proceedings of the Cornell International Symposium and Workshop on the Hydrogen Economy*, (1975), p. 95.

(17) Romero, J.B., Smith, D.W. and Dod, R.E., "Thermal Analysis and Optimization of Cryogenic Tanks for Lunar Storage", Ref. (6) above, Vol. 11 (1966), p. 231.

(18) Liebenberg, D.H., Stokes, R.W. and Edeskuty, F.J., "Chilldown and Storage Losses of Large Liquid Hydrogen Storage Dewars", Ref. (6) above, Vol. 11 (1960), p. 554.

(19) Kreith, F. and Dean, J.W., "Cooldown and Warmup of Large Powder-Insulated Dewars", Ref. (6) above, Vol. 8 (1968), p. 536.

(20) Woerner, R.C., "A Study of Liquid Cylinder Filling Techniques", Ref. (6) above, Vol. 13 (1968), p. 237.

(21) Jacobs, R.B., Richards, R.J. and Schwartz, S.B., "The Transfer of Liquefied Gases", Ref. (6) above, Vol. 1 (1960), p. 87.

(22) Scott, R.B., *Cryogenic Engineering,* D. Van Nostrand Company, Inc., Princeton, N.J., 1959, p. 253.

(23) Richards, R.J., Steward, W.G. and Jacobs, R.B., "Transfer of Liquid Hydrogen Through Uninsulated Lines", Ref. (6) above, Vol. 5, (1960), p. 103.

(24) Martinelli, R.C. and Nelson, D.B., "Prediction of Pressure Drop During Forced Circulation Boiling of Water", *Trans. ASME*, Vol. 70 (1948), p. 695.

(25) Haettinger, G.C., "Considerations in the Design, Selection and Use of Vacuum Insulated Pipe", Ref. (6) above, Vol. 11 (1966), p. 98.

(26) Flieder, W.G., Smith, W.J. and Wetmore, K.R., "Flexibility Considerations for the Design of Cryogenic Transfer Lines", Ref. (6) above, Vol 5 (1960), p. 111.

(27) Randall, R. and Sullivan, A, "A New Vacuum-Insulated Cryogenic Coupling", Ref. (6) above, Vol. 10 (1965), p. 451.

(28) Moore, R.W., Fowle, A.A., Bailey, B.M., Ruccia, F.E. and Reid, R.C., "Gas-Pressurized Transfer of Liquid Hydrogen", Ref. (6) above, Vol. 5 (1960), p. 450.

(29) Burke, J.C., Byrnes, W.R., Post, A.H. and Ruccia, F.E., "Pressurized Cooldown of Cryogenic Transfer Lines", Ref. (6) above, Vol. 4 (1960), p. 378.

(30) Drake, E.M., Ruccia, F.E. and Ruder, J.M., "Pressurized Cooldown of a Cryogenic Liquid Transfer System Containing Vertical Sections", Ref. (6) above, Vol. 6 (1961), p. 323.
(31) Bronson, J.C. et al, "Problems in Cooldown of Cryogenic Systems", Ref. (6) above, Vol. 7 (1962), p. 198.
(32) Jacobs, R.B., "Liquid Requirements for the Cooldown of Cryogenic Equipment", Ref. (6) above, Vol. 8 (1963), p. 529.
(33) Steward, W.G., "Transfer Line Surge", Ref. (6) above, Vol. 10 (1965), p. 313.
(34) Chi, J.W.H., "Cooldown Temperatures and Cooldown Time During Mist Flow", Ref. (6) above, Vol. 10 (1965), p. 330.
(35) Steward, W.G. , Smith, R.V. and Brennan, J.A., "Cooldown Transients in Cryogenic Transfer Lines", Ref. (6) above, Vol. 15 (1970), p. 354.
(36) Novak, J.K., "Cooldown Flow Rate Limits Imposed by Thermal Stresses in Liquid Hydrogen or Nitrogen Pipelines", Ref. (6) above, Vol. 15 (1970), p. 346.
(37) Cowley, S.C., Timson, W.J. and Sawdye, J.A., "A Method for Improving Heat Transfer to a Cryogenic Fluid", Ref. (6) above, Vol. 7 (1961), p. 385.
(38) Maddox, J.P. and Frederking, T.H.K., "Cooldown of Insulated Metal Tubes to Cryogenic Temperatures", Ref. (6) above, Vol. 11 (1966), p. 536.
(39) Allen, L.D., "A Method of Increasing Heat Transfer to Space Chamber Cryopanels", Ref. (6) above, Vol. 11 (1966), p. 547.
(40) Leonhard, K.E., Getty, R.C., and Franks, E.E., "A Comparison of Cooldown Time Between Internally Coated and Uncoated Propellant Lines", Ref. (6) above, Vol. 12 (1967), p. 331.
(41) Edeskuty, F.J. and Williamson, Jr., K.D., "Storage and Handling of Cryogens", Ref. (6) above, Vol. 17 (1972), p. 331.
(42) Andonian, M.D., "A Liquid Hydrogen Dewar to Supply Gas to Balloons in Flight", Ref. (6) above, Vol. 2 (1960), p. 83.
(43) Johnson, J.E., "Storage and Transportation of Synthetic Fuels", A Report to the Synthetic Fuels Panel, ORNL-TM-4307, Sept. 1972.
(44) "Liquid Hydrogen Systems 1973", Standard for Liquefied Hydrogen Systems at Consumer Sites, NFPA No. 50B-1973.
(45) Martin, F.A., "The Safe Distribution and Handling of Hydrogen for Commercial Application", *7th Intersociety Energy Conversion Engineering Conference*, San Diego, Cal. Oct. 1972, p. 1335.
(46) Ordin, P.M., "Review of Hydrogen Accidents and Incidents in NASA Operations", *9th Intersociety Energy Conversion Engineering Conference*, San Francisco, Cal., Aug. 1974, p. 442.
(47) Cassutt, L.H., Maddocks, F.E. and Sawyer, W.A., "A Study of the Hazards in the Storage and Handling of Liquid Hydrogen", Ref. (6) above, Vol. 5 (1960), p. 55.
(48) Zabetakis, M.G., "Hazards in the Handling of Cryogenic Fluids", Ref. (6) above, Vol. 8 (1963), p. 236.
(49) Zabetakis, M.G., Furno, A.L. and Martindill, G.H., "Explosion Hazards of Liquid Hydrogen", Ref. (6) above, Vol. 6 (1961), p. 185.
(50) Connolly, W.W., "A Practical Safety Standard for Commercial Handling of Liquefied Hydrogen", Ref. (6) above, Vol. 12 (1967), p. 192.
(51) Von Elbe, G. and Scott, T.H., Jr., "Hazards of Liquid Hydrogen in Research and Development Facilities", Atlantic Research Corp. Technical Report, ASD-TDR-62-1027, December 1962.
(52) Moore, C.S. et al, *NASA–Lewis Hydrogen Safety Manual,* National Aeronautics and Space Administration, December 10, 1959.
(53) Reider, R., Otway, H. and Knight, H.T., "An Unconfined Large Volume Hydrogen-Air Explosion", Paper presented at the National Meeting of American Institute of Chemical Engineers, May 1964.
(54) Moore, H.G., et al, "Space Shuttle Orbiter/Booster Airbreathing Engine Study", NASA CR-120861, December 1971.
(55) Pastuhor, A. and Ruccia, F., "Total Energy Transfer System for LNG and Electrical Power", *Cryogenics and Industrial Gases,* Vol. 4, n. 3, p. 14-18, March 1969.

(56) Belanger, B.C., and Jefferies, M.J., "Cryogenic Power Cable: Design and Test", *Cryogenics and Industrial Gases,* Vol. 7, n. 2, p. 15-22, March-April 1972.
(57) "Super Conductive Underground Electrical Transmission: One Step Closer", *Cryogenics and Industrial Gases,* Vol. 4, n. 7, p. 22-24, July, 1969.
(58) "Lab Spurs Cable R & D", *Cryogenics and Industrial Gases,* Vol. 8, n. 1, p. 29, January-February 1973.
(59) Minnic, S. and Fox, G.R., "Cryogenic Power Transmission", *Cryogenics,* Vol. 9, n. 3, p. 1965, June, 1969.
(60) Sindt, C., "A Summary of the Characterization Study of Slush Hydrogen", *Cryogenics,* Vol. 10, n. 4, p. 372, October 1970.
(61) Los Alamos Scientific Laboratory Report LASL P-94, Los Alamos, New Mexico, 1972.
(62) Cryogenic Materials Data Handbook, Air Force Materials Lab Report AFML-TDR-64-280, Wright Patterson Air Force Base, Ohio 1970.
(63) Nachtigall, A.J., NASA report TN D-7532, Lewis Research Center, Cleveland, Ohio, 1974.
(64) Keeton, S.C., Sandia Laboratories Report SAND 74-8217, Livermore, California, 1974.
(65) Pati, S.R.,and M. Cohen, *Acta Met.,* 1971, Vol. 19, p. 1327-32.
(66) Zackay, V.F., *Microstructure and Design of Alloys,* Institute of Metals, London, 1973, p. 591.
(67) Montano, J.W., NASA Report TM X-64791, Marshall Space Flight Center, Huntsville, Alabama, 1973.
(68) Los Alamos Scientific Laboratory Report LA-5472-PR, Los Alamos, New Mexico, 1973.
(69) Hallet, N.C., "Study, Cost and System Analysis of Liquid Hydrogen", NASA CR 73-226, June 1968.
(70) "Pipeline Economics", *The Oil and Gas Journal,* Vol. 69, n. 31, August 2, 1971.
(71) Gregory, D.P. and J. Wurm, "Production and Distribution of Hydrogen as a Universal Fuel," p. 1329-34 in *Proceedings of 1972 Intersociety Energy Conversion Engineering Conference*, San Diego, California, September 25-29, 1972.

SOLID HYDRIDES

The following reports were used as the sources of material for this chapter: N74-11728-735, AD-A016 256, BNL-18887, SAND-74-8219, BNL-18720, BNL-20162, BNL-50447 and BNL-18719. A complete bibliography starts on page 283.

MECHANISM OF HYDRIDE STORAGE

Metal hydrides have been proposed by two groups of investigators as a method of storing hydrogen: at the Brookhaven National Laboratories (1) and another at the Phillips Laboratory in the Netherlands (2).

Hydrides are metal-hydrogen compounds that have certain attributes that make them advantageous for storage of hydrogen. Unfortunately, there are also deficiencies that make hydrides unsuitable for certain types of hydrogen storage systems.

The hydride storage systems operate in the following manner. Hydrogen under pressure is pumped into a container filled with a suitable metal or alloy which absorbs hydrogen and ultimately forms a metal hydride compound. Since all of the useful metal-hydrogen compounds for storage are exothermic in nature, the formation of the metal hydride is accompanied by the release of the heat of formation. This heat must be removed from the system or the hydride forming reaction will stop and hydrogen will cease to be stored, an important point to remember when hydrides are proposed for practical applications, such as automotive fuel tanks. Each of the metal hydride systems will absorb hydrogen up to an equilibrium pressure called the dissociation pressure. The dissociation pressure depends mainly on temperature and the nature of the metal (3).

When the hydride is reheated, such as with waste heat from an engine, the hydrogen gas is evolved and builds to the equilibrium pressure inside the container. If gas is removed from the system the pressure is lowered and more hydride decomposes.

For greatest convenience of operation of a hydride storage stystem, one desires these dissociation pressures and temperatures close to atmospheric conditions. The typical behavior of metallic hydrides is most conveniently displayed by means of pressure-composition isotherms (Figure 4.1).

In this figure is shown the dissociation pressure as a function of the hydrogen-to-metal atom ratio for various temperatures. Beyond the curve labelled A', the true hydride appears, the reaction occurs at constant pressure, and the heat of formation, ΔH_f, must be either supplied or taken away, depending on which way the reaction is proceeding. It should be noted that it is desirable to have a low heat of formation in order to minimize the energy requirements to release hydrogen during the dehydriding part of the cycle.

FIGURE 4.1: PRESSURE-COMPOSITION ISOTHERMS

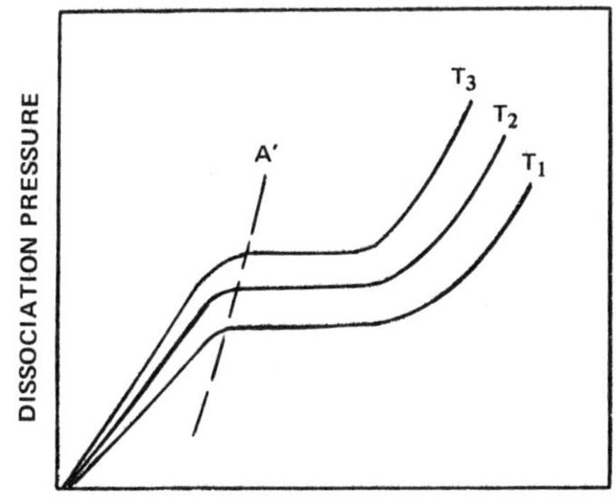

Source: AD-A016 256

Hydride heats of formation, ΔH_f, can be determined either by calorimetry or by determining the temperature-dependence of decomposition pressure. Table 4.1 gives values of ΔH_f for a number of hydrides. Hydrides which have a high decomposition pressure at low temperatures generally have a relatively small value of ΔH_f.

TABLE 4.1: HYDRIDE HEATS OF FORMATION

Initial and Final Compositions*	ΔH_f, kcal/mol H_2
$FeTiH_{0.1}$–$FeTiH_{1.95}$	~-7.5
$VH_{0.95}$–$VH_{2.0}$	-9.6
$Mg_2NiH_{0.3}$–$Mg_2NiH_{4.2}$	-15.4
Mg–MgH_2	-17.8
K–KH	-27.2
U–UH_3	-30.4
Zr–ZrH_2	-39.7

(continued)

TABLE 4.1: (continued)

Initial and Final Compositions*	ΔH_f, kcal/mol H_2
$Ca-CaH_2$	-41.7
$Li-LiH$	-43.3
$Ce-CeH_2$	-49.2

*When a metal or alloy is indicated as the initial phase, it is actually a hydrogen-saturated metal phase.

Source: BNL-18887

Figure 4.2 displays the range of dissociation pressures as a function of temperature for commonly utilized hydrides.

FIGURE 4.2: DISSOCIATION PRESSURES OF METAL HYDRIDES

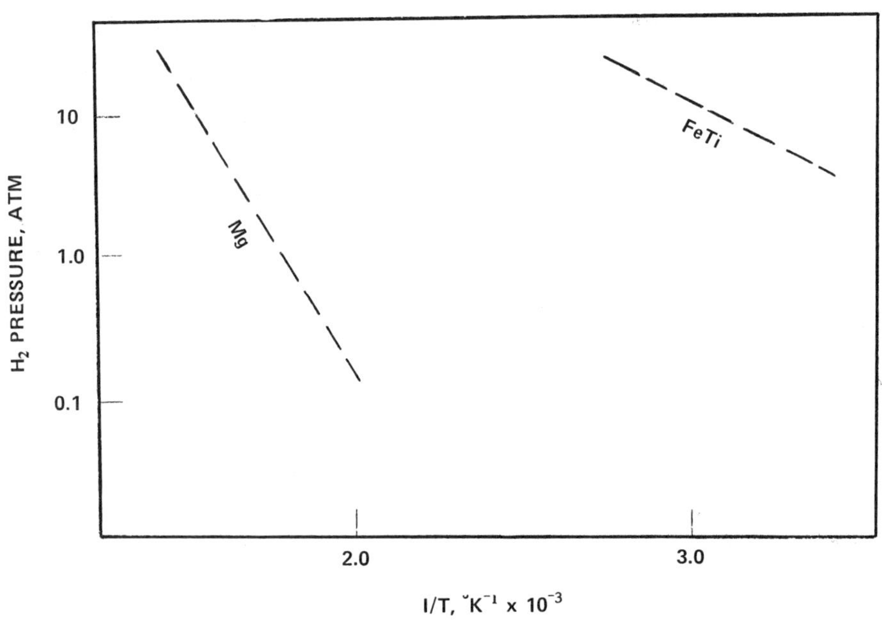

Source: AD-A016 256

Aside from considerations of heat of formation and pressures and temperatures involved in hydriding and dehydriding, it is further desirable to have a low weight of the storage system, especially for a transportation application. Similarly, one would like high density of storage so that a given weight of metal or alloy absorbs as much hydrogen as possible. The intermetallic compound, iron titanium hydride ($FeTiH_{1.5}$) satisfies the criteria of reasonably low heat of formation, and dissociation pressures and temperatures close to atmospheric conditions. But the use of this hydride requires almost 900 pounds of material to store a million Btu's in the form of hydrogen (approximately 14 pounds of hydrogen). These figures characterize the storage system designed by Brookhaven National Laboratories for Public Service Electric and Gas Company. Magnesium nickel hydride Mg_2NiH_4 is capable of storing an equivalent amount of hydrogen in half the weight as can be seen in the table below.

TABLE 4.2: COMPARISON (WT) OF $FeTiH_{1.5}$ AND Mg_2NiH_4

Hydride	Density	Available H_2
$FeTiH_{1.5}$	375 lb/ft^3	5.1 lb/ft^3
Mg_2NiH_4	160 lb/ft^3	5.3 lb/ft^3

Source: AD–A016 256

However, referring back to Figure 4.2 will show that a magnesium hydride system will operate at temperatures much higher than the iron titanium hydride system; a disadvantage. So at present, these lighter hydrides suffer the drawback of higher heat of formation (Table 4.1) and higher operating temperatures and pressures.

The most proven system to date is the Public Service Electric and Gas hydride-fuel-cell system with electrolytically produced hydrogen. This system operates at 500 psig with hydriding cooling water temperature of 59°F and dehydriding heating water temperature of 131°F. This is a stationary application. Transferred into a vehicular application, this would amount to 65 watt hours per pound of hydride which would give a range of 70 miles with a 1,000 pound tank at 30% energy conversion efficiency (4). Figure 4.3 is the pressure-composition isotherm for iron titanium hydride as used in the Public Service Electric and Gas system.

FIGURE 4.3: PRESSURE-COMPOSITION ISOTHERM FOR $FeTiH_{1.5}$ SYSTEM

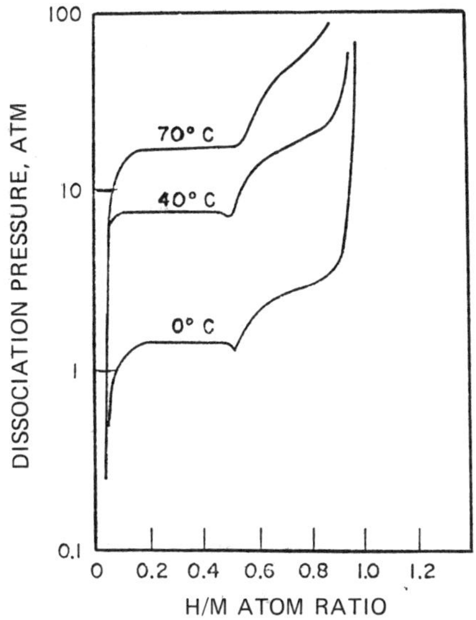

Source: AD-A016 256

TECHNICAL PROBLEMS OF HYDRIDE STORAGE

The technical problems which have been uncovered in connection with the investigation of hydride storage to date lie in the areas of heat transfer, deterioration of the hydride bed material, safety, embrittlement of materials used for the containers, low energy density of storage resulting in weight penalties, and the need to develop reliable hydrogen compressors. None of these appear to be serious problems, although in the area of deterioration of the hydride material, there is a lack of information needed to make an accurate determination of the extent of the problem. In the area of low energy density of storage, this appears to be limiting the application of hydride storage to stationary uses at present.

Heat Transfer

The operating temperatures and pressures of the hydride storage bed during the processes of hydriding and dehydriding as discussed in the previous section are related by the dependence of the so-called plateau pressure on temperature as shown in Figure 4.4 for iron-titanium.

FIGURE 4.4: PLATEAU PRESSURE-TEMPERATURE DEPENDENCE FOR IRON-TITANIUM HYDRIDE

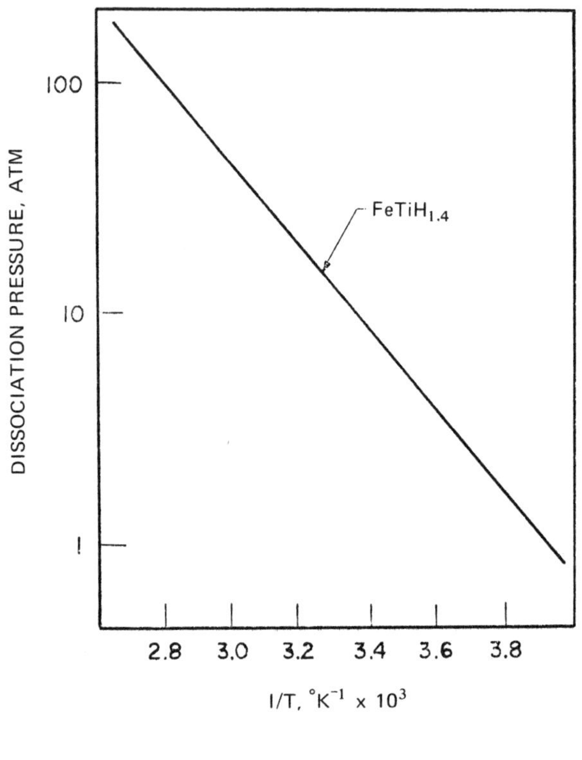

Source: AD-A016 256

The rates of hydrogen release and hydrogen absorption during the dehydriding and hydriding cycles are dependent upon the overall heat-transfer coefficient in the hydride bed and the temperatures and pressures of operation. This area does not appear to contain any serious limiting problems. Recent studies of the application of hydride storage systems to a bus (5) and a metal-hydride compact involving the use of a porous microstructure deal with various phases of the heat-transfer problem with a view towards practical applications of the hydride concept for transportation and improvement of the rate of hydrogen release.

Deterioration

The hydriding and dehydriding cycle with accompanying heat rejection and heat addition appears to cause crumbling of the metallic hydride particles due to the associated volume changes. With time, this tends to reduce the average granular size of the bed and may be a limiting factor in applying new approaches.

It may also contribute to the safety problem to be discussed next. The extent to which this phenomenon occurs is unknown at this time. Accumulated experience with the bed designed and constructed by Brookhaven National Laboratories and being tested at the Public Service Electric and Gas Company, Maplewood Laboratory, will provide additional information relating to this problem.

Safety

Recent information concerning the pyrophoric property of finely divided metallic hydride particles raises a concern that the operation of a hydride bed may become more hazardous as time goes on due to this phenomenon. This may also serve to limit design approaches involving higher operating temperatures once more suitable materials become available in an attempt to increase the energy density of storage in a hydride bed.

Steel Problems

Hydrogen embrittlement of high alloy steels is a common problem in hydrogen technology. The less severe ranges of application of alloys used in hydride work may tend to reduce the severity of the problem in this area. The basic difficulty is the fact that high-strength alloys which designers turn to for fabrication of highly stressed parts are the most apparently susceptible to hydrogen embrittlement. The data shown in Table 4.3 (6) illustrate this point.

TABLE 4.3: RELATIVE DEGREE OF HYDROGEN-ENVIRONMENT EMBRITTLEMENT OF CLASSES OF METALS

Degree of Embrittlement	Materials	Characteristics
Extreme	High strength steels Nickel-base alloys	Large decrease in notch strength and notched and unnotched ductility. Some decrease in unnotched strength. Propagation of surface cracks.

(continued)

TABLE 4.3: (continued)

Degree of Embrittlement	Materials	Characteristics
Severe	Ductile, lower-strength steels Pure nickel Titanium alloys	Considerable reduction of notch strength and unnotched and notched ductility. No reduction of unnotched strength. Propagation of surface cracks.
Slight	Metastable 300 series stainless steels Beryllium-copper Pure titanium	Small decrease in notched strength and unnotched ductility. Failure of unnotched specimens from within.
Negligible	Aluminum alloys Stable austenitic stainless steels Copper	Essentially unembrittled with no surface cracks.

Source: AD-A016 256

A study of hydrogen compatibility problems in hydride storage was done at Sandia Laboratories (SAND-74-8219). A description of the work done by these investigators follows.

The procedure used for temporary storage of hydrogen in hydrides is to react a metal or alloy with gaseous hydrogen at a relatively low temperature, then release the hydrogen when needed by heating the hydride to a higher temperature. The structural materials in the reactor may be exposed to high pressures, high temperatures, and cycling of both temperature and pressure. Thus, there is a possibility of hydrogen damage from all the mechanisms mentioned above. It is necessary to examine details of hardware design before a judgment can be made regarding potential hydrogen damage.

As an example, consider the reservoir built by Brookhaven National Laboratory for using iron titanium hydride ($FeTiH_{1.6}$) as a medium for storing off-peak electrical energy (7). The pressure vessel was made from a section of type 316 stainless steel pipe, 30.5 cm (12 inches) in diameter and 0.64 cm (0.25 inch) wall thickness. The vessel must contain hydrogen at 3.45 MPa (500 psia) pressure at ambient temperature. The highest temperature the vessel reaches in the hydride-dehydride cycle is approximately 60°C. This temperature is not high enough to cause blistering from methane or hydrogen gas in voids, so only the thermal cycling and crack growth mechanisms need be examined.

The rate of uptake of hydrogen in stainless steels is known to be decreased by surface oxide films. However, it is known that lithium hydride in contact with stainless steel at a slightly elevated temperature will reduce the oxide film and produce a relatively clean surface. Free energy change calculations show that $FeTiH_{1.6}$ should do the same, TiO_2 being a product of the reaction. Therefore, if the dissociation rate of H_2 molecules on the surface is assumed to be fast, the hydrogen uptake should be predictable from solubility (S) and diffusion coefficient (D) data.

Using these data, a computer program incorporating a technique called over-relaxation was written to compute hydrogen concentration profiles in the steel as a function of time in the hydriding-dehydriding cycle. An approximation was made that the profiles would be the same as in a flat plate of the same thickness with hydrogen present on one side only. Possible effects of surface oxide films and trapping of hydrogen at defects were ignored.

The results of calculations show that, for repeated 12 hour cycles, only the first 10 μm below the surface changes significantly in hydrogen concentration in response to changes in temperature and hydrogen pressure in the reactor. Even at a depth of 10 μm, there is a significant time lag in responding to temperature and pressure changes because of the low diffusion coefficient in austenitic steels at these temperatures. In the bulk of the material, there is a gradual increase in hydrogen concentration. Calculations show that a nearly infinite length of time is required for the hydrogen concentration to reach a few ppm near the outer surface of the reservoir.

If the same reservoir is made from a low carbon steel with a 1 cm (0.393 inch) wall thickness, the results change drastically because the hydrogen solubility is much lower and the diffusivity is much higher in ferritic steels.

(S) at 33.6°C can be shown to be 4.3×10^{-3} ppm, compared to 7 ppm for stainless steel. (D) at 33.6°C is 8×10^{-6} cm^2/s, compared to 3×10^{-12} for stainless steel.

Hydrogen concentration profiles were computed for the time, temperature, and pressure conditions during the cycle given in Table 4.4, which was obtained from a figure from reference (7). After the third sequential cycle, it was found that a condition developed in which the concentration for each grid point from the table was essentially the same as in the preceding cycle for the same time and distance from the surface. A set of 3 concentration-versus-time plots are shown in Figure 4.5 for distances of 0.1, 0.3 and 0.8 cm below the surface in contact with the gas. The hydrogen concentrations are relative in the sense that the concentration at a given grid point and time is normalized to the concentration at that grid point when t = 0 (start of the cycle).

TABLE 4.4: TIME, TEMPERATURE AND PRESSURE CONDITIONS EXISTING IN THE FeTi HYDRIDE STORAGE SYSTEM DURING ONE COMPLETE CYCLE

Time (hr)	Temperature (°C)	Pressure (MPa)
0.00	20	3.55
0.15	30	3.55
0.30	40	3.55
0.45	50	3.55
0.60	58	3.55
0.80	55	3.55
1.00	50	3.55
2.00	38	3.55
3.00	26	3.55
4.00	20	3.55
5.00	16	3.55
6.00	16	3.93
7.00	26	0.689
11.00	33	0.124
11.50	43	0.103
12.00	43	0.103

Source: SAND-74-8219

FIGURE 4.5: RELATIVE HYDROGEN CONCENTRATIONS (NORMALIZED TO HYDROGEN CONCENTRATIONS AT t = 0) AT SELECTED DEPTHS IN A CARBON STEEL PIPE VERSUS TIME DURING COMPLETE HYDROGEN CHARGING AND DISCHARGING CYCLE

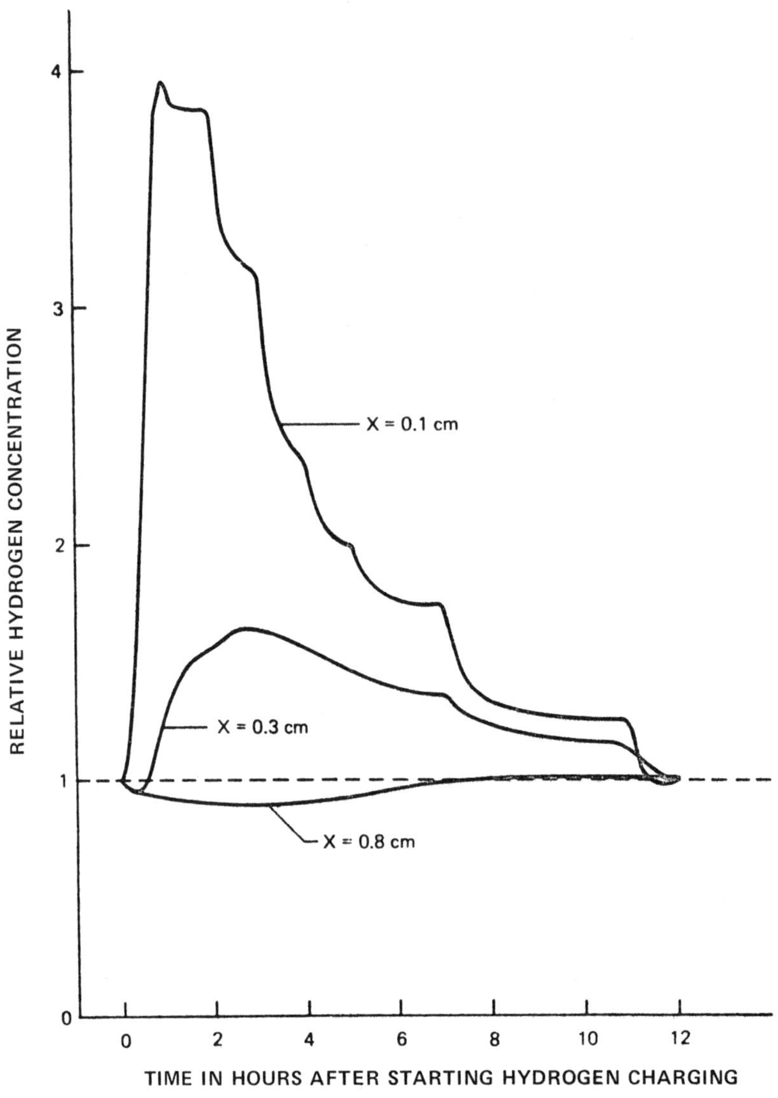

Source: SAND-74-8219

Note in Figure 4.5 that at a depth of 0.1 cm, the concentration changes with the pressure and temperature fluctuations, but with a characteristic lag time. At 0.3 cm, the changes are small, and the time lag is a sizeable fraction of the total cycle time. At 0.8 cm below the surface, the fluctuations are nearly zero.

In principle, thermal cycling such as described above could result in a gradual buildup of defects in the material because of hydrogen precipitation. For this set of conditions, however, problems should not arise with either a stainless or a low carbon steel. For cycles involving higher temperatures and shorter times, degradation of the material could occur.

The possibility of embrittlement due to the crack growth mechanism should also be examined for the Brookhaven reactor. A stress calculation was performed using a finite element stress code and neglecting residual stresses in the material. The minimum radius of curvature allowed by ASTM Standards for a torispherical schedule 40 pressure vessel cap was used in the calculation.

The maximum tensile stress encountered was about 165 MPa (24,000 psia) for the maximum gas pressure of 3.93 MPa (570 psia). The maximum tensile stress encountered in the weld between the schedule 20 cylinder and the cap is approximately 107 MPa (15,500 psi). If the yield strength of the steel is taken to be 207 MPa (30,000 psi) the safety factor in design based on yielding is 1.25 for the most critical area.

In assessing the possibility of a failure due to crack growth in hydrogen, it is assumed that the combined effects of the applied stress, residual stresses, and metallurgical defects will lead to some localized yielding and surface cracks. The question which needs to be answered is how large the flaws must be before they will propagate and result in a failure.

It is impossible at present to answer this question satisfactorily because threshold stress intensity (K_{ISCC}) values for hydrogen embrittled stainless and low alloy steels have not been measured. The most relevant data available (8) are fracture toughness (K_{IC}) values for notched bars of 347 stainless steel in helium and hydrogen under a pressure of 34 MPa (5,000 psia). A value of K_{IC} of 33 $MN/m^{3/2}$ was reported, which might correspond to a K_{ISCC} half as large as this value for the Brookhaven reactor as a worst case.

The fracture mechanics expression for a flaw in an infinite sheet (5) which can be used for an approximate calculation is $K = \sigma\sqrt{\pi a}$ where a is the crack length and σ is the tensile stress of interest (165 MPa or 24,000 psia). For a value of $K = 16.5\ Mn/m^{3/2}$, a is found to be 0.32 cm or 0.12 in. This flaw size is nearly half the cylinder wall thickness. It is doubtful that such a flaw would be introduced during manufacture and it could be detected rather easily. Thus even with a worst case calculation, it appears that the reactor is designed safely with respect to crack growth in hydrogen at 3.45 MPa (500 psi) pressure.

From this analysis of the design of Brookhaven's 316 stainless steel reservoir for hydrogen storage, it is concluded that the design is conservative. Although the chances of uncovering a problem are small, two inspections or tests should be performed. The welded sections should be examined to be sure that no cracks or hardened areas exist. Also, some 316 stainless specimens should be exposed to $FeTiH_{1.6}$ and 3.45 MPa (500 psi) hydrogen to be sure that the uptake of atomic hydrogen from $FeTiH_{1.6}$ does not degrade mechanical properties significantly.

Since the environment is not very severe and the design conservative, there is a good chance that a carbon or low alloy steel could be used for the pressure vessel. Qualification tests on the material selected under the conditions to be used in service would be required. The cost saving would be large if the process were scaled up in size, or if a large number of units were made.

This recommendation does not apply if hydride compounds which dissociate at a higher temperature are substituted for $FeTiH_{1.6}$. For example, if TiH_2 or MgH_2 were used, the maximum operating temperatures would be approximately 600° and 400°C, respectively. At these temperatures, hydrogen damage could be a problem with either a stainless or a low alloy steel.

Compressor

The nature of compressor problems associated with hydride storage systems will depend on the size of the system. For small rates of flow, as typified by the PSE&G/BNL hydride reservoir, diaphragm compressors offer a suitable design solution (these may also be suitable injection systems involving gaseous hydrogen fuel in diesel engines). Larger system designs, as for example the Brookhaven National Laboratories proposed 26 Mw hydride/fuel cell concept, will run into the design problems relating to compressors.

METAL HYDRIDE SCREENING EXPERIMENTS

It has been the goal of the Brookhaven National Laboratory research program to develop hydrides, or, rather, hydrogen-metal systems, which will have a high hydrogen content and which will meet certain requirements imposed by their use in conjunction with devices that use hydrogen for the production of energy.

A description of BNL's screening procedure and criteria for evaluation follows. Rates of hydrogen-metal reactions vary widely, and depend on several factors. One is the intrinsic nature of the system. Thus, certain titanium-cobalt alloys react, even when in large chunks, almost as fast as hydrogen can be supplied; while pure magnesium powder reacts very slowly. Another factor is cleanliness of metal surface; an oxide film will often result in a long induction period before a good rate is attained.

Still another factor is the state of subdivision of the metal. This can usually be greatly increased by subjecting a sample to a series of hydriding-dehydriding cycles. Absolute surface areas as high as 2 m^2/g have been obtained in this way, and the product was highly active toward hydrogen. Finally, it is possible to increase the rate of the combination reaction by the addition of small quantities of solid catalysts. Thus, the formation of MgH_2 is accelerated by the presence of nickel (or more accurately Mg_2Ni).

There are a number of criteria by which one may judge the suitability of a metal-hydrogen system for energy storage. For example, the hydrides are exothermic and energy must be supplied for their decomposition. This need not, however, be a source of inefficiency. All energy producing devices, whether fuel cells or combustion engines, produce waste heat and it should be possible to utilize this heat for the decomposition of the hydride. Therefore a balance is required between the heat produced and that required, both as to quantity and quality (temperature).

In other words, the hydride should have an appreciable decomposition pressure (at least one atmosphere) at the temperature of operation of the energy-producing device. There are, of course, additional criteria by which one can judge the worth of a material according to one's particular needs. For example, a unique set of criteria has been developed by which any candidate metal-hydrogen system can be measured against the EPA program goals. These criteria are listed in Table 4.5 and were used for screening a large number of alloys which are listed in Table 4.6.

All of these materials, when judged by this particular set of criteria score lower than catalyzed magnesium hydride and can be eliminated as hydrogen storage media. It should be noted that the criteria, as presently constituted, include reversibility. This specification, when applied in the usual sense, eliminates certain complex hydrides [e.g., $Mg(AlH_4)_2$], which otherwise may be very attractive as hydrogen storage media.

Such compounds can only be made indirectly using wet chemical techniques. Future work will examine the possibility of simplifying the synthesis of such compounds through the use of catalysts and/or alternate wet chemical reactions. Thus, if simple syntheses are possible, such systems, even if not directly reversible, could then be considered as practical hydrogen storage media.

TABLE 4.5: CRITERIA FOR EVALUATION OF A CANDIDATE METAL HYDRIDE

Key:

x - Rejects candidate
o - Unknown
1 to 5 - Rating scale

Criteria:

Reversibility –
- x - Series reactions required to constitute hydride and process is not simple or cheap.
- 1 - Special apparatus and extreme conditions are necessary
- 5 - Reaction takes place in situ with heating, cooling or pressurization

Hydrogen Content (by weight) –
- x - Contains less than 5% hydrogen
- 1 - Contains 5 to 10%
- 3 - Contains 10 to 20%
- 5 - Contains more than 20%

Hydrogen Content (by volume) –
- x - Contains less than 4 x 10^{22} atoms H/ml of hydride
- 1 - Contains 4 to 6 x 10^{22} atoms H/ml of hydride
- 2 - Contains 6 to 8 x 10^{22} atoms H/ml of hydride
- 3 - Contains 8 to 10 x 10^{22} atoms H/ml of hydride
- 5 - Contains more than 10 x 10^{22} atoms H/ml of hydride

(continued)

TABLE 4.5: (continued)

Pressure/Temperature Relation –

x - $P > 2{,}000$ psia at 150°F
$T > 570$°F at 15 psia

1 - P = 1,000 to 2,000 psia at 150°F
T = 390° to 570°F at 15 psia

2 - P = 500 to 1,000 psia at 150°F
T = 300° to 390°F at 15 psia

3 - P = 200 to 500 psia at 150°F
T = 5° to 300°F at 15 psia

5 - $P < 200$ psia at 150°F
$T < 5$°F at 15 psia

Heat of Dissociation –

x - Heat of Dissociation $(H_D) > 0.4$ Heat of Combustion (H_C) of hydrogen
1 - $0.2 < H_D/H_C < 0.4$
3 - $0.1 < H_D/H_C < 0.2$
5 - $H_D/H_C < 0.1$

Safety (Hydrided and Dehydrided) –

x - More hazardous than gasoline
1 - Same overall degree of hazard as gasoline
3 - Somewhat less hazardous than gasoline
5 - Significantly less hazardous than gasoline

Cyclic Stability - physical –

x - Effective for <100 cycles
1 - Effective for 100 to 300 cycles
3 - Effective for 300 to 500 cycles
5 - Effective for more than 500 cycles

Cyclic Stability - contamination –

x - Less than 300 cycles with high purity gas (HPG)
Less than 100 cycles with normal purity gas (NPG)

1 - 300 to 500 cycles with HPG
100 to 300 cycles with NPG
<100 cycles with impure gas (IG)

3 - >500 cycles with HPG
>300 cycles with NPG
100 to 300 cycles with IG

5 - >500 cycles with NPG
>300 cycles with IG

Cost/Availability –

x - More than $1,000 (20 gal gasoline equivalent) near or long term
Long range availability unlikely

1 - $500 to $1,000 near and long term
Long range availability in doubt

3 - $200 to $500 near and long term
Availability reasonably certain

5 - <$200 near and long term
Availability certain

Physical Properties –

x - Low melting point. Volatile in operating range. Corrosive, etc.
1 - Properties marginally acceptable
5 - Acceptable

Reaction Kinetics –

x - Rate inadequate regardless of equipment design
1 - Rate adequate but equipment complex
5 - Rate adequate in simple equipment

(continued)

TABLE 4.5: (continued)

Thermal Conductivity –
- x - Limits hydrogen availability regardless of equipment
- 1 - Acceptable for use but with complex heat exchange equipment
- 5 - Rate adequate in simple equipment

Source: BNL-18887

TABLE 4.6: METAL HYDRIDE SCREENING EXPERIMENTS

----Alloy Composition----		---------- Max. H ----------	
Wt %	Formula	Content Wt %	Remarks
39 RE 61 Mg	Mg_9RE^*	4.1	
39 Ce 61 Mg	Mg_9Ce	4.4	Pyrophoric
25.8 Mg 74.2 RE	Mg_2RE^*	2.1	
39.9 Ba 60.1 Mg	Mg_2Ba_{17}	3.9	
90 Mg 10 Mn	$MgMn_{0.05}$	0.5	
57.8 Mg 42.2 Pd	Mg_6Pd	0.8	
62.5 Al 37.5 Mg	$MgAl_{1.5}$	0.7	
53.6 Zn 46.4 Mg	Mg_7Zn_3	0.9	
63.4 Mg 36.6 Si	Mg_2Si	0.1	
56.1 Mg 43.9 Al	$MgAl_{0.7}$	1.7	
58 Al 42 Mg	$MgAl_{1.2}$	1.4	
54.8 Mg 45.2 Ca	Mg_2Ca	1.4	
68 Pb 32 Mg	Mg_2Pb	1.1	
46.7 Mg 26.3 Mn 27.0 N	$MgNi_{0.5}Mn_{0.5}$	~3.5	
49.0 Mg 36.7 Ti 14.3 Ni	$MgTi_{0.38}Ni_{0.12}$	2.08	
46.6 Mg 26.7 Ti 26.7 Cu	$MgTi_{0.29}Cu_{0.22}$	3.8	
49.0 Mg 30.6 Ti 20.4 Mn	$MgTi_{0.39}Mn_{0.23}$	1.7	
39.5 Mg 30.8 Zr 30.7 Cu	$MgCu_{0.30}Zr_{0.21}$	2.7	

(continued)

TABLE 4.6: (continued)

--- Alloy Composition --- Wt %	Formula	--- Max. H --- Content Wt %	Remarks
50.0 Mg 39.4 Zn 10.6 Ti	$MgZn_{0.29}Ti_{0.11}$	2.1	
95 Mg 5 Pb	Solid solution	-	No apparent reaction
95 Mg 5 Sn	Solid solution	-	No apparent reaction
95 Mg 5 Zn	Solid solution	-	No apparent reaction
77 Sb 23 Mg	Mg_3Sb_2	-	No apparent reaction
60.6 Zn 24.8 Al 14.6 Mg	$Mg_{32}Al_{49}Zn_{49}$	-	No apparent reaction
	$Al_5Cu_6Mg_8Si_6$	-	No apparent reaction
	$Al_2Cu_2Mg_2Zn_2$	-	No apparent reaction
	AlCuMg	-	No apparent reaction
	Al_5Mo	-	No apparent reaction
100	Al (amorphous)	-	No apparent reaction
100	Al fines	-	No apparent reaction
96 Mg 2 Al 2 Li	Solid solution	0.4	
95 Mg 5 Li	Solid solution	0.4	
65 Mg 35 Ba	$BaMg_{10.5}$	~5.1	4.2% available H

*RE = Mischmetal.

Source: BNL-18887

IRON-TITANIUM-MANGANESE ALLOY FOR HYDROGEN STORAGE

After screening, several hydrides were found to be of interest but the material nearest to practical application is iron-titanium hydride, which can be synthesized through the direct union of hydrogen with the intermetallic compound FeTi.

U.S. Patents 3,508,414 and 3,516,623 disclose methods and apparatus for utilizing iron-titanium alloys to store hydrogen by the formation of hydrides. One difficulty which has been discovered in the use of iron-titanium alloys for hydrogen storage is the effect of the presence of oxygen in the alloys in small amounts. For example, it has been discovered that the presence of oxygen in the amount of 7,000 ppm in commercially available iron-titanium reduced substantially the maximum hydrogen that could be stored and the equilibrium dissociation pressure was increased. This had the effect of increasing the costs involved in storing hydrogen by the use of these alloys.

A process by *J.J. Reilly and R.H. Wiswall, Jr.; U.S. Patent 3,922,872; Dec. 2, 1975;*

assigned to U.S. Energy Research and Development Administration discloses that the addition of manganese to the intermetallic alloy FeTi in certain specific amounts not only increases the amount of H_2 which can be stored and at a lower pressure but also has the effect of compensating to a significant extent for the presence of oxygen, permitting significant increases in the amounts of hydrogen which can be stored under more convenient and economical pressures.

In accordance with the process, there is provided a three component alloy capable of reversible sorption of hydrogen having the chemical formula $TiFe_{1-x}Mn_x$ where x is in the range of about 0.02 to 0.5. There is also provided a method of storing hydrogen comprising contacting gaseous hydrogen with a solid alloy of $TiFe_{1-x}Mn_x$ where x is in the range of about 0.02 to 0.5.

An alloy in accordance with this process may be prepared by melting granules or small ingots of Fe, Ti, and Mn in an arc or induction furnace within an inert atmosphere followed by cooling. The cooled alloy, in order to be utilized for the storage of hydrogen is comminuted or granulated and then activated by outgassing at high temperature (300°C) and exposing to H_2 for a short time followed by outgassing again and cooling under hydrogen with about 1 atmosphere pressure. In order to form the hydride the activated alloy is exposed to H_2 at a pressure usually 10 atmospheres above dissociation pressure at that temperature, due to hysteresis type effects. The hydriding pressure should for best results be at least twice the dissociation pressure because of the already mentioned hysteresis effect.

Figures 4.6a and 4.6b show curves illustrating the H_2 storage characteristics of alloys incorporating the principles of this process and comparing them with similar alloys not incorporating this process.

FIGURE 4.6: DISSOCIATION PRESSURE-COMPOSITION ISOTHERMS FOR IRON-TITANIUM-MANGANESE ALLOYS

a.

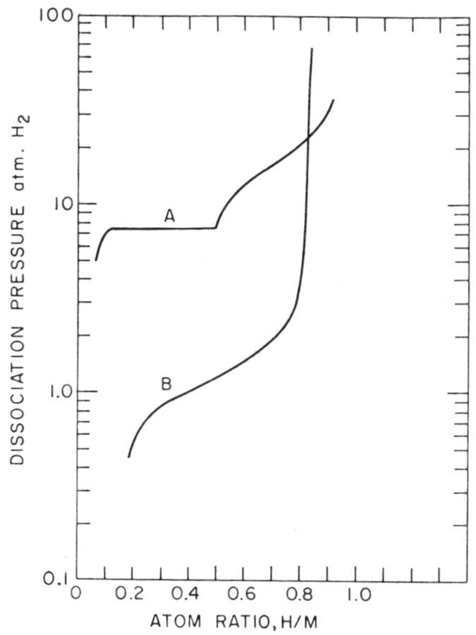

(continued)

FIGURE 4.6: (continued)

b.

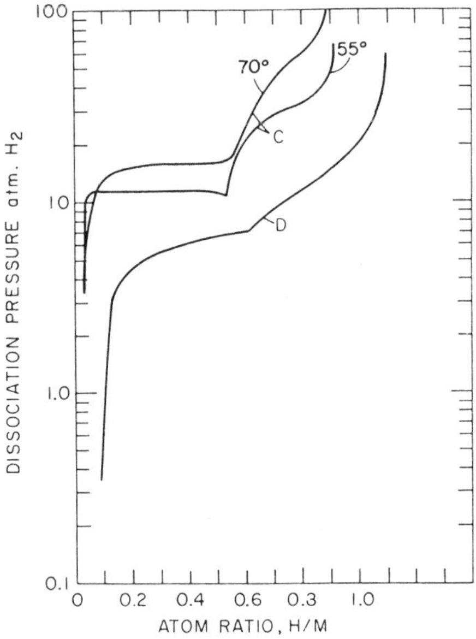

Source: U.S. Patent 3,922,872

An alloy was prepared with the composition (A) of FeTi and the dissociation pressure-composition isotherms for this alloy are shown in Figure 4.6a. The H_2 dissociation pressure of this alloy can be seen from the curve at 40°C to be at least 7.2 atmospheres and reaches 25 atmospheres at the maximum H_2 concentration. A similar alloy (B) was prepared in which some of the iron was displaced by Mn and had the formula $TiFe_{0.7}Mn_{0.3}$.

The dissociation pressures for this alloy at the same temperature, as shown in Figure 4.6a, range from 0.42 to 9 atmospheres for the same amount of stored H_2 as in alloy (A). In the drawing, the atom ratio, H/M is defined as the ratio of atoms of hydrogen to total atoms of metal.

It was found that for other temperature conditions the presence of Mn displacing some of the iron additionally made it possible to increase the amount of H_2 which could be stored as well as reducing the dissociation pressure. Curves **C** in Figure 4.6b shows isotherms for a FeTi alloy at 55° and 70°C while curve **D** shows the isotherm at 61°C for the composition $TiFe_{0.8}Mn_{0.2}$. Not only does alloy D have a lower dissociation pressure but in addition H_2 storage capacity was increased by about 10% by weight. This is shown by the upper limits of the curve.

MODELLING STUDIES OF FIXED-BED METAL HYDRIDE STORAGE SYSTEMS

A study was conducted at Brookhaven National Laboratory (Report BNL-18720) to examine the dynamic behavior of two metal hydride storage systems. The metals chosen for the system were alloys—magnesium-nickel and iron-titanium.

The hydride formation reactions are all exothermic, but to varying degrees. Thus, it is necessary to remove or supply heat from or to a hydride storage system during the hydriding or dehydriding process. Such a storage system must provide for the transfer of this heat. There are two general categories of heat exchange, direct and indirect. Direct exchange involves physical contacting of two media exchanging heat; indirect exchange occurs across a boundary impermeable to the exchanging media.

This section will discuss one direct exchange scheme termed a convection bed and one indirect scheme termed a conduction bed. The models are focused on the dynamics of hydrogen release, the reverse process of hydride formation; application to the hydrogen sorption process should be a straightforward extension of release models.

Often, the proposed source of heat for the hydrogen release process is waste heat from the hydrogen combustion process. For example, in automobiles the source of this heat could be the radiator coolant or exhaust gases. In such systems, the storage facility might be closely coupled to the dynamics of the combustion load. However, for the purposes of this study, the source of heat to the storage facility is assumed independent of the release process, and the storage system is examined as an isolated unit.

The ultimate success of this work depends on the knowledge of the kinetics of hydride formation and decomposition. At this time, the study of the dynamic behavior of promising hydrides is limited. The selection of the two working hydrides is mainly based on their use in practical applications of immediate interest. Magnesium-nickel is a lightweight hydride considered as a good candidate for hydrogen storage in automobiles. Iron-titanium is suitable for stationary storage due to its low reaction temperature and low heat content.

The thermal data for hydrides are relatively more abundant and the thermal properties of the two hydrides which are discussed can be found in references (9) (10) (11) and (12). The conductivity of a hydride is usually lower than its metal alloy. In a conduction model, variations in the thermal conductivity can play an important role. The only available data (11) for these two hydrides were used in the analysis.

Convection Bed Model

It is assumed that the hydride will be stored as a granular solid in a cylindrical pressure vessel. Since the material is a solid, it is likely that the simplicity of a fixed bed scheme would be preferred. The granular form of the solid would be necessary from internal mass transfer rate consideration.

This model is a direct heat exchange model; the heat of reaction necessary to liberate hydrogen is supplied by circulating hot hydrogen, which is heated externally, through a charged hydride bed. The advantages of such a system would be that first, heat transfer to the bed would not be limited by poor hydride thermal conductivity as it might be in conduction beds and second, a much simpler container design would be possible.

Because the reaction is endothermic and heat is being supplied from one end of the bed, the reaction will travel from the hot end to the cool end. Whether this reaction zone is broad and covers the entire bed or is a narrow band moving from one end to another depends on the relative rates of heat and mass transfer and chemical kinetics. The conditions are considered to be as follows:

(1) Slug gas flow
(2) Insulated wall
(3) Constant gas and solid physical properties
(4) Intraparticle heat conduction fast relative to surface heat transfer ($hD_p/kp \ll 1$), where h is the heat transfer coefficient, D_p is the diameter of the spherical particle in feet and k_p is the thermal conductivity of the particles
(5) The particle sees an external environment of uniform temperature passing at a uniform mass flux.

The calculations involved in the convection bed are described in detail in the original report and are not included here. However, they were based on physical properties given in references (9)-(11) and the conditions are in Table 4.7.

TABLE 4.7: CONDITIONS FOR CONVECTION BED MODEL

	Mg_2NiH_4	FeTiH
Particle diameter, ft	0.00273	0.00273
Bed void fraction	0.4	0.4
Bed pressure, atm	1.0	3.0
Inlet gas temperature, °F	840.0	160.0
Inlet bed temperature, °F	60.0	60.0
Gas inlet flow rate, lb/hr-ft²	50.0	50.0
Gas heat capacity, Btu/lb-°F	3.5*	3.4*
Gas viscosity, lb/hr-ft	0.0338*	0.0227*
Gas thermal conductivity, Btu/hr-°F-ft	0.15*	0.1*
Heat transfer coefficient	**	**

*Reference (13). **References (14) and (15).

Source: BNL-18720

Conduction Bed Model

The conduction bed model is of considerable practical interest in automotive applications and is analyzed using as a basis, fluid flow through an inner passage to provide or to release heat during the dehydriding or hydriding process. A constant-property solid loosely packed in the annulus passage is assumed. On the tube side, it is assumed that the constant-property fluid flow will provide a constant heat flux to the solid hydride through the wall. The conditions considered are as follows:

(1) Steady state slug flow in the inner tube
(2) Constant inlet and outlet fluid temperatures
(3) Heat transfer from the inner wall only

(4) Insulated outer wall
(5) Heat flux arbitrarily varying in the axial direction
(6) Hydrogen will be released at the end of the bed or supplied at the entrance of the bed.

Table 4.8 lists the conditions for the conduction bed model which were used for the calculations.

TABLE 4.8: CONDITIONS FOR CONDUCTION BED MODEL

	Mg_2NiH_4	FeTiH
Particle diameter, ft	0.00273	0.00273
Bed void fraction	0.4	0.4
Bed pressure, atm	1.0	3.0
Inlet fluid temperature, °F	840.0	160.0
Outlet fluid temperature, °F	740.0	120.0
Initial bed temperature, °F	480.0	60.0
Gas inlet flow rate, lb/hr-ft^2	4,000.0	4,000.0
Gas heat capacity, Btu/lb-°F	3.5*	3.4*
Gas viscosity, lb/hr-ft	0.0338*	0.0227*
Gas thermal conductivity, Btu/hr-°F-ft	0.15*	0.1*
Heat transfer coefficient	**	**

*References (13). **See reference (16).

Source: BNL-18720

Results of Studies

What is immediately apparent is that the reaction zone is an extremely narrow band moving down the bed at a constant rate. This implies a process which is limited by the hot hydrogen flow rate. Because of the high inlet gas temperature and low hydrogen pressures chosen, the chemical rate is very high, and the heat supply rate is controlling.

Since the available kinetics data were for fairly small particles, the model was applied to cases employing comparably sized particles about 20 mesh (0.00273 ft), and assumed these to be free of internal particle mass transfer effects over the whole range of temperatures. The pressure dependence of the chemical rate was modelled as a function of $(P_{dissociation}-P_{bed})/(P_{dissociation})$. The hot hydrogen feed rate was determined by energy considerations in obtaining a hydrogen release rate of 10 lb/hr out of a two foot diameter bed, as required for automotive applications.

The total bed pressure drop under these conditions is only about 1 psi/ft. The temperature plateau occurs at a solid temperature where the hydride dissociation pressure equals the bed pressure and further reaction is prevented by the equilibrium relationship. Any heat remaining in the gas below this temperature can only be utilized for sensible heating of the bed ahead of the reaction zone.

There are several important conclusions to be drawn from these results. As far as the storage bed alone is concerned, the rate at which hydrogen is released is only a function of the flow rate of hot hydrogen (even for the extremely high demand automotive applications). The rate limitation in a real convective storage system will probably occur in transferring heat to the hydrogen which is fed to the bed. Another important result is that the system becomes rapidly self-supporting, allaying earlier fears that a high temperature magnesium alloy based system would require a large starter heat source.

Dynamic load variations (sudden changes in hydrogen demand levels) are handled by the area of the bed under the temperature plateau. By permitting the bed pressure to drop during a period of increased demand, the sensible heat stored in this section of the bed becomes available for the heat of reaction, and additional hydrogen becomes available without waiting for additional external heat to be supplied.

The most serious shortcoming of the study is the available kinetics data. These data lack pressure dependence information, lack sufficient high temperature information, and lack rate versus particle diameter information in internal mass transfer controlling regimes (if any exist).

A sample calculation using the simplified model, was done for an iron-titanium system which is similar to the operating system of the Public Service Electric and Gas Co. bed. The bed was built at Brookhaven National Laboratory for the PSE&G Co. It was found that the release rate is roughly constant, at 0.036 lb/hr-ft. In terms of the PSE&G bed which has nine U tubes of length, six feet per leg, under these conditions the model predicts a maximum release rate of 3.0 lb/hr.

HIGH EFFICIENCY POWER CYCLES USING METAL HYDRIDE COMPRESSORS

Closed cycle gas turbines using helium are a power conversion option for high temperature heat sources. In such a conventional cycle, high coolant temperatures (e.g., about 850°C) are necessary for reasonable thermal efficiencies (e.g., 40%), with the helium gas being compressed mechanically. If helium is replaced with hydrogen gas, the hydrogen can be compressed by an absorption/desorption cycle in a bed of metal hydride particles.

The absorption process releases heat and the desorption process absorbs heat. In a practical compressor, the desorption can be carried out at a relatively low temperature (T about 100°C), and the heat can be obtained from low grade geothermal or solar sources. Since the compressor work is supplied by the auxiliary low grade thermal energy, the conversion of high temperature heat to electricity can be much higher. Thus, 80 to 90% of the high grade heat can be converted to electricity, instead of 40% with conventional cycles.

Wiswall and Winsche (unpublished data) originally conceived of a cycle similar in nature to that described. In their original hydride engine scheme hydrogen was driven off a metal hydride bed by heat from a low temperature heat source, e.g., 65°C. The hydrogen was expanded isothermally and then adiabatically to do work and then subsequently collected on a metal hydride bed at a lower temperature, e.g., 30°C. For vanadium hydride, they estimated an efficiency of 9.4% which is 90% of the Carnot value.

Work has been done at Brookhaven National Laboratory *(Report BNL-20162)* on a high efficiency power cycle using a metal hydride compressor. This cycle has several significant advantages over conventional power cycles:

(1) The high temperature heat input per unit electrical output can be reduced by a factor of 2 to 3. This conserves scarce resources used to produce high grade heat, and greatly reduces the cost of the high temperature heat per unit electrical output.
(2) Although the low grade heat used in this cycle (T about 120°C) could be converted to electricity in a conventional cycle, the efficiency would be very low ($<10\%$). In the proposed cycle, low grade heat is used efficiently.
(3) Significant overall cost savings may be obtainable, compared to cycles that generate electricity using only a high temperature heat source.

The intermetallic compound of iron and titanium appears a promising material for a hydride compressor. Its projected price is low, it can store large amounts of H_2, and has rapid H_2 absorption/desorption rates. Although better materials may be found, for this study only iron-titanium is considered.

Iron-titanium hydride is being investigated at Brookhaven National Laboratory (BNL) as a material for hydrogen storage for electrical peak shaving applications (17)(18)(19). Several iron-titanium storage beds have been built, with the largest bed holding 14 lb of H_2. Considerable measurements of physical properties have also been made.

The controlling step for a hydride compressor is probably H_2 absorption. Kinetics are least favorable at this point, since bed temperature is low and overpressure small. For this study, rates of absorption as a function of pressure and composition were determined in small iron-titanium tests beds containing commercial material. H_2 absorption was studied at pressures of 150, 200, 250, and 300 psia, with desorption at 600 psia (limited by the available steam temperature). The results of these tests show that even this relatively nonoptimal material is adequate for a practical hydrogen compressor.

Operation of an Iron-Titanium Bed as a Hydrogen Compressor

Iron titanium alloy (50-50 mol percent) can readily adsorb hydrogen at ambient temperatures according to the general reactions

$$FeTi + \tfrac{1}{2}H_2 \rightleftharpoons FeTiH$$

$$FeTiH + \tfrac{1}{2}H_2 \rightleftharpoons FeTiH_2$$

The overall reaction which forms the hydride ($FeTiH_x$) is exothermic (about 7,000 cal/g mol H_2). Upon decomposition of $FeTiH_x$ the same amount of heat must be added to the system. If heat is not added, the hydride will cool down and decomposition will cease. Thus, the design of a hydrogen compressor using FeTi involves heat removal during the charging process and heat addition during the discharge process.

A simple design for a compressor would be a shell and tube heat exchanger, with the $FeTiH_x$ compound on the shell side and a coolant/heating bundle penetrating the granular bed. Hydrogen could be injected and removed from the bed through small diameter porous metal tubes dispersed in the bed and connected to a common header. Cooling would be by ambient temperature cooling water circulating through the tubes.

During H_2 desorption hot water (or steam) from the low temperature heat source would circulate through the tubes. Figure 4.7 shows a conceptable design of an iron-titanium hydride compressor.

The pressure composition isotherms of the $FeTiH_x$ system generally exhibit some hysteresis. This results in some inefficiency in the cycle; the reaction is not completely reversible since the equilibrium charging pressure is slightly higher than the equilibrium discharge pressure. However, it is shown later that with available material this inefficiency is tolerable. Experimental work to further reduce hysteresis could be of value.

A practical embodiment of a hydride compressor contains about 1,000 pounds of FeTi; it was designed primarily as a hydrogen storage vessel and is currently being used by the Public Service Electric and Gas Company of New Jersey. The storage reservoir is a shell and tube heat exchanger (FeTi on shell side) where the tube bundle assembly is combined with six porous tubes running the length of the assembly and connected to a central header.

FIGURE 4.7: CONCEPTUAL DESIGN OF AN $FeTiH_x$ ABSORPTION RESERVOIR FOR COMPRESSOR APPLICATIONS

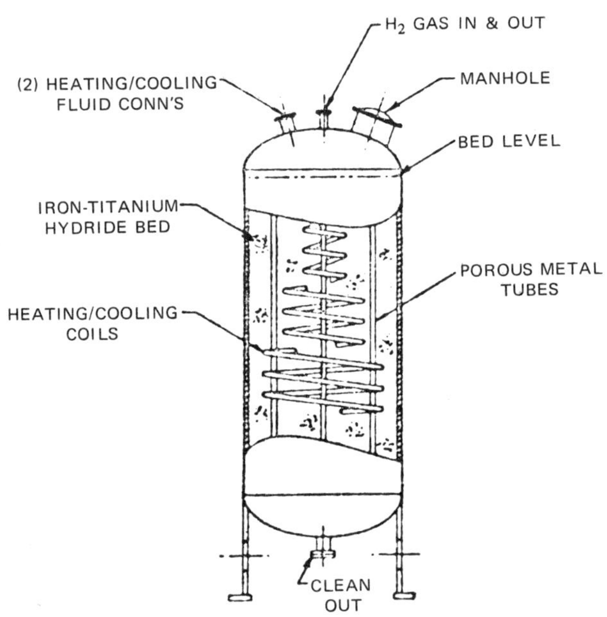

Source: BNL-20162

An important practical concern in the design of an $FeTiH_x$ compressor is the behavior of the material after thousands of repetitive cycles.

In the development program at Brookhaven, $FeTiH_x$ beds have been cycled 1,200 times with no evidence of significant deterioration in kinetics or hydrogen capacity. The experiments were terminated at 1,200 cycles in order to test alternative materials. There is no known reason why $FeTiH_x$ cannot undergo an unlimited number of cycles. The predominant physical change after many cycles is a reduction in FeTi particle size and the creation of some fines. In the shell and tube compressor design this is not a serious design problem.

In the design of an $FeTiH_x$ compressor or storage reservoir, the important parameters are the charge and discharge rates and the pressure composition isotherms. Kinetic and equilibrium experiments have been carried out in a small test compressor, a one inch i.d. stainless steel tube, ten inches long, with an external jacket through which steam or cold water can be circulated.

Hydriding kinetic data were obtained at H_2 pressures of 150, 200, 250, and 300 psia starting with an initial composition of $FeTiH_{0.1}$, with coolant water at about 16°C. At 150 and 300 psi the bed composition reaches $FeTiH_{0.5}$ in 8 and 1 minutes, respectively. (Discharge pressure was 600 psia, which was limited by available steam temperature.) Dehydriding rates are considerably faster than hydriding rates.

Thus if kinetics are limiting, system design will probably be limited by hydriding rather than the dehydriding. However, present data and experience suggest that the reaction rate for large systems will be heat transfer limited. Models have been developed for the hydriding/dehydriding processes based on this assumption, for various vessel configurations and coolants.

High and Low Temperature Heat Sources

Many high temperature heat sources are potential candidates for high efficiency power cycles. These include nuclear reactors (LWR's, HTGR's, LMFBR's, etc.), fusion reactors, fossil fuel fired boilers (coal, oil, and gas), and high temperature solar and geothermal sources. At present, only LWR's, HTGR's, fossil fuel fired boilers, and high temperature geothermal sources are sufficiently developed and low enough in cost to be practical.

Reactors can be coupled relatively easily to the high efficiency cycle, since they use closed-cycle coolants and the heat can be supplied to the high efficiency cycle without having to recover sensible heat from exhaust combustion gases, as must be done with fossil fuel fired heat sources to avoid wasting high temperature heat and reducing cycle efficiency. Incoming air is preheated in a regenerator.

The heat of combustion is then effectively supplied at a single source temperature to the high efficiency cycle. Such preheating is standard in MHD cycles and can be carried out relatively easily to temperatures well above 1300°K (using pebble-bed regenerators). With fossil fuel fired boilers the source temperature available to heat the hydrogen working fluid will depend primarily on materials limitations, since heat must be transferred from the combustion gases across metal tubes to the hydrogen. A source temperature of about 1000°K should be achievable with alloys similar to those used in standard methane reformers.

Pressure limits the source temperature available from LWR's to about 600°K. Source temperatures achievable with present HTGR's are considerably higher (about 1000°K), and advanced HTGR's (either pebble-bed or prismatic) should achieve about 1200°K. High temperature efficiency (η_{HT}) will be greater for HTGR's than for LWR's, but not significantly.

For the proposed cycle, the real performance gain of HTGR's over LWR's is in the increased efficiency in the conversion of low temperature heat (η_{LT}). The value for HTGR's will be almost twice that for LWR's because of the higher source temperature. If the costs of low temperature heat and the hydride compressor are not a significant part of the overall power cost, there would be no economic reason for preferring one reactor to the other. If these costs are a substantial part of overall cost, then HTGR's would be preferred.

High temperature geothermal heat is an interesting possibility for the high temperature heat source. In the proposed high efficiency cycle, geothermal heat at 600°K can be transformed into electricity with an efficiency of about 80% instead of the 25% possible. Since the high temperature geothermal resources are relatively limited or extractable only by costly deep drilling, the threefold increase in efficiency with use of high temperature geothermal heat would be an important factor in evaluating cost and conservation considerations.

Rex (20) estimates the geothermal energy ($T > 400°K$) in the region of the Jemez volcano to be 8.5 x 10^{20} cal to a depth of 6 km. The energy above 600°K is about 6 x 10^{20} cal. At 80% average conversion efficiency this corresponds to generation of 640,000 megawatt (Mw) centuries of electricity if 100% of the stored energy is extracted.

For a more reasonable extraction of only 10% of the stored energy, 64,000 Mw centuries would be generated. There are a number of such areas in the United States, and it appears that the total high grade geothermal energy could supply a large fraction of U.S. needs if used in high efficiency cycles.

Fusion and high temperature solar energy are attractive possibilities for future high temperature heat sources. Each is virtually unlimited in terms of resources, and each should have a very small environmental impact. One of the consequences of using a high efficiency cycle is that differences in capital cost between alternative high temperature sources are reduced, as compared with those using a conventional cycle.

For example, suppose that the cost difference between sources A and B is $300/kw(e) and that both use a conventional cycle with 35% thermal efficiency. If both use a high efficiency cycle with 90% efficiency, the difference decreases to $116/kw(e). The more expensive heat source, which is also more environmentally acceptable, becomes much more attractive economically if a high efficiency cycle is used.

At this point, low-grade solar and geothermal energy seem to be the only practical low temperature heat sources for the proposed high efficiency cycle. Waste heat from condenser water or cooling towers in conventional power plants is too low in temperature to be useful for a hydride compressor. The required temperatures could be obtained by extracting steam at an intermediate stage of turbine expansion; this would reduce the efficiency of the steam cycle, but the overall plant thermal efficiency should increase. The net gain would probably be small, however, and may not be of significant value.

Ocean thermal energy is an intriguing possibility for a low temperature heat source, but to make effective use of the small temperature differential (about 20°C) between upper and lower ocean layers requires a hydride compressor with a relatively small overpressure to accommodate hysteresis and kinetics. Such a compressor may be possible, but more research on the properties of metal hydrides is needed before a definite conclusion can be reached about the feasibility of using ocean thermal energy for the proposed cycle.

While sources of high grade geothermal energy ($T \geqslant 300°C$) are relatively limited or difficult to reach, low grade geothermal energy (T about 100°C) is much more plentiful and accessible. For example, Blackwell (21) has statistically analyzed numerous heat-flow and temperature measurements and concludes that the geothermal provinces in the western U.S. (2×10^6 km^2) could supply 200,000 Mw centuries of electricity, assuming exploitation to a depth of 3 km and extraction as electricity of 1% of all stored energy above 100°C.

However, most of this energy is relatively low grade (temperatures of about 200°C), and conventional cycles would yield at most about 10% thermal efficiency. If it is assumed that this low grade energy is used for the hydride compressors in a high efficiency cycle with $\eta_{LT} = 0.3$, and that 10% of the available low-grade energy in the 3 km depth is extracted, 600,000 Mw centuries of electrical generation could be supplied from the western U.S. Exploitation to a depth of 5 km would increase the generation to about 6,000,000 Mw centuries, again assuming that 10% of the available energy is extracted.

The normal temperature gradient in the Eastern U.S. is about 30°C/km so that useful temperatures for hydride compressors could be reached at depths below about 3 km. If 10% of the low-grade energy between 3 and 5 km were to be extracted for input to the hydride compressors, at $T \geqslant 120°C$, 5,000,000 Mw centuries of electricity could be generated in the Eastern U.S. assuming an effective area of 4×10^6 km^2. Practical exploitation of this low-grade heat source depends on development of techniques for extraction of energy from hot dry rock sources.

On the basis of average well costs (1973 dollars) given as a function of depth by Altseimer (22) and assuming that a production and reinjection well pair produces 50 Mw(t), geothermal energy for the hydride compressors would require an investment of $12/kw(t) for 3-km-deep wells and $40/kw(t) for 5-km-deep wells.

By way of comparison, Austin (23) estimates a cost of $5/kw(t) for geothermal energy in the Salton Sea area. The wells would be 1.6 km deep, with 100 Mw(t) extracted per pair of production and reinjection wells. The hot brine/steam mixture would exit from the wells at 230°C. The energy content between the exit condition and saturated liquid at T = 120°C could be used in hydride compressors.

In summary, the amount of geothermal energy available for use in hydride compressors in a high efficiency cycle seems to be virtually unlimited. Furthermore, the cost of extraction should be a relatively small part of the total power cost, even in the Eastern U.S., where it is probably too high for geothermal energy to be considered for conventional power cycles.

Solar collectors could also be used to supply low-grade heat for hydride compressors. The necessary temperatures can be readily achieved by using flat-plate solar collectors with selective coatings to provide high absorptivity in the visible range, low in the infrared. Mirror systems, which are used to achieve the temperatures necessary for conventional cycles, are not needed in this application, but a low-concentration, low-cost, nonsteerable mirror system combined with a flat-plate collector may be desirable.

For solar collectors with an average rating (averaged diurnally and annually) of 200 w/m^2, one could afford to pay \$7/m^2 for the solar collector, assuming a capital investment of \$35/kw(t) for solar heat. There may be a problem in achieving this cost, and the cost of thermal storage must also be included in any detailed assessment.

If overpressure in hydride compressors can be minimized, solar ponds with working temperatures of about 70°C could be used as heat sources. Low-grade heat from solar ponds should cost less than that from flat-plate collectors.

A relatively large area would be required if solar flat-plate collectors or ponds were used. For a 2,000 Mw(e) installation, with η_{LT} = 0.3 and a 50% coverage of land by solar collectors or ponds, with an average rating of 200 w/m^2, a collection area of 66 km^2 would be required, corresponding to a circle of 4.6-km radius around the installation.

The transmission costs for H_2, as discussed previously, would be quite small for a collection area of this size. The land might be difficult to acquire or too expensive in some locations, although in general Western locations should present no problem. Acquisition costs of several thousand dollars per acre would be economically acceptable.

Converting existing high temperature heat sources, either nuclear or fossil, to use a high efficiency cycle is attractive and should be possible for many locations. The principal expense would be for the low temperature heat source and a hydride compressor. The incremental cost of modifications could be substantially less that the cost of building new generating plants. The electrical output of such a converted plant should be more than doubled.

Transmission of Hydrogen from Compressors to High Temperature Source

The proposed cycle requires several thousand Mw(t) of low grade solar or geothermal energy to operate the hydride compressors for a large power plant. This low grade energy is located at a number of distributed sources. The low grade heat could be transmitted by hot water in pipelines to the high temperature heat source and hydride compressors.

A better choice is to locate the hydride compressors at the low grade heat sources; H_2 is then compressed and transmitted at high pressure to the high temperature heat source, and the low pressure H_2 is returned to the low grade heat sources. The mass flow rate of H_2 gas is much smaller (typically two orders of magnitude) than the water flow rate required with the first alternative.

Also, the H_2 pipes need not be thermally insulated, and the large quantities of waste heat discharged by the hydride compressors are distributed over a wide area, avoiding localized thermal pollution.

The high temperature heat source is then served by a net of high and low pressure H_2 lines. Detailed costs of such a net depend on the specific locations of the hydride compressors, flow quantities, etc., and will be site dependent.

Implications of High Efficiency Power Cycle

The proposed cycle has several important implications:

(1) Scarce resources (oil, gas, high grade uranium ore, and high grade geothermal) can be greatly extended with the proposed high efficiency cycle. Their consumption per kwh output can be cut almost by a factor of 3.

(2) The proposed high efficiency cycle should result in substantial cost and environmental benefits. The capital cost and fuel cost per Kw(e) of the high temperature heat source should be cut by almost a factor of 3. Considerable reductions in environmental impact associated with electric generation are possible with the high efficiency cycle. These include reductions in land disruption (strip and underground mining), water and air pollution (acid mine drainage, gaseous and particulate emissions), and radioactive waste storage and disposal requirements.

(3) Existing power plants could be retrofitted with the high efficiency cycle where an adequate supply of low temperature heat is available. The capital cost of retrofitting an existing power plant should be substantially less than that of constructing a new plant for additional capacity, and fuel costs would also be lowered.

(4) For power plant sites with suitable low-grade energy sources, the proposed cycle would convert a high temperature heat to electricity with higher efficiency than any alternative cycle. For example, an advanced topping cycle (MHD, thermionic, potassium rankine) combined with a conventional steam cycle would only convert 55 to 60% of the high temperature heat to electricity vs. 90% with the proposed cycle. Similarly, using electrolyzers the proposed cycle could convert 70% of high temperature heat to H_2 chemical energy, which is considerably better than any thermochemical H_2 process.

METAL HYDRIDES FOR GROUND TRANSPORT

Among the hydrides studied at Brookhaven (Report BNL-18719), two are of special interest for storage use. Magnesium hydride formed from the metal and containing a small percentage of nickel as a catalyst has been found to be a lightweight material with a potential for substituting for gasoline in conventional vehicle designs. The second material $FeTiH_2$, has ideal properties for use as a storage medium but is a heavy material which incurs a serious weight penalty when considered for vehicle use. The study of other hydride materials is continuing.

Groups of hydrides showing a potential for automotive use as a result of the initial review are magnesium and aluminum binary and complex hydrides and magnesium and aluminum alloy hydrides.

For hydrogen fuel storage and particularly in transportation systems, a simple, direct, and reversible reaction is desired. For example, a magnesium-nickel alloy can be reacted with hydrogen under specific conditions of temperature and pressure to produce Mg_2NiH_4. Reduction of the pressure or an increase in the temperature will then cause the hydride to dissociate and release hydrogen gas. An example of this is shown in Figure 4.8.

FIGURE 4.8: PRESSURE COMPOSITION ISOTHERM OF Mg_2Ni-H_2 SYSTEM

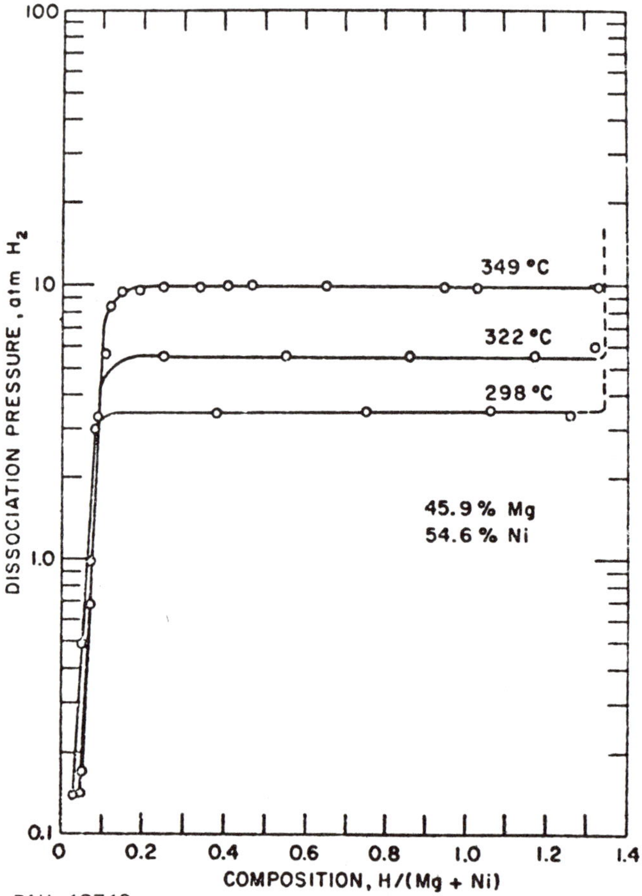

Source: BNL-18719

The equilibrium relationship for Mg_2NiH_4 is presented in this plot of pressure vs. composition. As the pressure over Mg_2Ni alloy is increased, hydrogen is taken up as a solid solution of hydrogen in the alloy. A pressure plateau then appears as the hydride starts to form. This effect continues until all of the Mg_2Ni has been converted to the hydride at a composition of Mg_2NiH_4. It is the plateau region which is of interest in storage applications since a large adsorption of hydrogen can be accomplished at a relatively constant pressure.

In practice, an hysteresis effect is present depending on whether the hydriding or dehydriding reaction is taking place. This phenomenon is shown clearly in the case of FeTi, Figure 4.9.

FIGURE 4.9: HYSTERESIS IN THE FeTi-H SYSTEM (40°C)

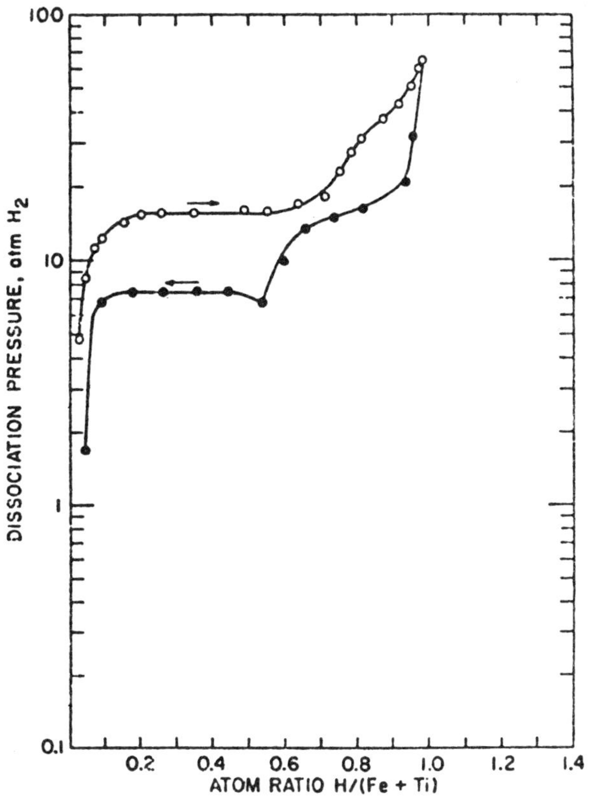

Source: BNL-18719

The hydrides available suffer varying weight disadvantages when compared with gasoline use in vehicles. However, a comparison of this type is unrealistic today when environmental and fuel source restraints are so important. By incorporating new vehicle developments at the same time conversion of vehicles to hydrogen fuel is being accomplished, hydrides can probably be included without penalty.

For example, higher engine efficiencies realized with hydrogen reduce the weight of hydrogen required for a given performance rating. A saving of up to 1,000 pounds in vehicle weight has been suggested as a benefit derived from the use of the Wankel engine in standard automobiles (24). By the use of judicious trade-offs and development efforts aimed at taking advantage of hydrides and hydrogen, it is reasonable to believe that a vehicle approaching the performance of today's automobile can be produced.

A part of the Brookhaven effort is directed toward studies which will allow the evaluation of specific hydrides coupled to engines with specific characteristics and result in an optimum selection of operating conditions and components. The initial effort involves the characterization of storage bed behavior and is a prerequisite to any serious study of a complete vehicle system. Until the response of the bed to load change and heat input is determined, only steady state estimates of vehicle behavior are reasonable.

Two hydride bed concepts are under consideration. One scheme, Figure 4.10, is based on flowing hydrogen through the packed bed of hydride and using this hydrogen to transport heat directly to or from the particles depending on whether hydride dissociation or formation is in progress. The hydrogen is recirculated through the bed after passing through an external heat exchanger. This type of storage bed is referred to as a convection bed. Hydrogen is charged to or withdrawn from the circulating hydrogen loop.

FIGURE 4.10: HYDROGEN FUELED ENGINE—CONVECTIVE HYDRIDE STORAGE UNIT

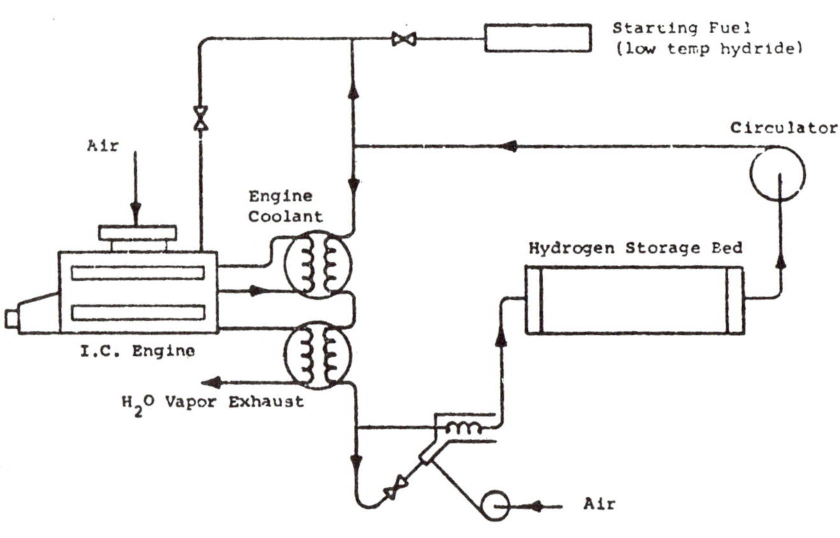

Source: BNL-18719

A second bed design being investigated, Figure 4.11, uses the conduction process to transfer heat from the engine coolant or exhaust to the hydride through heat exchange surface built into the bed. No external exchanger or circulator is required.

The choice of bed type depends to a large extent on the hydride in use. Nickel-catalyzed magnesium hydride has a small particle size which tends to pack, and unless it could be bonded into relatively large particles or fluidized, its use in a convection bed would result in a prohibitive pressure drop. This fact would probably lead to the use of a conduction bed design with internal heat exchange for this material. On the other hand, a hydride with a large or controllable particle size such as $FeTiH_2$ can be used in either configuration. Higher operating pressures characteristic of this hydride would make a convection bed with its simpler construction a strong favorite.

Modelling studies described earlier indicate that the necessary hydrogen rates can be achieved with these beds. In the case of a material like $FeTiH_2$, with a low heat requirement and a low dissociation temperature, there is little difficulty in getting the necessary heat to the hydride particles. In the case of magnesium hydride, the heat of reaction is about a third of the heat of combustion of the hydrogen released and the temperature required is approximately 600°F.

The amount of heat available as waste heat from a conventionally designed engine is limited. Engine coolant could supply some of the heat required to raise the bed to operating temperature but exhaust gas heat would have to provide the heat of reaction.

In fact, some hydrogen would have to be burned in order to provide enough heat at a suitable temperature. About 10% of the hydride would be used for this purpose.

FIGURE 4.11: HYDROGEN STORAGE SYSTEM—LOW TEMPERATURE HYDRIDE

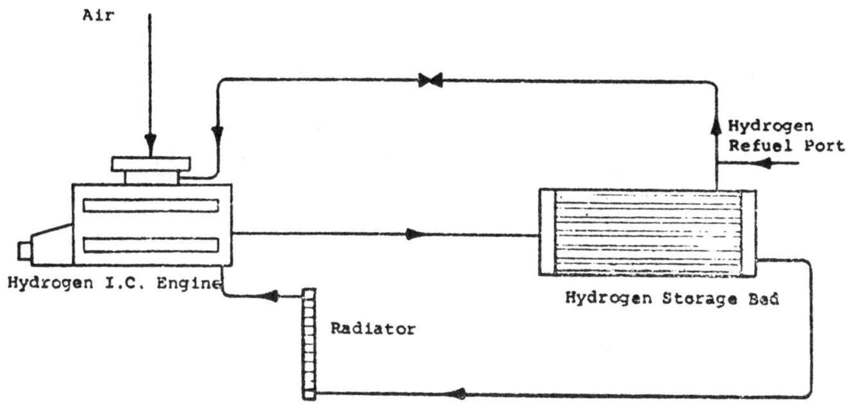

Source: BNL-18719

The charge of hydride and fuel required to match gasoline fuel performance (20 gallon supply) would be some 650 pounds if supplemental heat is considered and an engine efficiency of 23% is used. It, in fact, engine efficiency can be increased as suggested using hydrogen, the hydride and fuel load would be about 500 pounds.

Startup of a bed of MgH_2 would require a small quantity of hydrogen stored in a low temperature hydride. Models of both types of bed show that the hydrogen release occurs across a narrow band of hydride so that sensible heat is only required to heat material in the volume from the heat wall or gas inlet to the reaction zone.

As the bed is depleted, and depending on how much the bed is allowed to cool down, more heat will be required to start the hydrogen release. Compartmental operation could be used to minimize the amount of depleted material involved by isolating the depleted material from the unused hydride. A trade-off between an increase in complexity and the use of more hydride to provide extra heat would probably be required in handling the restart question.

The above discussion was directed to the use of hydrides in a conventional automobile where MgH_2 or a similar lightweight hydride is necessary. Use of this size vehicle implies the possible use of this storage technique in gasoline or diesel fueled vehicles of any size including trucks and buses. Hydrogen and hydrides should not be limited only to uses where they substitute directly and equivalently for gasoline. This approach would ignore several special situations. Examples include vehicles used in enclosed areas where hydrocarbon emissions would rule out gasoline or where hydride could substitute for other forms of energy storage such as batteries.

Superficially, $FeTiH_2$ would seem to be too heavy to consider for vehicle use. However, examination of its properties and some preliminary energy balances show it to have a surprising potential for application. Low equilibrium temperatures and a low heat of dissociation make it very simple to use in a vehicle where engine coolant can easily provide the heat and where ambient temperatures are adequate for starting purposes.

The hydride weight ultimately limits the range or performance of the vehicle by limiting the quantity of fuel carried. The use of $FeTiH_2$ results in a vehicle which will produce an engine output of 65 watt hours per pound of hydride after factoring in an efficiency of 30%, assuming waste heat utilization and limiting the charging time. The vehicle would perform reasonably like a standard auto and would have a range of about 70 miles with a 1,000 pound storage unit. The car would be an ideal urban car, comparable in purpose to electric vehicles under development. Experience with $FeTiH_2$ indicates that it would probably be safer than gasoline. Refueling times of 15 to 30 minutes are anticipated.

Another hydride application which could be considered is as a source of hydrogen for injection into gasoline fueled I.C. engines, the concept under development at the Jet Propulsion Laboratory for NASA (25). The use of hydrides in the form of either of the hydrides discussed would eliminate many of the complexities involved in the use of a reformer and not add to the hydrocarbon emission problem. Hydrides would of course add complexity to the refueling operation.

The development and introduction of hydrides for vehicular use is best done in terms of fleet vehicle systems. A taxi or bus fleet with a central garage operation would provide the proper degree of control over fuel storage and charging operations. Regular operating and maintenance schedules would be relatively certain. To reach this stage requires the development of efficient and safe storage bed designs, the selection and testing of one or more engines, the design of the vehicle and the integration of the fuel storage and engine systems.

REFERENCES

(1) Hoffman, K.C. et al, "Metal Hydrides as a Source of Fuel for Vehicular Propulsion," paper SAE 690232 presented at the International Automotive Engineering Conference, Detroit, January 13-17, 1969.

(2) van Vucht, J.H.N., et al, "Reversible Room Temperature Absorption of Large Quantities of Hydrogen by Intermetallic Compounds," Philips Research Reports 25, 133-40, Eindhoven, The Netherlands, Phillips Research Laboratories, April 1970.

(3) Wiswell, R.H. Jr. and Reilly, J.J., "Metal Hydrides for Energy Storage," Proceedings of 7th Intersociety Energy Conversion Eng. Conf. 1972, San Diego, Calif., Sept. 25-29, 1972.

(4) Buchner, H. and Saufferer, H., "The Short and Medium Term Development and Practical Application of Hydrogen Powered Vehicles," presented at NATO Conference, Dusseldorf, 1974.

(5) Wall, T. and Feeney, R., "An Engineering Study of the Storage of Hydrogen on a Bus," ME 44 Report, Mech. Eng. Dept., Stevens Inst. of Tech., May, 1975. (R.F. McAlevy, advisor).

(6) Hord, J. (ed.), "Selected Topics on Hydrogen Fuel," Report NBS IR-75-803, U.S. Nat. Bur. of Stds., Jan. 1975.

(7) Strickland, G., J.J. Reilly and R.H. Wiswall, *Proceedings of the Hydrogen Economy Miami Energy Conference,* University of Miami, Miami Beach, Florida, March 1974, p. S4-10.

(8) Harris, J.A. and M.C. Van Wanderham, Pratt and Whitney Aircraft Report NASA CR-124394, July 1973.

(9) Reilly, J.J. and Wiswall, R.H., "The Formation and Properties of Iron-Titanium Hydride," BNL-17711, January 1973.

(10) Reilly, J.J. and Wiswall, R.H., "The Reaction of Hydrogen with Alloys of Magnesium and Nickel and the Formation of Mg_2NiH_4," *Inorganic Chemistry,* p. 2254 -2256, #11, November 1968.

(11) Wirsing, E., Brookhaven National Laboratory, January 1974.

(12) Reilly, J.J., Wiswall, R.H. and Hoffman, K.C., "Metal Hydrides as a Source of Hydrogen Fuel," Pres. Division of Fuel Chemistry, Am. Chem. Soc., Chicago, September 1970.

(13) Eckert, R.G. and Drake, R.M., *Heat and Mass Transfer,* McGraw-Hill Book Company, Inc., New York (1969).

(14) Gupta, A.S. and Thodos, G., "Direct Analogy Between Mass and Heat Transfer to Beds of Spheres," *Journal American Institute of Chemical Engineers* 9, 751-754 (1963).

(15) Gupta, A.S. and Thodos, G., "Transitional Behavior for the Simultaneous Mass and Heat Transfer of Gases Flowing Through Packed and Distended Beds of Spheres," *Ind. Eng. Chem. Fundamentals* 3, 218-220 (1964).

(16) McAdams, W.H., *Heat Transmission,* p. 219, McGraw-Hill Book Company, Inc., New York (1964).

(17) Salzano, F.J., ed. "Hydrogen Storage and Production in Utility Systems," First Annual Report, BNL-19249 (July 1974).

(18) Strickland, G., and Reilly, J.J., "Operating Manual for the PSE&G Hydrogen Reservoir Containing Iron Titanium Hydride," BNL-50421 (February 1974).

(19) Burger, J.W. et al, "Energy Storage for Utilities via Hydrogen Systems," Proceedings, 9th IECEC Conference, p. 428, San Francisco, California (Aug. 26-30, 1974).

(20) Rex, R.W. and Howell, D.J., "Assessment of Geothermal Resources," in *Geothermal Energy,* Chap. 3, P. Kruger and C. Otte, Editors, Stanford University Press 1973.
(21) Blackwell, D.D., "Geothermal Resources of Marysville, Montana," Proceedings, 9th IECEC Conference, San Francisco, California (Aug. 26-30 1974).
(22) Alteseimer, J.H., "Geothermal Well Technology and Potential Applications of Subterrene Devices, A Status Review," LA-5689-MS (1974).
(23) Austin, A.L., Higgins, G.H., and Howard, J.H., "The Total Flow Concept for Recovery of Energy from Geothermal Hot Brine Deposits," UCRL-51366 (1973).
(24) Cole, D.E., "Update on the Wankel Rotary Engine," EPA-First Symposium on Low Pollution Power Systems Development, Ann Arbor, Michigan, Oct. 14-19, 1973.
(25) Breshears, R., Cotrill, H., and Rupe, J., "Partial Hydrogen Injection into Internal Combustion Engines – Effect on Emissions and Fuel Economy," EPA-First Symposium on Low Pollution Power Systems Development, Ann Arbor, Michigan, Oct. 14-19, 1973.

USAGE OF HYDROGEN FOR FUEL AND ENERGY STORAGE

The following reports were the sources of material for this chapter: N74-11728-735, N74-16800, AD-787 484, BNL-18721, BNL-19266, BNL-19249, ORNL-TM-4305, LAUR-73715 and UCRL-51228. A complete bibliography starts on page 283.

HYDROGEN UTILIZATION DEVICES

Water-Modified H_2-O_2 (Aphodid) Burner

When a hydrogen combustion process incorporates the use of oxygen rather than air as the oxidizer, no nitrogen compounds or other undesirable gases will be present in the exhaust. Equation [1] represents the theoretical hydrogen-oxygen reaction to water vapor products, while equation [2] illustrates the same reaction with air.

$$[1] \quad 2H_2 + O_2 \longrightarrow 2H_2O$$

$$[2] \quad 2H_2 + O_2 + 3.76N_2 \longrightarrow 2H_2O + 3.76N_2$$

Note that in the H_2-air reaction 3.76 mols of nitrogen are involved with the reaction but do not contribute to that reaction. In fact, since the nitrogen is in contact with the burning hydrogen-oxygen mixture it absorbs a portion of the combustion energy which results in a lower theoretical flame temperature for the overall mixture. The accepted value of theoretical flame temperature for the reaction shown in equation [1] is 3080°K, while that for equation [2] is 2400°K. If equation [2] is written so as to include excess air, the following equation can result.

$$[3] \quad 2H_2 + xO_2 + 3.76xN_2 \longrightarrow 2H_2O + yN_2 + zNO_X + O_2$$

Note here that at the reaction temperature at least part of the excess oxygen has reacted with the excess nitrogen to form various oxides of nitrogen (NO_X).

The NO_x products can later combine with hydrocarbons and sunlight in an urban environment to form photochemical smog. In addition to the undesirable polluting effect of the reaction shown in equation [3], the additional quantity of products drives the combustion temperature still lower. Thermodynamically, the maximum possible efficiency of a heat engine is represented by the Carnot cycle efficiency

$$[4] \qquad \eta = \frac{T_H - T_L}{T_H} \times 100$$

where T_H represents the temperature of heat supply (combustion) to the engine and T_L represents the temperature of heat rejection from the engine. This relationship illustrates that as the combustion temperature of an engine is reduced the theoretical efficiency will also be reduced. Obviously, the reaction represented by equation [1] is preferable to those of equations [2] or [3], since the engine operating with this reaction, equation [1], would not only tend to be more efficient but also would be nonpolluting.

The aphodid burner is a device that uses hydrogen and oxygen as in equation [1] to produce high temperature steam that can later be used by any one of several types of steam engines or processes. This burner also uses an additional water input to lower the combustor exit temperature to metallurgically acceptable values.

Ideally the water spray could be eliminated and the resulting reaction, equation [1], would produce the theoretical flame temperature of 3080°K. This temperature would in turn produce a Carnot cycle efficiency of

$$[5] \qquad \eta = \frac{(3{,}080 - 294)(100)}{3{,}080} = 90\%$$

if the heat rejection temperature is taken to be T_L = 294°K. This efficiency then becomes the limiting efficiency in a cycle using an aphodid burner. Assuming a water spray input to reduce aphodid burner outlet temperature to 2480°K, in a steam turbine which uses cooled turbine blades capable of withstanding these combustion gas temperatures, and assuming reasonable mechanical and electrical efficiencies, an overall cycle efficiency of 60% seems likely. In fact, efficiencies of up to 70% have been predicted by the General Electric Corporation (1).

Present steam power plants use fossil fuels and air at combustion temperatures of about 1875°K to produce steam in a heat exchanger at an approximate temperature of 860°K. The resulting efficiency for such a modern plant operation under these conditions will be approximately 44%. This cycle efficiency is not expected to increase significantly as long as the present type of boiler is used.

The same type of plant, however, using an aphodid burner operating on hydrogen and oxygen should be able to increase this efficiency by sixteen percentage points to 60%, or an increase of 36% over a similar fossil fueled plant. In power plant type applications the aphodid burner actually takes the place of the conventional boiler. Its greatest advantage, however, is its smaller size. Studies at Rockwell International indicate that this burner may replace boilers that are 500 times larger in physical size (2). They estimate that the aphodid burner will be smaller than an average living room while a comparable fossil fuel boiler would be nearly as large as a six story building. Conventional power plants emit many or all of the following atmospheric contaminants depending on the type of fuel used: ash, soot, carbon dioxide, carbon monoxide, sulfur dioxide, nitrogen oxides, water vapor, and others. An aphodid burner plant in contrast would emit only pure water and this would probably be in the liquid form that would not affect the atmosphere.

The aphodid burner not only promises a significant breakthrough in steam cycle efficiencies, but also provides a means of generating power without objectionable air pollution. This burner is still in the experimental stage but there appears to be no serious obstacle to either its development or the development of its associated turbine equipment.

Hydrogen Fueled Internal Combustion Engine

Today's internal-combustion (I-C) engine traditionally burns fuel with air to produce shaft power. This fuel is usually liquid or gaseous hydrocarbons derived from petroleum, and may be brought to the engine in pipelines (stationary engines) or carried with the engine (mobile power plant). The engine generally falls into one of the following categories: carbureted reciprocating, injected reciprocating, rotary trochoid (Wankel), rotary turbine (gas turbine or jet engine), and rocket.

In all but the last of the above engines the fuel acts on the engine's working fluid (air) to increase its energy content so that the engine can extract a portion of this energy as useful work. In theory, the performance of an engine is unaffected by the type of fuel. In reality, however, the performance and emissions of an engine are directly related to the type of fuel burned.

Engine combustion is extremely complex and any analysis should consider both the engine and the fuel as a system, forming a basis for comparison such that the same engine operating on two different fuels forms two separate systems. In theory, I-C engines will utilize a hydrocarbon fuel (isooctane, C_8H_{18}) in an air oxidizer as described by the following simplified equation.

$$C_8H_{18} + (12.5)O_2 + (12.5)(3.76)N_2 \longrightarrow 8CO_2 + 9H_2O + 47N_2$$

In reality, however, the reaction seldom if ever progresses to this ideal end state. In addition, most fossil fuels contain impurities such as sulfur and metal compounds. More realistic reaction products, therefore, will include carbon dioxide, water, nitrogen, carbon monoxide, nitrogen oxides, hydrocarbons, sulfur compounds, metallic compounds, alcohols, aldehydes, and ketones; all but the first three of which are undesirable air pollutants. In contrast to the reaction described in the above equation, an engine fueled with a hydrogen-air mixture would theoretically yield the following products.

$$2H_2 + O_2 + 3.76N_2 \longrightarrow 2H_2O + 3.76N_2$$

In an actual engine with excess air entering the reaction, one would expect the following reaction.

$$2H_2 + xO_2 + (3.76x)N_2 \longrightarrow 2H_2O + yNO_X + zN_2 + O_2$$

The only expected atmospheric pollution from a hydrogen fueled engine would be nitrogen oxide compounds. The combustion and emission characteristics of hydrogen appear so favorable that research into hydrogen engine technology has expanded at a rapid rate, especially in the last few years. Results have been so favorable that all of the previously defined engine types have been operated on hydrogen with relative success.

Extensive tests relating emissions and performance have been performed by R.G. Murray and R.J. Schoeppel at Oklahoma State University (OSU) using an injected reciprocating single cylinder engine. Wide-open-throttle nitric oxide emission characteristics of this engine and similar fossil fueled engines are shown in Figure 5.1.

FIGURE 5.1: EMISSION CHARACTERISTICS OF OKLAHOMA STATE UNIVERSITY HYDROGEN ENGINE

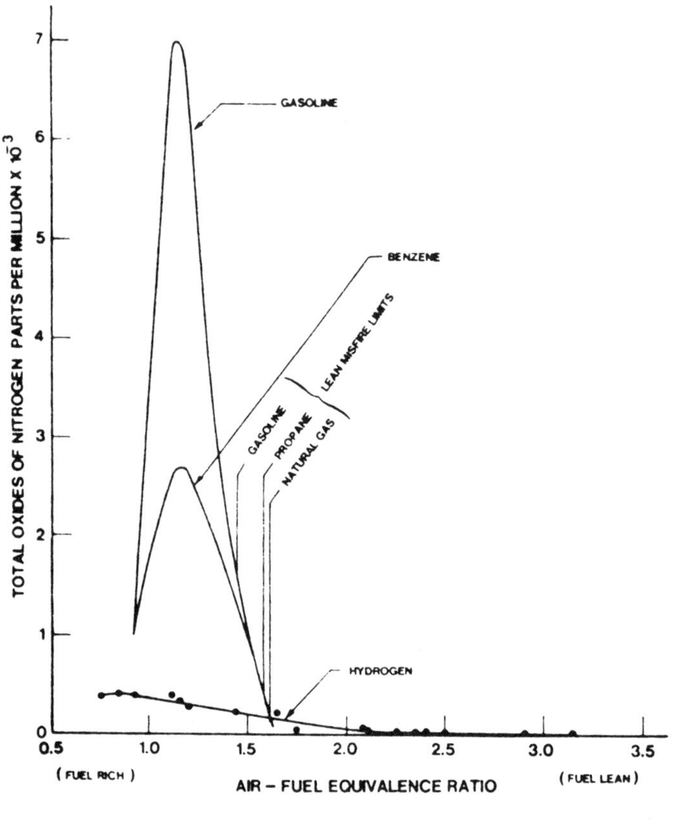

Source: N74-11728-735

The top curve of Figure 5.1 represents an expected maximum emission from a carbureted gasoline fueled engine, while the middle curve represents a reasonable minimum for a carbureted fossil fueled spark ignition (S-I) engine. The lower curve represents measured NO_x emissions from the OSU engine, which is less than one-fifth of the expected emissions shown by the other two curves.

The injection of hydrogen at or near proper combustion time eliminated any tendency for knock or crankcase explosion, provided smooth engine operation throughout a wide range of air-fuel ratios and engine speeds, and eliminated hydrogen displacement of air in the induction system which yielded an improved volumetric efficiency.

This engine was designed specifically to test emission and performance characteristics. No attempt was made to improve emissions or adapt the engine to a practical use. Operational characteristics, however, indicated that further development could have improved performance and adapted the engine to a practical use.

The work of Murray and Schoeppel was directed primarily towards correlating engine emissions to performance. Previous to their effort, extensive engine testing had been performed in Germany by Rudolph Erren and was later continued in England by Erren and Campbell (3). Erren's work led to successful vehicle operation in Germany before World War II. These vehicles used either hydrogen or hydrogen-hydrocarbon mixtures for fuel. Unfortunately, most of the documentation of this work was destroyed in World War II.

R.O. King and coworkers experimented with hydrogen fueled engines but were unable to operate engines for extended (more than 12 hours) periods of time without experiencing severe knock. The knock encountered by King was caused by carbon or ash deposits from burned lubricating oil. This tendency towards knock may limit the usefulness of the carbureted hydrogen engine in some utilization areas. Recently several automobiles have operated for limited periods of time on hydrogen, notable among these are:

M.R. Swain and R.R. Adt of the University of Miami, who have modified a 1,600 cc Toyota engine with a 9 to 1 compression ratio. This engine uses a modified injection technique.

A UCLA student-faculty group has operated several single and multicylinder carbureted engines on hydrogen as well as an 8 cylinder Ford engine-equipped Gremlin automobile.

An Illinois Institute of Technology student-faculty group ran a 1972 Vega with a propane carburetor (4).

Roger Billings and others at Brigham Young University entered a carbureted Volkswagen in the 1972 urban vehicle competition and won the emissions category in a field of over 60 competing vehicles.

The peak emissions reported by Billings are about ten times higher than those reported by Murray and Schoeppel. The carbureted technique used by Billings may account for this marked increase in pollution level. The injection principle undoubtedly leads to charge stratification with a resulting reduction in NO_x emission. All of the above investigators except Erren have limited their investigation to spark ignition (S-I) reciprocating engines. Erren worked with both S-I and compression ignition (C-I) engines successfully. Investigators at the Brookhaven National Laboratory have operated a Wankel (rotary trochoid) engine on mixtures of hydrogen and air. Successful operation was achieved; however, no emissions data are available.

The first reported gas turbine to operate on hydrogen was a converted J-57 Pratt and Whitney engine (see Figure 5.2). Gaseous hydrogen was supplied to J-57 axial tube single-can burners by an auxiliary centrifugal pump and gasifier system. The stable operation of the J-57 confirmed that conventional gas turbines could be readily adapted to hydrogen fuel (5).

FIGURE 5.2: SCHEMATIC DIAGRAM OF THE MODIFIED J-57 ENGINE

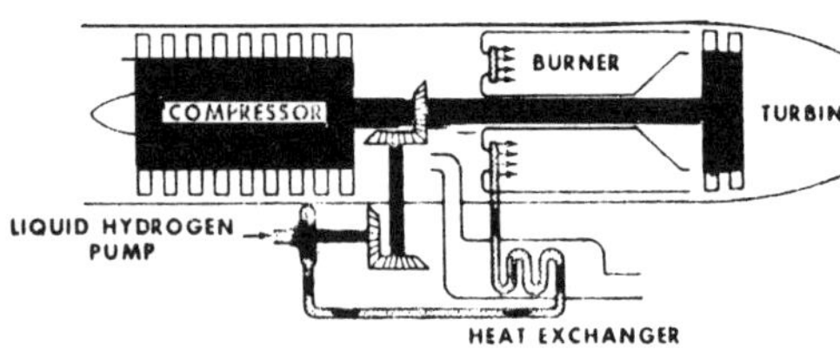

Source: N74-11728-735

A second turbine, the Pratt and Whitney 304 (see Figure 5.3), was specifically designed to utilize the expansion of liquid hydrogen to provide the work for the compressor. The combustion energy was utilized to provide thrust. All performance predictions for this engine were confirmed within the first 25 test hours when the project was terminated in 1958.

FIGURE 5.3: SCHEMATIC DIAGRAM OF THE PRATT AND WHITNEY 304 ENGINE

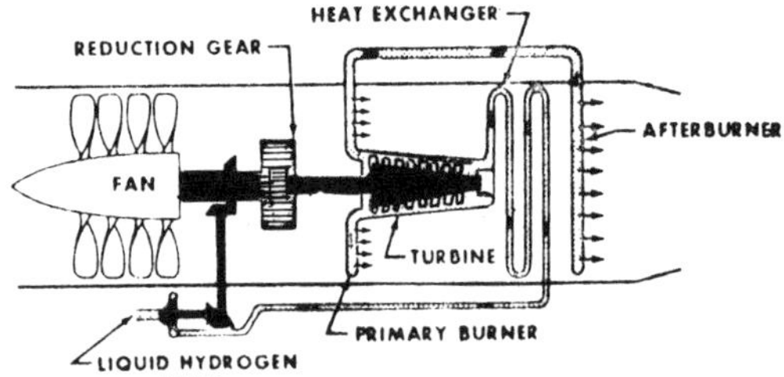

Source: N74-11728-735

The last reported hydrogen gas turbine operation occurred when the National Advisory Committee for Aeronautics (NACA) converted one engine of a B-57 bomber to hydrogen. Testing started with the airplane at cruise whereupon the left engine was switched to liquid hydrogen from the left tiptank. No performance or emission data are reported for any of these engines; however, all were reported to have operated excellently. (These turbine engines are discussed further in the section, Fuel for Transportation Systems.)

Hydrogen-oxygen fueled rocket engines have been a spectacular success in recent years. The Saturn series of rockets has utilized H_2-O_2 engines in upper stages for numerous successful earth and moon orbits. Liquid hydrogen is pumped to 1,000 psi and directed to the thrust chamber cooling jacket where its temperature is increased sharply. It then expands through a turbine which provides the pressurization work for both the hydrogen and the oxygen. These engines have undergone over 100 firings for a total time of 6 hours without repair.

Hydrogen fueled engine research and development is far from complete; however, the preliminary efforts performed to date indicate that all existing engine types can be operated with hydrogen. All of these engines appear to be capable of excellent performance characteristics at very low levels of pollution.

Catalytic Burners

A particular property of hydrogen that gives it a marked advantage over other fuels is its ability to react with the oxygen in air when in contact with a catalyst. This property has allowed the fabrication of unusual flameless heating devices. Research by the Institute of Gas Technology and the American Gas Association has produced several types of hydrogen fueled catalytic burners that have been successfully demonstrated to the public (4).

Detailed information is not available on these burners; however, it is known that a very thin catalytic film, probably platinum or palladium, is suspended on a substrate (ceramic). When hydrogen is passed over this surface, flameless combustion takes place. The rate of combustion and the resulting temperature are functions of the rate of fuel supply and the exposed surface area of the catalyst.

An important advantage of this burner is that its operating temperature is under 1750°K, which is below the temperature of formation of significant quantities of nitrogen oxides. Another advantage of a catalytic burner is its inherent safety. An inoperative ignition source or a burner accidentally left on poses a serious problem in a natural gas appliance. A hydrogen catalytic heater, on the other hand, would only produce unwanted heat if accidentally left on.

Developmental efforts on catalytic burners are far from complete; however, it appears that such a device could be used for water heating, cooking, process heating and space heating. The only likely technical obstacles to development of practical systems are cost and availability of catalytic materials. These materials must be not only inexpensive, but impervious to poisoning by fuel additives and impurities.

Flame Burners

Hydrogen in the gaseous form can be used as fuel for all combustion applications that use natural gas today. Hydrogen combustion characteristics and the effect they have on existing natural gas burners are discussed in this section. The changes that will need to be made in burner design, valves, and metering devices are also studied.

There are essentially two types of combustors used today for heating and cooking, the catalytic combustor described previously and the gas phase or open flame combustor that is examined in this section. The most common type of burner today, the flame burner, works on the principle that a jet of fuel entrains air and the flammable mixture is burned downstream in one port or a series of ports.

The satisfactory performance of a burner depends upon flame stability, which is bounded by the flashback and blow-off tendency. When fuel velocity is higher than flame velocity, blow-off occurs (the flame advances away from the burner) and the flame is extinguished. If fuel flow velocity is slower than flame velocity, flame enters the burner or flashback occurs. Both of these are affected by the fuel/air ratio and gas temperature.

The combustion characteristics of hydrogen are quite different than those of methane, which is the chief constituent of natural gas. The characteristics important to flame burners will be reviewed as follows:

Hydrogen burns in air with a nonluminous almost invisible flame (7).

The temperature of hydrogen-air flames is hotter than methane-air flames at stoichiometric conditions (2400°K compared to 2190°K).

Flame velocity for hydrogen is very high, about 10 times the flame velocity of methane (7).

Limits of flammability for hydrogen and methane are similar at the lower limit but vary greatly at the upper limit. The limit of flammability of hydrogen in air at ambient conditions is 4 to 75%, methane in air is 5 to 15 volume percent.

The quenching distance, the maximum size opening through which a flame will not pass is approximately 0.06 cm for hydrogen as compared to 0.23 cm for methane (6)(8).

Ignition qualities of hydrogen (6) require much less energy than that required for methane.

All of these characteristics must be considered when designing or using flame-type burners.

Hydrogen burners will be different than natural gas burners in the following ways. To produce the same amount of heat, about three times the volume of hydrogen, as compared to methane, is required. But since the flow rate of hydrogen through a given area is approximately three times as high as methane at the same pressure, existing fuel jets should be sufficient. Since hydrogen is more prone to flashback but has a much larger flame stability range, burner port size can be made much smaller, thus increasing the quenching effect (8).

Pilot lights or some electrical ignition devices should work well because of hydrogen's low ignition energy. Hydrogen burns at about 200°K above methane flame temperature. This increase could cause existing burners to become too hot;

therefore, the selection of suitable burner materials is very important. Converting to a hydrogen fuel will necessitate the modification or replacement of all natural gas burners for the following reasons: port size will have to be reduced (flashback control), air mixture devices changed because of different fuel volumes, and the burner head itself may need replacement in order to avoid material problems due to high temperatures.

Fuel Cells

In principle, fuel cells are much like batteries; they are devices to transform chemical energy directly to electrical energy without using an engine as an intermediate step. Fuel cells differ from a battery in that the chemical energy input to the cell can be continuously replenished while a battery has a limited supply and once this supply is exhausted the battery is useless. The simplest fuel cell is a hydrogen-oxygen fuel cell as shown in Figure 5.4.

FIGURE 5.4: CROSS SECTION OF SIMPLE HYDROGEN-OXYGEN FUEL CELL

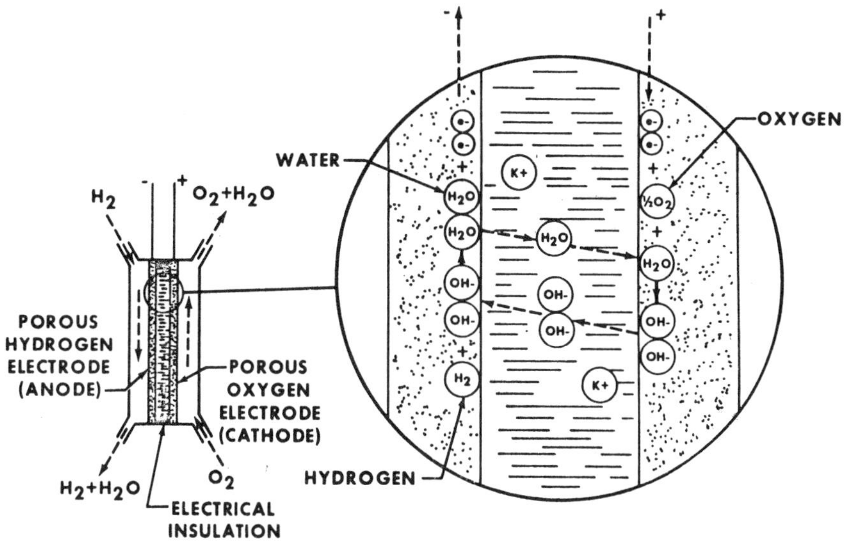

Source: N74-11728-735

Hydrogen flows to the anode where it breaks up into hydrogen ions and electrons according to the following equation.

$$\text{Anode} + H_2 + 2OH^- \longrightarrow (\text{Anode} + 2e^-) + 2H_2O$$

Oxygen, at the same time, flows to the cathode and reacts as illustrated by the following equation (shown on the following page).

$$(\text{Cathode} + 2e^-) + \tfrac{1}{2}O_2 + H_2O \longrightarrow \text{Cathode} + 2OH$$

The net result of these two reactions on opposite sides of the cell is

$$H_2 + \tfrac{1}{2}O_2 \longrightarrow H_2O$$

where the hydroxyl ion is transported from the cathode to the anode through an electrolyte and the electrons are transported from the anode to the cathode via an external circuit. This electron flow is the useful electrical output of the fuel cell. If the external circuit is opened, the cell stops functioning automatically. Electrolytes can be acids, bases, or molten salts. For hydrogen, however, potassium hydroxide (KOH) is usually preferred to acids since it is less corrosive and since under comparable conditions, the oxygen electrode gives about 0.1 v more potential.

Air can also be used as the oxidant, however, it must continuously be circulated so the excess nitrogen and other gases can be removed. If this is not done, the nitrogen will significantly interfere with oxygen-electrode contact, seriously limiting output. In addition, contaminants in the air can poison catalysts contained by the electrode, reducing the life of the cell.

Catalysts are needed at the reaction interface to assist the breakdown of hydrogen and oxygen into ions. Precious metals, such as platinum, palladium, or rhodium are the best catalysts; however, nickel can be used with hydrogen in KOH electrolytes (9). The most promising electrode technology uses plastic-bonded electrodes. Here small particles of Teflon are mixed with small particles of platinum black.

The resulting paste is pressed, heated and sintered into an electrode that is thin, flexible, and strong, yet one that remains porous and resists flooding due to the hydrophobic properties of Teflon. The electrical output of each cell is direct current electricity at a peak theoretical potential of 1.23 volts. Voltage, however, drops with increased current until the usable voltage of each cell may be approximately 0.8 volt. Cells must be stacked in series to get higher voltage.

Actual current outputs of H_2-O_2 cells usually range from 40 to 155 amp/ft^2 of projected electrode surface. Power outputs of typical H_2-O_2 fuel cells range from 50 to 2,500 watts. Cell life of up to 3,000 hours has been reported. The most spectacular success for fuel cell application has been its use in Gemini and Apollo space vehicles. When weight, required useful life and power output are all considered, the fuel cell was judged to be the best power source for these spacecraft.

Main advantages for the fuel cell are high efficiency, absence of moving parts, and lack of noise. Unlike heat engines, the fuel cell is not limited by the Carnot cycle theory. Thermal efficiences of about 83% (coulombic efficiency of nearly 100%) can be achieved at low currents.

Practical cells can run at thermal efficiencies of 60% with present technology. This efficiency can be quite attractive when compared to present steam power plant efficiencies of 55% maximum. Disadvantages are high cost, and scarcity of electrode materials. The direct current electrical output may be a disadvantage depending on application. If alternating current is required, an inverter must be used which adds to the cost and weight and reduces overall efficiency.

Fuel cells function excellently in some applications particularly in spacecraft. Advances in catalyst and electrode technology, as well as cost, will have to be made, however, before fuel cells can be considered seriously for general use in the residential, commercial, transportation or industrial market.

Other Devices

Magnetohydrodynamic (MHD) electric generation plants have received some attention in recent years as possible high efficiency generation systems. Problems with materials, and low magnetic field density however, eliminate this scheme as a practical power generation device in the near future.

The Stirling cycle engine has been in limited use for years, particularly in Europe (10)(11). This includes use both as an engine and a refrigerator. Since the Stirling engine can use heat transfer between substances, as well as combustion, to produce power, it has the capability of deriving power from the vaporization of cryogenic hydrogen. Characteristically this cycle has a higher thermal efficiency, but, further development will be needed to overcome weight and endurance problems.

H_2-O_2 COMBUSTION POWERED STEAM-MHD CENTRAL POWER SYSTEMS

Central power stations with magnetohydrodynamic (MHD) generator topping cycles have the potential for very high efficiency. Because MHD generators have no moving parts and use volume rather than surface forces to extract power, they have the ability to operate at much higher temperatures than turbines. The combustion temperature and pressure usable in MHD generators are limited only by the maximum acceptable local heat transfer rate.

High-temperature high-efficiency MHD topped power plants can be constructed with negligible efflux pollution if oxygen is used in conjunction with a clean fuel. Use of oxygen, however, requires an oxygen plant. Hydrogen as a fuel tends to minimize the difficulties associated with using oxygen since hydrogen requires a minimum of oxygen per unit of heat release.

Combustion with pure oxygen requires larger MHD generator expansion ratios (inlet/exit pressure) to extract a given fraction of the enthalpy in the flow than combustion with air. However, if large expansion ratios can be achieved, higher fractions (over 45%) of the steam enthalpy can be extracted with reasonable power density in fuel-oxygen generators. This results from the expansion process being more nearly isothermal in generators using oxygen rather than air.

For cooled-wall combustors and generators the maximum allowable combustion pressure is limited by the allowable maximum local heat transfer rates. H_2-O_2 combustion at pressures up to 30 atm does not result in heat transfer rates that are excessive for water cooled combustors. For limited inlet pressures, very high expansion ratio generators require subatmospheric generator exit pressures. Steam MHD generators can be operated with less than $^1/_{10}$ atm exit pressures since the steam can ultimately be condensed to water and pumped with little work to atmospheric pressure.

Steam MHD System Performance

The H_2-O_2 steam MHD generator can be easily integrated into cycles using steam turbines. Figure 5.5 shows one such cycle. If the H_2 and O_2 are preheated, the cycle efficiency is increased more by recycling water through the high pressure recuperative-boiler and turbine than by increasing the combustion temperature. The gross efficiency of the high MHD expansion ratio cycle is 70% (using the higher heating value) for a preheat temperature of 1500°K. Thus, this steam MHD cycle has the potential for obtaining a net efficiency of over 60% after subtraction of the power for O_2 production. Even without preheat the high expansion ratio steam MHD topped cycles offer the potential of over 50% efficiency.

FIGURE 5.5: H_2-O_2 COMBUSTION STEAM MHD POWER PLANT

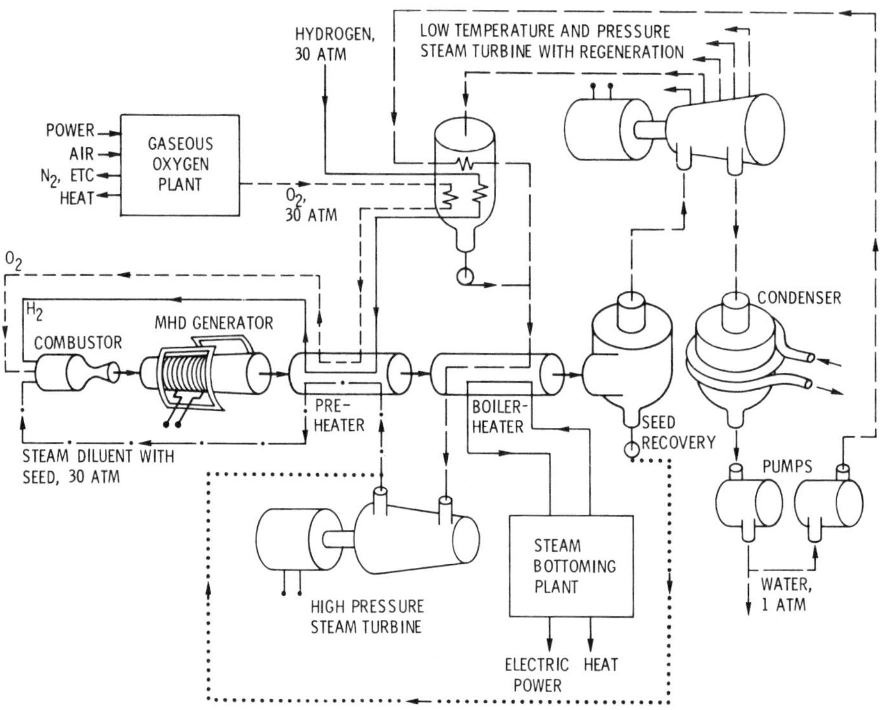

Source: N74-16800

Four options of the power plant shown in Figure 5.5 are examined herein. In option 1, the H_2, O_2, and the steam diluent to the combustor are preheated to the steam bottoming plant temperature (839°K), and in the MHD generator the flow is expanded to a total pressure of 1 atm. In options 2, 3, and 4, 1500°K preheat is assumed and the expansion is to a total pressure of 1, 0.5, and 0.1 atm respectively. All options except option 4 use a low pressure turbine to expand the flow to a pressure of 0.1 atm before the steam is condensed. In all options the combustor is operated stoichiometrically at 30 atm and 3468°K. To maintain this combustion temperature for the two different preheat temperatures the steam diluent flow is varied from 17 to 30% of the total MHD hydrogen and oxygen mass flow. Although high and low pressure steam turbines are schematically shown as distinct components in Figure 5.5, they would in fact be integral parts of the steam bottoming plant. The preheater in practice would be subdivided into low temperature and high temperature preheaters. The low temperature preheater would be an integral part of the bottoming plant boiler heater and would provide a temperature of 839°K. Separate sets of high temperature refractory pebble bed type regenerators are required to preheat the O_2 and H_2 to 1500°K. The steam diluent can be effectively used as a regenerator purge at the end of each regenerator thermal cycle.

The MHD generator is assumed to be seeded with cesium hydroxide. The cesium seed in the MHD generator exhaust can conceptually be easily recovered as liquid cesium hydroxide which would start to condense to droplets at a temperature of approximately 600°K. Since the cesium seed will condense as cesium hydroxide independent of how it was injected, it is also desirable to inject it in this form, thereby simplifying the seed recovery and generator chemistry.

For each option an MHD channel was designed with an area variation so as to maintain a constant Mach number. This constraint automatically avoids choking of the flow and adverse pressure gradients. The generator seed fraction and Mach number are optimized to maximize the average generator power density. The generator performance is calculated assuming one dimensional flow and using a multispecies equilibrium chemistry program (12) to determine the species concentration and plasma conductivity. An ideal segmented Faraday generator is assumed, but power density is decreased inversely with the Hall parameter for values greater than one to account for experimental departures from ideal performance at high Hall parameters.

Various cycle parameters and conditions for the four steam MHD options are shown in Table 5.1. Table 5.1 also contains, for comparison, similar parameters for the coal-air MHD system studies by Bergman et al (13) with lowest power cost.

TABLE 5.1: MHD POWER PLANT CHARACTERISTICS AND COSTS*

	H_2-O_2 STEAM MHD				COAL-AIR MHD
	OPTION 1	OPTION 2	OPTION 3	OPTION 4	TOPPED STEAM
Combustion Chamber					
Temperature, K	3468	3468	3468	3468	2836
Pressure, Atm	30	30	30	30	8
Preheat Temp, K	839	1500	1500	1500	1473
MHD Generator					
Expansion Ratio	30	30	60	300	6.7
Max. Magnetic Field, T	6	6	6	6	6
Enthalpy Extraction, %	31	31	37	47	23
Av. Power Density, MW/M^3	76	76	40	8.4	18

(continued)

TABLE 5.1: (continued)

	H_2-O_2 STEAM MHD				COAL-AIR MHD TOPPED STEAM
	OPTION 1	OPTION 2	OPTION 3	OPTION 4	
MHD Power/Net Power	.74	.82	.89	1.04	.63
Steam Power/Net Power	.48	.38	.30	.12	.56
O_2 Plant Power/Net Power	.21	.20	.18	.17	-
Compressor Power/Net Power	-	-	-	-	.19
High Temp Preheater Thermal Power/Net Power	0	.31	.27	.26	.52
NET EFFICIENCY, %	48	51	55	60	50
MAJOR MHD COMPONENTS, $/kW$_{MHD}$					
MHD Channel	.53	.53	.63	2.1	1.5
MHD Magnet	4.74	4.74	8.96	50.8	14.7
Inverter	15	15	15	15	15.4
High Temp Preheater	0	.92	1.79	8.85	15.2
Compressor	-	-	-	-	10.3
Sum, $/kW$_{MHD}$	20.3	21.4	26.4	76.8	57.1
CONSTRUCTION, $/kW$_e$					
MHD Plant	26.8	31.2	42.3	144	63.9
Steam Plant	44.4	35.1	27.4	11.6	54.0
Total, $/kW$_e$	71.2	66.2	69.7	156	118
Power Plant Cost, $/kW$_e$	90.4	84.1	88.5	198	150
Capital Cost, $/kW$_e$	128	119	125	280	214
Power Cost neglecting Fuel Cost, mills/kW-hr$_e$					
Power Plant	4.68	4.35	4.58	10.23	8.03
O_2 Plant Other than Elec.	1.03	.98	.91	.83	-
Total	5.71	5.33	5.49	11.06	8.03

*Based on 1974 dollars.

Source: N74-16800

Capital and Operating Costs for Steam MHD Systems

Estimates of the cost of steam MHD systems are made in a manner consistent with previous cost estimates for coal-air MHD systems by Bergman et al (13). Analysis of their study indicates that MHD system cost can be well estimated from an assessment of the cost of the major MHD plant components, the steam plant cost, and the relative power outputs of the MHD and steam plants. The major cost MHD components are the MHD generator channel and magnet, the inverters required to convert the DC output to AC, and the high temperature preheater.

The resulting component costs and the capital cost for the four steam MHD plants are shown in Table 5.1. MHD channel cost was estimated on the basis of wall area using a cost of \$13,000/m^2. The magnet cost was scaled on the basis of the total magnetic energy in the channel assuming \$0.025/J.

These costs as well as the high temperature preheater cost are based on scaling the cost of Bergman et al (13). For the clean fuel heated preheater their option 2 was scaled. It was based on previous studies by Heywood and Womach (14). For this study the preheater structure cost was scaled by the ratio of the maximum pressures and the resulting total cost was scaled on the basis of the required heat transfer surface assuming turbulent heat transfer and equal fractional pressure drops.

As shown in Table 5.1, steam MHD options 1, 2, and 3 all yield low capital cost. Option 4, however, has significantly higher cost. The large cost increase in option 4 is primarily caused by the magnet cost associated with the lower average MHD power density operation. Only in option 4 does the high temperature preheater cost become significant because of the low pressure in the hot gas flow.

In the other three options this cost is small compared to the coal-air MHD system because of the following: [1] clean gas heated heat exchangers are significantly less expensive than those heated by steams containing coal ash; [2] the steam MHD systems preheater heat flows are a much smaller fraction of the electric power; and [3] the steam MHD system preheaters operate with higher T than the coal-air MHD system.

Table 5.1 also shows the cost of power neglecting the fuel cost for the four steam MHD options. This is estimated on the basis of 7,000 hr/yr plant operation with the annual power production cost being 25.6% of the power plant capital cost plus the oxygen plant production cost for other than electric power. The power costs minus fuel costs of the steam MHD options 2 and 3 (5.3 and 5.5 mills/kwh_e respectively) compare favorably with similar costs for either the coal-air MHD (8.0 mills/kwh_e) or conventional coal fired steam plants (6.7 mills per kwh_e with no SO_x removal, 8.7 ± 0.5 mills/kwh_e with SO_x removal).

Comparison of Power Systems

Table 5.2 shows representative average power cost for conventional steam plants using various fuels and for both coal-air MHD and steam MHD systems. In this table, energy and/or power transportation cost is neglected and only east/midwest surface mined coal is considered. Fuel costs are based on 1985 projections.

Table 5.2 shows that the option 3 is the most attractive steam MHD system. This 55% efficient steam MHD system is also more attractive than all alternative clean fuel conventional steam power plants, particularly if they are fueled from domestic shale, coal, or crude. Table 5.2 also shows that the direct coal fired power plants can produce power at the lowest average cost, the lowest cost system being a conventional coal fired steam plant with no SO_x removal. It is 0.9 mills/kwh_e below the coal-air MHD topped steam plant which would remove SO_x.

The power production cost will be only one major factor in determining future power plant choices; additional aspects of the cost of power in a city as well as the environmental impact of the power plant must be considered. In Figure 5.6 the cost of power in a city is evaluated for various power plant concepts. The city is assumed to be either or both 100 miles from east or midwest surface coal mines or 1,500 miles from western surface coal, syngas, syncrude (shale or coal), and domestic crude.

FIGURE 5.6: COMPARISON OF CITY POWER COSTS FOR VARIOUS POWER PLANT CONCEPTS

ELECTRIC PLANT LOCATION AND TYPE	PLANT EFFICIENCY, %	ENERGY/POWER TRANSPORTATION
I. IN-CITY PLANTS A. COAL-AIR 1. CONVENTIONAL STEAM A. NO SO_x REMOVAL B. WITH SO_x REMOVAL 2. MHD TOPPED STEAM	 40 40 50	EAST/MIDWEST COAL 100 MI BY RAIL WEST COAL 1500 MI BY EITHER UNIT OR INTEGRAL TRAIN
B. CONVENTIONAL PLANTS CLEAN FUEL-AIR 1. OIL 2. GAS (SYNGAS) 3. HYDROGEN	 40 40 40	1500 MI PIPE TRANSMISSION EXCEPT IMPORTED AND H_2 H_2 AS BELOW
C. H_2-O_2 STEAM MHD (INTEGRAL O_2 PLANT GASSIFIER 100 MI FROM CITY)	55	EAST/MIDWEST GASSIFIER AT COAL MINE WEST COAL 1500 MI TO GASSIFIER 100 MI GH_2 PIPE TRANSMISSION
II. COAL PLANTS 100 MI FROM CITY A. OVERHEAD ELECTRIC TRANS 1. STEAM SO_x REMOVAL 2. MHD TOPPED STEAM B. UNDERGROUND ELECTRIC TRANS 1. STEAM SO_x REMOVAL 2. MHD TOPPED STEAM	 40 50 40 50	EAST/MIDWEST PLANT AT COAL MINE WEST COAL 1500 MI BY RAIL 100 MI ELECTRIC TRANS
III. WESTERN COAL MINE PLANTS A. STEAM SO_x REMOVAL B. MHD TOPPED STEAM	 40 50	1500 MI OVERHEAD ELECTRIC TRANS

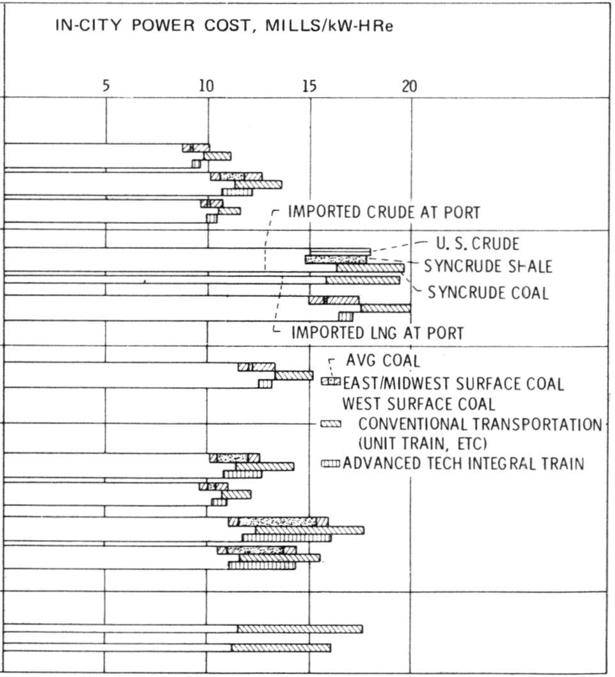

Source: N74-16800

TABLE 5.2: REPRESENTATIVE POWER COST (ENERGY/POWER TRANSPORTATION NEGLECTED)

TYPE PLANT	EFFICIENCY %	AV. FUEL COST* mills/kW-hr$_e$	AV. POWER COST mills/kW-hr$_e$
CONVENTIONAL STEAM			
East/Midwest Coal			
No SO_x Removal	40	2.2	8.9
With SO_x Removal	40	2.2	10.9
Oil			
Domestic Crude	40	8.9	15.5
Imported Crude	40	6.2	12.8
Syncrude - Shale	40	8.1	14.7
Syncrude - Coal	40	10.4	17.0
Gas			
Syngas - West Coal	40	8.7	15.4
Imported LNG	40	7.1	13.7
Hydrogen			
East/Midwest Coal	40	8.8	15.5
COAL-AIR MHD TOPPED STEAM			
East/Midwest Coal	50	1.8	9.8
H_2-O_2 STEAM MHD			
East/Midwest Coal			
Option 1	48	7.3	13.0
Option 2	51	6.9	12.2
Option 3	55	6.4	11.9
Option 4	60	5.9	16.9

*Based on 1985 projections.

Source: N74-16800

In-city power costs are calculated including energy/power transportation costs for in-city power plants using coal and clean fuels, for coal fired power plants 100 miles from the city with either overhead or underground electric transmission. For the hydrogen fueled power plants the hydrogen is assumed to be piped to the city from a gasifier located 100 miles outside the city.

Figure 5.6 shows that a 55% efficient steam MHD power plant has an attractive potential compared to domestic fueled clean fuel conventional steam power plants. This type plant is used in many cities in the United States. Oil and gas fired power plants account for approximately 25% of the U.S. electric power production. Figure 5.6 also shows that steam MHD power cost will be higher than in-city coal fired power cost but may have lower cost potential if the coal fired plant is required to be located 100 miles outside the city (the assumed gasifier location) and if electric power transmission into the city is required to be underground like the hydrogen transmission.

Advantages and Disadvantages

H_2-O_2 combustion powered steam-MHD central power systems offer the potential of producing power without air pollution (except for the emission of CO_2 at a remote gasifier), the potential of low in-city thermal pollution power plants, the potential of lowering the cost of power from clean fuel power plants, and the potential of lowering the economic penalties associated with requiring coal consumption to be remote from cities and elimination of overhead electric power transmission. Steam MHD systems should, in addition, be the least difficult MHD technology to develop because of their clean fuel and relatively simple chemistry.

The steam MHD systems examined have not been fully optimized but should be representative of the general potential of such systems. The cost estimates, herein, for these systems are of course speculative in nature as are many other aspects of a comparison of future power plants including the prospects for satisfactory SO_x removal for conventional coal-steam plants, the feasibility of developing the difficult technology required for direct coal-air MHD systems, and future fuel and transmission costs.

The major potential disadvantage of the proposed steam MHD power plant system is not associated with the MHD plant but with the hydrogen gasifier. For this study the gasifier efficiency for conversion of coal Btu's to hydrogen upper heating value Btu's is only 54%. Thus, even though the MHD plant has a high 55% efficiency, the overall efficiency of converting coal to electricity is lower than in conventional steam plants.

However, having noted this point, it should also be recognized that if significantly higher efficiency hydrogen gasifiers can be realized then large potential reductions in the estimated power costs could also be realized since these steam MHD power plants are low capital cost and high fuel systems. In a similar way improvements in oxygen plant technology could also lead to significant improvements in these proposed steam MHD systems.

OPERATING EXPERIENCES WITH HYDROGEN FUELED RECIPROCATING ENGINES

The widespread current use of reciprocating engines, both mobile and stationary, warrants attention as a potential application for hydrogen fuel. The "recip" has been developed extensively into a low cost, reliable power plant, competitive in power and efficiency with other hydrocarbon-fueled engines. While not nearly so developed as its gasoline-fueled counterparts, the hydrogen-fueled recip has already shown enough promise so as to be the most investigated of hydrogen-fueled power plants.

Owing to its similar thermodynamics and application, the rotary combustion (Wankel) engine is treated in this section along with other types of true reciprocating engines. Each of these categories has been investigated experimentally, but most attention has been given to the hydrogen/air, spark-ignition engine.

Operating experiences with hydrogen-fueled reciprocating engines can be readily categorized by way of the reactants used, hydrogen-air, hydrogen + other fuel-air, and hydrogen-oxygen; or by way of engine type, spark ignition and compression ignition; or by way of induction method, naturally-aspirated and supercharged. Reciprocating heat engines (for example, Stirling and Feher-cycle engines) which do not use hydrogen combustion products as a working fluid are not dealt with in this section.

Hydrogen/Air Engines

As mentioned above, the spark-ignition hydrogen/air engine has been most widely investigated, in fact, since the 1920s. Compression-ignition hydrogen engines were used in the 1920s and 1930s but no operating experience appears to have been reported since World War II.

Until about 1970, published data on hydrogen operation of reciprocating engines has been almost exclusively from single-cylinder, naturally-aspirated, noncylinder fuel injected (non-CFI), test-bed engines. While some early multicylinder operating experience is recognized, quantitative data and operational details are very incomplete.

Similarly, reports of fuel-injection at either low or high pressure have been very limited. Expectedly, it did not prove difficult for engines designed primarily for gasoline to be operated on hydrogen, as anticipated even by an early science fiction writer (15). No major accidents have occurred during such operations, though a few engine failures which were attributable to hydrogen use have been reported (16).

The major operational problem exposed to date is that of performance. As is shown below, all available quantitative test data indicate, for example, a substantial decrease in peak power available from a naturally aspirated spark-ignition engine operated with hydrogen rather than gasoline. This is consistent with the fundamental considerations and calculations. In contrast, however, spark-ignition supercharged (including CFI) engines and compression-ignition engines have not been investigated enough to allow such statements.

The summaries following draw performance information from such operational data as are available. Performance variables such as: power output, specific power output (power/displacement, power/piston area), mean effective pressure (MEP), thermal efficiency and exhaust emissions are emphasized in relation to operating variables such as: compression ratio, equivalence ratio (fuel/air mass ratio relative to stoichiometrically correct), and speed (of rotation and of piston displacement).

As is seen below, a comprehensive view of prior experimental data of this type allows a useful quantitative characterization of hydrogen-fuel operations, at least for the naturally-aspirated spark-ignition engine.

Spark-Ignition Engines—Overall: The historical trend of experience with hydrogen-fueled spark-ignition engines is apparent in Tables 5.3 and 5.4. These tables of engine specifications summarize prior conditions for operating both unmodified "Otto-cycle" engines and those recent engines which have been modified to provide for exhaust-gas recirculation ("EGR") and water-injection ("WI"). "Quality control" (fuel/air ratio variation), rather than "quantity control" (throttling), has been used with rare exception.

It is noteworthy that virtually all these engines have been used for relatively modest test programs using hydrogen as a fuel. Often very specific reasons for testing limit the data obtained, for example, the knock-limit tests of Downs et al (17) or Anzilotti et al (18)(19). In other cases, testing was for demonstration purposes, for example, for the 1972 Urban Vehicle Design Contest (20)(21). In Table 5.4, MEP stands for mean effective pressure.

TABLE 5.3: SUMMARY OF HYDROGEN/AIR ENGINE EXPERIENCE—ENGINE SPECIFICATIONS AND POWER OUTPUT

INVESTIGATOR	NO. OF CYLINDERS	DISPLACEMENT (IN^3)	COMPRESSION RATIO	MAX. ENGINE SPEED (RPM)	MAX. PISTON SPEED (FT/MIN)	MAX. REPORTED POWER (HP)	MAX. POWER / DISPLACEMENT (HP/IN^3)	MAX. POWER / PISTON AREA (HP/IN^2)
Tizard (55)	1	127	-	-	-	-	-	-
Ricardo	1	127	5.45/1	1500	2000	28	.22	1.76
(29)	"	"	7/1	"	"	36	.28	2.26
(38)	1	127	5.45/1	1500	2000	-	-	-
Burstall (30)	1	127	5/1	1000	1333	15.5	.12	0.97
	"	"	7/1	"	"	18.7	.15	1.18
Erren (15)	-	-	8/1	-	-	-	-	-
Egerton (54)	-	-	-	-	-	-	-	-
Oemichen (36)	-	-	12/1	1500	-	-	-	-
King (16)(28)(31)(32)	1	37	4/1-20/1	1800	1350	12	.32	1.44
Downs (17)	1	31	8/1-20/1	1500	-	-	-	-
Anzilloti (18)(19)	1	37	8/1-12/1	900	625	-	-	-
Holvenstot (27)	8	3850	8/1	550	825	217	.056	.51

(continued)

TABLE 5.3: (continued)

INVESTIGATOR	NO. OF CYLINDERS	DISPLACEMENT (IN^3)	COMPRESSION RATIO	MAX. ENGINE SPEED (RPM)	MAX. PISTON SPEED (FT/MIN)	MAX. REPORTED POWER (HP)	MAX. POWER / DISPLACEMENT (HP/IN^3)	MAX. POWER / PISTON AREA (HP/IN^2)
Karim (34)	1	37	6/1-16/1	900	675	-	-	-
(35)	1	37	14/1-20/1	900	675	-	-	-
Swain (25)	4	98	9/1	2600	1200	16*	.16*	-
Billings (23)	1	6.65	5.5/1	2900	725	1.64	.25	.370
Adt (52)	4	197	10/1	1400	875	6.9*	.035*	.13*
Finegold (20)	8	351	8.9/1	-	-	-	-	-
Lynch (33)	Various, few details given							
Adt (53)	8	197	10/1	1450	906	16.8*	.085*	.33*
Breshears (50)	8	-	-	2000	-	-	-	-
Finegold (22)	8	350	8.5/1	1500	870	43	.12*	.43*
Stebar (24)	1	37	8/1	1200	900	1.3	.035	.16
Thomas (37)	4	140	8/1	2500	1500	12.*	.086*	.31*

NOTES:
* based on brake rather than indicated horsepower

Source: AD-787 484

TABLE 5.4: SUMMARY OF HYDROGEN ENGINE EXPERIENCE–PARTIAL PERFORMANCE DATA

INVESTIGATOR	AT MAX. REPORTED POWER				AT MAX. REPORTED EFFIC.				PERFORM-ANCE CURVES?
	EQUIV. RATIO, Φ	THERMAL EFFIC., η (%)	MEP (PSI)	EXHAUST NO_x (PPM)	EQUIV. RATIO, Φ	THERMAL EFFIC., η (%)	MEP (PSI)	EXHAUST NO_x (PPM)	
Tizard (55)	-	-	-	-	-	-	-	-	-
Ricardo	sl.>1.0	31	115	-	.38	38	60	-	Yes
(29)	-	43^{***}	74^{***}	-	-	-	-	-	No
(38)	$.9^{\dagger}$	$33^{\dagger}$	-	-	$.9^{\dagger}$	$40^{\dagger}$	-	-	No
Burstall	.77	31	94	-	.38	35	53	-	Yes
(30)	.85	37	118		.50	38	77		
Erren (3)	-	45	-	"none observed"	-	45	-	"none observed"	No
Oemichen (36)	-	-	-	-	-	52	-	-	No
King (16)(28)(31)(32)	1.0	38 (CR = 14/1)	143	-	.39	52 (CR = 20/1)	60	-	Yes
Downs (17)	-	-	-	-	-	-	-	-	No
Anzilloti (18)(19)	-	-	-	-	-	-	-	-	No
Holvenstot (27)	≅1.0	29^{*}	82^{*}	-	-	-	-	-	Yes
Karim (34)	-	-	-	-	-	-	-	-	No
(35)	-	-	-	-	-	-	-	-	No

(continued)

TABLE 5.4: (continued)

INVESTIGATOR	AT MAX. REPORTED POWER				AT MAX. REPORTED EFFIC.				PERFORMANCE CURVES?
	EQUIV. RATIO, Φ	THERMAL EFFIC., η (%)	MEP (PSI)	EXHAUST NO_x (PPM)	EQUIV. RATIO, Φ	THERMAL EFFIC., η (%)	MEP (PSI)	EXHAUST NO_x (PPM)	
Swain (25)	-	-	50*	1900	-	-	-	-	No
Billings (23)	.8-.9	33-23	66	7000	.5	34	53	2.5	Yes
Adt (52)	.28	17*	19*	2.2	-	-	-	-	Yes
Finegold (20)	-	-	-	-	-	-	-	-	No
Lynch (33)	-	-	-	-	-	-	-	-	No
Adt (53)	.47	25*	47*	-	.37	28*	38*	-	
Breshears (50)	-	-	-	-	.35	39	-	.06 gm/ihp-hr	Yes
Finegold (22)	≅.6	31†*,***	65*,***	12	-	-	-	-	Yes
Stebar (24)	.2**	40** (Lean-limit-operation)	23**	13** gm/ihp-hr	-	-	-	-	Yes
Thomas (37)	-	23*, **	27*, **	-	-	-	-	-	No

NOTES:
* based on brake rather than indicated horsepower
** not at maximum power; only one operating point reported
*** limited by preignition and/or backfire
† varying % heat rejection reported at varying % full load

Source: AD-787 484

More recently, specific concerns for part-load efficiency (22) and pollutant emissions (23)(24) have yielded data that are interesting new explorations. However, again, these are compromised in utility owing to the limited test aims, programs, (and presumably funding) involved. Several current investigations are reportedly of somewhat more comprehensive scope, but data have not yet been reported from these. Typical of presently available data is a lack of indicated (rather than the less useful brake) horsepower data at commonly high piston speeds over a range of speeds and fuel/air ratios.

Reported performance is summarized in Table 5.3 and 5.4 for unmodified and modified engines. Cases for which performance curves of any type are available are identified. However, it was judged sufficient in summarizing briefly to characterize just two key operating points, whenever possible: "maximum" power output (P_{max}) and "maximum" thermal efficiency (η_{max}). In each case, the term "maximum" implies simply the conditions of the largest reported value for either performance variable; the values are not necessarily the maxima which might have been observed using a different or more complete test plan, etc.

Unmodified Naturally-Aspirated Spark-Ignition Engines: In this section only those engine operations are considered which involve the admission of hydrogen and air together into the engine cylinder. Port injection (25) is counted in this class. Direct injection of fuel into the engine cylinder (CFI) is considered below under the separate heading of Supercharged Spark-Ignition Engines; the effects of such hydrogen injection hinder direct comparison between operation with natural aspiration and with injection.

While not explicitly recognized in Reference (26), direct injection of gaseous hydrogen into the closed cylinder in effect yields variable supercharging as equivalence ratio is varied. Though not highly significant in the case of high-molecular-weight hydrocarbon fuels, this effect is substantial with (low-molecular weight) hydrogen. Data from operations involving EGR and WI are considered in the next section.

Collected thermodynamic performance data are shown in Figures 5.7 and 5.8 where thermal efficiencies (η) and mean effective pressures (MEP) are plotted as functions of equivalence ratio (Φ). The points shown represent values calculated from experimental data in a way which allows rough comparison despite the differences in compression ratios at which the experimental data were obtained. The method used was that of scaling each reported efficiency to the corresponding efficiency at a common reference CR of 10/1.

A detailed discussion of the data analysis will not be provided here, but can be found in the source report, AD-787 484.

In summary, actual efficiencies of hydrogen engines appear to be approximately the same fraction or somewhat higher (80 to 90%) than the fraction of fuel/air cycle efficiencies in gasoline engines.

Actual volumetric efficiencies are comparable (approximately 75 to 90%) with gasoline and with hydrogen operation. This suggests that the relative performance of gasoline and hydrogen engines can reasonably be assessed from the relative performance indicated by fuel/air cycle calculations for the two fuels.

FIGURE 5.7: COLLECTED EXPERIMENTAL THERMAL EFFICIENCIES OF H_2/AIR, NON-CFI ENGINES*

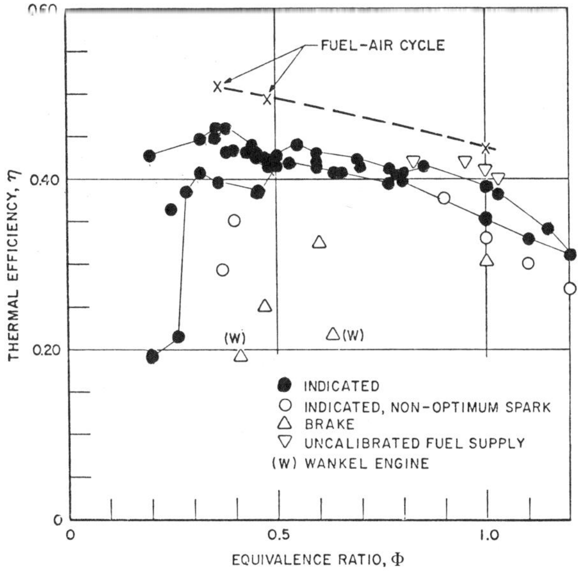

*Data scaled to common CR = 10/1

FIGURE 5.8: COLLECTED EXPERIMENTAL MEAN EFFECTIVE PRESSURES OF H_2/AIR, NON-CFI ENGINES*

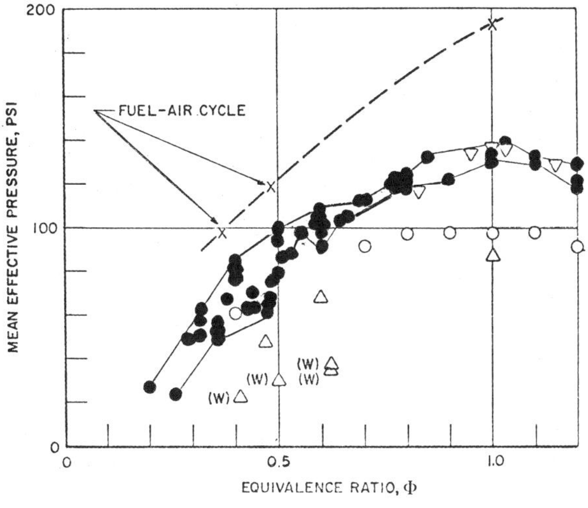

*Data scaled to common CR = 10/1

Source: AD-787 484

Modified, Naturally-Aspirated, Spark-Ignition Engines: A small number of data largely at single operating conditions are available describing the performance of hydrogen engines modified with exhaust-gas recirculation (EGR) (20)(22)(23) or water injection (WI) (22)(23). In gasoline engines and in hydrogen engines free of backfire, both of these modifications tend to decrease NO_x emissions (mostly due to decreased combustion temperatures). This occurs at the expense of decreased efficiency and power output (due to dilution). Small amounts of injected water may, however, increase efficiency and power output slightly (23).

A major additional effect of EGR and WI in hydrogen engines is the elimination of backfire. As a result of backfire suppression by substantial water injection (M_{H_2O}/M_{H_2} = 2), Finegold has reported from throttled hydrogen/air operation the possibility of increasing power output in such a quantity controlled situation by decreased throttling before the occurrence of backfire. This can be done reportedly without a major concurrent effect on efficiency, at least at $\Phi = 1$.

In a quality controlled operation, low levels of water injection, for example, M_{H_2O}/M_{H_2} = 0.3, have been reported to eliminate backfire and simultaneously to increase power and efficiency slightly, at least for the given engine operating conditions (2,900 rpm, $\Phi = 0.625$). In a quality controlled hydrogen engine, these small gains presumably can add to observed increases in power provided by operation at a higher Φ with backfire.

In contrast, the elimination of backfire by EGR has not been shown to allow even slight intrinsic increases in power or efficiency at the same operating conditions. In at least one given operating condition, EGR elimination of backfire (25% EGR) has led to significant efficiency and power decrease (for example, 5%). This is in accord with operating experiences with gasoline engines using EGR.

Unfortunately, the data presently available (both theoretical and experimental) are insufficient to clearly show any merits for EGR and WI in hydrogen engines aside from NO_x suppression. While NO_x suppression is apparent, the extent to which this benefit trades off with efficiency and power output losses cannot presently be considered clear.

Supercharged Spark-Ignition Engines: Supercharging is defined to include not only the use of above atmospheric manifold pressures but also cylinder fuel injection (CFI), the injection of gaseous hydrogen into the engine cylinder after closing of the intake valve. As does conventional manifold pressure supercharging, CFI hydrogen operation can allow engine operation with substantially greater air rates and therefore, higher MEP and power than otherwise is possible.

The low density of hydrogen severely decreases the air rate of a carburetted or port-injected reciprocating engine compared with that of a gasoline or even a hydrocarbon-gas engine. Direct injection into the cylinder allows the induction of pure air rather than a mixture during the intake stroke of a four-stroke engine. This allows the air rate of a low density hydrogen engine to be essentially that of a carburetted engine operated on a higher density fuel such as gasoline.

Supercharging via CFI operation is the subject of patents by R.A. Erren in 1929, and Erren's pioneering work, summarized in Reference (3), is apparently the earliest reported use of hydrogen-engine supercharging of any kind.

TABLE 5.5: SUMMARY OF CFI HYDROGEN/AIR ENGINE EXPERIENCE–ENGINE SPECIFICATIONS AND POWER OUTPUT

INVESTIGATOR	NO. OF CYLINDERS	DISPLACEMENT (IN^3)	COMPRESSION RATIO	MAX. ENGINE SPEED (RPM)	MAX. PISTON SPEED (FT/MIN)	MAX. REPORTED POWER (HP)	MAX. POWER / DISPLACEMENT (HP/IN^3)	MAX. POWER / PISTON AREA (HP/IN^2)
Erren (15)	-	-	8/1	-	-	-	-	-
Murray (26)(39)	1	8.3	6.5/1	4000	1250	2.8	.34*	.63*

NOTES:
*based on brake rather than indicated horsepower

TABLE 5.6: SUMMARY OF CFI HYDROGEN/AIR ENGINE EXPERIENCE–PARTIAL PERFORMANCE DATA

INVESTIGATOR	AT MAX. REPORTED POWER				AT MAX. REPORTED EFFIC.				PERFORMANCE CURVES?
	EQUIV. RATIO, Φ	THERMAL EFFIC., η (%)	MEP (PSI)	EXHAUST NO_x	EQUIV. RATIO, Φ	THERMAL EFFIC., η (%)	MEP (PSI)	EXHAUST NO_x (PPM)	
Erren (15)	-	45	-	"none observed"	-	45	-	"none observed"	No
Murray (26)(39)	1.89	12	76	340	.78	17	63	752	Yes

Source: AD-787 484

Tables 5.5 and 5.6 designate those operating experiences which have involved hydrogen spark-ignition engines with CFI and for which at least a few data are available. There have apparently been no hydrogen-engine operations with conventional (manifold) supercharging, although plans for some have been made. As the tables indicate, scant operational data are available. This is due to the loss of Erren's records during World War II. The only substantial body of test data available today is that obtained by Murray and Schoeppel in 1971 on a modified, commercial, one-cylinder engine (26)(39).

Unfortunately, the major emphasis of the recent hydrogen CFI-engine work is on demonstrating feasibility and pollutant emission levels of an injected engine. Significant data are missing apparently for this reason; only brake horsepower and efficiency data are presented. Lack of friction loss and injection power use data preclude direct calculation of indicated horsepower, MEP, and efficiency from the reported brake data.

Particularly for the low horsepower, modified engine which was used it is highly speculative to attempt generalization of the results by consideration of indicated, rather than brake power output and efficiency. For example, brake MEP's and efficiencies appear low for the circumstances, but might arise simply from unreported frictional losses and injection power take off rather than from more fundamental effects of the fuel injection.

An even more fundamental shortcoming of Murray and Schoeppel's reported operating experience is that it remains to date unrelated to any fundamental theory dealing with the special feature of their experience—direct cylinder injection.

Rotary-Combustion Spark-Ignition Engine: The only reported application of hydrogen as a fuel for rotary-combustion (Wankel) engines is the demonstration project carried out at Brookhaven National Laboratory (40)(41). For a demonstration and development of metal-hydride hydrogen storage, the BNL workers have operated an 8 hp Wankel engine with hydrogen/air mixtures at several speeds (1,000 to 4,000 rpm) and loads. Performance data are sketchy owing to interest in the engine being peripheral to concern for metal-hydride storage. The BNL hydrogen-fueled Wankel engine operates smoothly and quietly from idle (approximately 900 rpm) to a speed of 4,700 rpm.

Compression Ignition Engines: The only data available concerning operating experience with pure-hydrogen-fueled compression ignition engines is apparently that of Helmore and Stokes (42). Gerrish and Foster (43) describe some of Helmore and Stokes' results. While most discussion of these results is postponed to a later section on combined-fuel operation, some of the combined-fuel results are relevant to pure-hydrogen compression-ignition operation.

Since these combined-fuel investigations apparently involved homogeneous rather than injected hydrogen/air mixtures, the following observations relate only to homogeneous-charge compression-ignition engines, not to common, injected, Diesel engines. There are apparently no data available for hydrogen-injection, compression-ignition engines. Helmore and Stokes operated with combined fuel (90% hydrogen/10% fuel oil) only after violent detonation resulted during an attempt to burn pure hydrogen at CR = 11.6/1. Unfortunately, Gerrish and Foster's citation is not clear on operating conditions for these tests. Detonation at nine-tenths full power appears to imply $\Phi = 0.9$ or perhaps slightly less. Helmore and Stokes detonation was presumably autoignition and, according to Karim's data on violently knocking autoignition, a maximum CR of about 13 is allowable without violent knock and autoignition. Especially since such phenomena are dependent on configuration and test conditions, the differences between Karim's limit of 13 and Helmore and Stokes' 11.6 (at high Φ) are probably inconsequential.

It appears that compression-ignition engines employing homogeneous mixtures are not practical owing to autoignition at least for Φ approaching 1. It must be noted at CR's exceeding approximately 12/1 that elevated intake temperatures may decrease this CR substantially, judging from Karim's data (34).

Combined-fuel operations also have implications regarding pure-hydrogen operations. At compression ratios of 13.4/1 and 15.6/1, Gerrish and Foster found that hydrogen equivalence ratios could not exceed about 0.5 and 0.4, respectively, without preignition or backfire when small pilot changes to fuel oil were injected (equivalent to $\Phi_{\text{fuel oil}} = 0.022$). This behavior may reflect the same phenomenon studied by King et al who found preignition and backfire attributable to solid carbon particles. However, at Φ of 0.5 and 0.4, homogeneous hydrogen/air mixtures would not compression-ignite without at least an exceedingly small pilot charge, for CR = 13.4/1 and 15.6/1, respectively.

The implication is that somewhere in the range of $0.5 < \Phi < 0.9$, homogeneous compression-ignition operation is impractical for CR above about 14 or 15/1. This observation is consistent with King's observation in spark-ignition engines that backfire and preignition could be eliminated at CR = 14/1 but occurred at $CR \geqslant 16/1$ for $\Phi \geqslant 0.8$. Owing to lack of operating experience data, no generalizations can be made concerning injected-hydrogen compression-ignition engines.

Hydrogen/Oxygen Engines

While a number of proposals for H_2/O_2 reciprocating engines have been made [for example, (44)(45)], operating experience is very limited and performance data sketchy. One of the most documented experiences is the development by Vickers, Inc. of an auxiliary power unit (APU) for space application (46)(47). An H_2/O_2 recip was considered promising for space application because of its high specific power (small volume) and low specific weight. Otherwise, operations have apparently been limited to exploratory experiments and demonstrations rather than substantial performance testing (48).

Table 5.7 summarizes the engine specifications and operating conditions for these H_2/O_2 engine operations. Also Table 5.7 summarizes the performance data reported. Since there is virtually no data available from the experiments of the Perris Smogless Automobile Association (48), further consideration is limited to the operation of the APU.

The Vickers APU was operated with O_2 and H_2 injection and vacuum exhaust (for space application) and shows high specific power (4.75 bhp; 1.75 bhp/in^3, max), brake thermal efficiency (0.32, max) and bmep (173 psi, max). Corrected by Vickers data for friction power, these data imply the maximum indicated power, specific power, efficiency, and MEP cited in the table.

TABLE 5.7: SUMMARY OF HYDROGEN/OXYGEN ENGINE EXPERIENCE–ENGINE SPECIFICATIONS AND POWER OUTPUT

INVESTIGATOR	NO. OF CYLINDERS	DISPLACEMENT (IN^3)	COMPRESSION RATIO	MAX. ENGINE SPEED (RPM)	MAX. PISTON SPEED (FT/MIN)	MAX. REPORTED POWER (HP)	MAX. POWER / DISPLACEMENT (HP/IN^3)	MAX. POWER / PISTON AREA (HP/IN^2)
Morgan (47)	1	2.72	12.7/1	4000	1025	5.15	1.9	3.0
Underwood (48)	Various, few details given							

TABLE 5.8: SUMMARY OF HYDROGEN/OXYGEN ENGINE EXPERIENCE–PARTIAL PERFORMANCE DATA

INVESTIGATOR	AT MAX. REPORTED POWER			AT MAX. REPORTED EFFIC.			PERFORMANCE CURVES?
	EQUIV. RATIO, Φ	THERMAL EFFIC., η (%)	MEP (PSI)	EQUIV. RATIO, Φ	THERMAL EFFIC., η (%)	MEP (PSI)	
Morgan (47)	6.	.32*	190**	Same as Max. Power			Yes
Underwood (48)	-	-	-	-	-	-	No

NOTES:
* Based on available O_2 in (rich) mixture and includes, as input, expansion work of H_2 supplied at 400 psig.
** Includes effect of expansion work by H_2 supplied at 400 psig, O_2 supplied at approx. 500 psig.

Source: AD-787 484

An additional correction is required to correct vacuum-exhaust operation to the more conventional ambient-exhaust conditions; with ambient exhaust, work is lost to compression of the exhaust residual by reactant injection. For the compression ratio used (12.7/1), however, this represents a negligibly small correction compared with that for the expansion work of the high-pressure injected reactants (which is included in the brake thermal efficiency calculations by Vickers).

The APU engine was operated hydrogen-rich ($1.4 \leqslant \Phi \leqslant 7.0$) and was run over a range of speeds (2,000 to 4,000 rpm), indicated power outputs (0.75 to 5.28 hp), indicated MEP's (55 to 192 psi), and indicated thermal efficiencies (0.125 to 0.356) without exceeding design limits on cylinder-head temperature (apparently approximately 1400°F at an unspecified location).

Heated hydrogen (500°F) was injected first at various supply pressures up to 400 pounds per square inch followed by ambient-temperature oxygen at unspecified pressure (apparently 500 to 600 psi). Injection timing was not optimized. It appears that Φ was determined in each run as the minimum ($\Phi > 1$) consistent with maximum allowable head temperature.

Mixed-Fuel Hydrogen Engines

Spark-ignition engines have been operated with mixtures of hydrogen and gasoline by several investigators (24)(29)(49)(50) and with mixtures of hydrogen and kerosene (29). Experiences with hydrogen and fuel oil in compression-ignition engines are also reported (43)(51). Mixtures of hydrogen with other gas fuels have also been tested (27)(42)(54)(56).

The most recent efforts have been with mixed-fuel spark-ignition engines and have been primarily concerned with extending hydrocarbon-fuel operation to lean mixture ratios (for example, $\Phi = 0.5$) where pure gasoline and fuel oils will not burn easily. The impetus toward lean mixtures derives from the decreased NO_x emissions expected with low equivalence ratios. This gain is, of course, concurrent with an inherent loss of MEP (that is, output torque and power) but with the likelihood of increased efficiency.

Ideally, one would compare performance data for the various mixed-fuel operations in a manner similar to that used above for hydrogen/air spark-ignition engines. Of course, an additional operating variable becomes influential, one expressing the mixed-fuel composition, for example, the hydrogen mass-fraction of the mixture. Unfortunately, in two major reports of operating experience, values are not specified either for this variable (50) or for some other significant variable [for example, Φ, (29); MEP or power, (50)]. Even in one case where hydrogen fractions are reported (24), the values are, at each value of Φ (for the mixture), the minimum hydrogen fractions observed to allow operation without misfire.

While these lean-limit data on mixed-fuel operation are useful, they do not allow ready comparison with other data for different hydrogen fractions or for engines with different lean limits for misfiring. Some of the missing data required for meaningful comparisons can be back-calculated (by making reasonable assumptions). However, the results of such calculations have not, to date, allowed a satisfactory comparison to be made; unresolved discrepancies exist.

Current efforts at the Jet Propulsion Laboratory (50) and one report from General

Motors Research Laboratories (24) as well as a much earlier report by Ricardo (29) corroborate the ability to operate hydrogen/gasoline/air spark-ignition engines at overall equivalence ratios as low as $\Phi = 0.1$. Below $\Phi = 0.6$, NO_x emissions are reduced to less than 0.4 g/ihp-hr, fulfilling the Environmental Protection Agency 1977 standard. CO emissions are low enough (for example, 4.5 g/ihp-hr) at even higher Φ. Unburned hydrocarbons are, however, high, exceeding those of the EPA standard (0.5 g/ihp-hr) for all Φ and exceeding the standard by a factor of 6 to 10 at $\Phi \leq 0.5$.

Indicated MEP's appear to be about the same or somewhat above those for hydrogen operation particularly at Φ's near 1.0 where the mixtures used contain little hydrogen. Indicated thermal efficiencies, on the other hand, appear to be lower than with pure hydrogen operation by perhaps 3 to 5 percentage points. The collected data currently available are not unambiguous, however, as regards both efficiency and MEP values, and further information and analysis are needed before a comprehensive view of mixed-fuel operations is available.

ENERGY STORAGE FOR UTILITIES–BROOKHAVEN OPTIMIZATION MODEL STUDIES

Elements and Cost of Hydrogen Storage Electric Peaking Plants

The elements of hydrogen production from off-peak electricity, storage of the hydrogen and reconversion of the hydrogen to electricity in a fuel cell or high temperature turbine can be used as components in an electric storage system. Production of hydrogen by water electrolysis is a commercial technology. Cost projections for advanced electrolytic systems are in the range of \$50/kw and efficiencies of 80% or greater are claimed.

A number of approaches exist for the storage of hydrogen, including the use of metal hydrides, liquid hydrogen, and high pressure gas storage. Cursory examination of these options suggests that the use of iron-titanium hydride for hydrogen storage will be the most efficient mode.

Hydrogen can be converted to electricity in fuel cells which are currently under development and are projected to achieve efficiencies up to 65%. Direct combustion of the stored hydrogen with pure oxygen to produce high pressure steam in very simple and inexpensive combustion devices, and the conversion of the enthalpy of the steam to electricity in high temperature turbines is presently feasible.

Hydrogen electric storage combines the elements listed in Table 5.9 to operate as an electric peaking plant. The assumed capital costs and the efficiency of the electrolyzer and fuel cell systems are also listed in Table 5.9.

It is assumed that the metal hydride storage facility operates on waste heat and losses due to cooling water and pumping power are negligible in comparison to electrolyzer and fuel cell losses. The total capital cost of the storage facility is assumed to be \$310/kw. This cost is based on very preliminary design work. It is possible that careful design and optimization will reduce this cost to a lower level; however, the analysis presented below is based on the above installed cost.

TABLE 5.9: TECHNICAL ELEMENTS AND COST OF HYDROGEN ELECTRIC STORAGE PLANTS

Elements of Hydrogen Storage Peaking Plants

AC Electricity Input
AC-DC Rectifier
Water Electrolyzer
Metal Hydride Hydrogen Storage
Fuel Cell
DC-AC Inverter

Assumed Capital Costs	$/kw
Power conditioning and electrolyzer	60
Metal hydride hydrogen storage	70
Fuel cell and inverter	180
	310

Assumed Efficiencies	
Power conditioning and electrolyzer	75%
Fuel cell and inverter	55%

Source: BNL-18721

As pointed out above, it is not possible to make broad generalizations regarding the application of energy storage devices in a specific utility system without detailed knowledge about the installed capacity, costs, load structure, and future growth, etc. However, it is possible to step back and examine broadly the role of a specific storage concept in the total United States energy delivery system.

The Brookhaven Energy System Optimization Model (OM) is a linear programming model of the national energy system (57)(58). The output of the model is a matrix consisting of 13 energy supply and 15 demand categories. Spare categories listed may be replaced by new supply or demand categories. The model selects the energy supply categories and energy flows to the various demand categories in such a way as to minimize the dollar cost of operating the total system.

All the energy demands in the model must be satisfied and the system must operate within realistic resource and capacity constraints as imposed in the real world. For this study, the energy system in the year 1985 is considered. Supply and demand projections are based on data from AET-8 (59) and the U.S. Department of the Interior (60).

Although other supply categories are included, the main categories in the model are nuclear, coal, oil and natural gas. These are assumed to be used after processing as the direct fuel forms such as gasoline, heating oil, natural gas, etc. or converted to electricity.

In all cases, processing, conversion and utilization efficiencies were taken to be the same as the average efficiency in the system today. Demands fall into four general categories: residential and commercial, including space conditioning and water heating; transportation, both public and private; industrial including petrochemicals, process heat demands, and iron and steel production; and miscellaneous electrical demands. Within the present technological limitations complete interfuel substitutability is allowed. The electric load duration structure incorporated into the model is consistent with the national average.

The results of the study presented below are not intended to be an exhaustive or complete study of electric storage in the United States energy system. They are presented to illustrate the utility of the model in assessing the role of a particular electric energy storage concept. However, the results obtained serve to illustrate not only the role of hydrogen storage for electric peaking, but also yield quantitative insights into the versatility, optimal capacity, and unique benefits of this particular electric storage concept.

This approach makes evident and quantifies the alternatives of satisfying other demands using hydrogen derived from off-peak electricity, and an optimal distribution of hydrogen between demands is developed for a given set of conditions.

Four runs with the OM are used for this assessment activity. The conditions and constraints for these runs are described below. Briefly, the study focuses on the availability of oil, the use of hydrogen generated from off-peak electricity to satisfy direct thermal demands, and the possible role of synthetic natural gas in fuel cell converters to satisfy peak demands as an alternative to electric energy storage.

Cases Considered

In order to examine the role of hydrogen energy storage for electric peaking applications it is assumed that the only competing options are pumped storage and gas turbines. It is assumed that the output from central station electric facilities to meet the peak demands does not exceed the output from pumped storage and gas turbine generators. The amount of pumped storage capacity available in the year 1985 is assumed to be limited so that only 0.215×10^{15} Btu_e of electricity would be produced from pumped storage facilities. The energy to satisfy peak demands is approximately 6% of all the electricity delivered in the United States energy system.

Based on the system described above the capital cost of a hydrogen storage electric peaking plant is taken to be \$310/kw which is appreciably more expensive than a pumped storage plant, or equivalent storage costing \$150/kw when both systems are assumed to have a 30 year life. However, there is a set of conditions under which the cost of operating the system would be minimized and hydrogen storage electric peaking plants would be an attractive option from the point of view of the national energy system, even at \$310/kw. This analysis considers four cases for which the OM was run. Each of the cases and run results is described below.

Case 1–Abundant Oil: In this case it is assumed that 55×10^{15} Btu_t of oil and $1,800 \times 10^{6}$ tons of coal would be available per year to the system, as well as up to 15×10^{15} Btu_t of nuclear fuel for Light Water Nuclear Reactor Central Station Electric Generating Plants (LWR). The results of this run were as follows. Gas turbines and pumped storage satisfied all of the peak demand and delivered 0.370×10^{15} Btu_e and 0.215×10^{15} Btu_e respectively in the projected year 1985. Some available off-peak electric energy, 0.540×10^{15} Btu_e was used for water-heating which was allowed in all runs as an off-peak demand up to a maximum level of 0.687 x 10^{15} Btu_e. If more pumped storage capacity had been allowed the system would have used it in place of gas turbines and would have shifted the remaining oil to satisfy the water heating demand. All the pumped storage capacity was charged with the LWR.

The system used almost all of the oil allowed, 54.16 x 10^{15} Btu_t, but only 706.5 x 10^{15} tons of coal per year out of the 1800 x 10^6 tons per year available. Approximately 30.6 x 10^{15} Btu_t of the oil used was imported. The cost of electricity delivered from hydrogen peaking plants, taken to be \$25.35/10^6 Btu_e, would have had to be reduced by \$8.79/10^6 Btu_e in order for hydrogen storage peaking plants to be used in the system.

If this had been the case off-peak power would have been shifted from water heating to hydrogen production. The average cost of electricity generated by peaking gas turbines in 1985 and delivered to point of end use was \$22/10^6 Btu_e. In contrast the cost of delivered LWR intermediate load electricity was \$6.39 per 10^6 Btu_e.

The total cost of operating the energy system was \$167.2 x 10^9, and the total installed electric generating capacity was 0.523 x 10^6 Mw_e. Case 1 reflects the state of the system if allowed to grow at the present rate with the same technology and available resources. A decrease in the level of gas turbine capacity would have occurred if the price of #2 fuel oil, taken to be \$0.88/10^6 Btu_t, were increased to \$2.23/10^6 Btu.

Case 2–Scarce Oil: The basic assumption in this run is that the amount of oil available is restricted and the system must make interfuel substitutions to compensate for the shortage. It is assumed that 46.1 x 10^{15} Btu_t of oil per year is available of which only 22.5 x 10^{15} Btu_t is imported, that is, a reduction of approximately 26% from Case 1. Sufficient oil was allowed to satisfy all the petrochemical and transportation demands, and 5.10 x 10^{15} Btu_t of oil per year was used in oil steam electric plants as in Case 1.

As in Case 1, the maximum allowed amount of nuclear capacity was installed; however, the amount of coal used annually in central station electric plants increased from 7.211 x 10^{15} Btu_t to 24.832 x 10^{15} Btu_t. The total consumption of coal is 1,414 x 10^6 tons per year, twice the level used in Case 1.

Pumped storage was used to the maximum extent allowed and delivered 0.215 x 10^{15} Btu_e per year as in the previous case. Gas turbines were no longer used as peaking devices, having been replaced by hydrogen storage peaking plants using off-peak electricity. Increased electrification in the industrial sector resulted in a high peak power demand, and some central station peaking capacity was installed.

Approximately half of the total hydrogen produced, 1.2 x 10^{15} Btu_t, was used to supply the demand for process heat. Again in this case all the other available off-peak power was used for electric water heating which increased to the maximum extent allowed, 0.687 x 10^{15} Btu_e, from the level of 0.540 x 10^{15} Btu_e in Case 1. The total cost of operating the system annually was \$187.6 x 10^9 compared to the previous cost of \$167.2 x 10^9. The increase was primarily due to the greater electrification necessitated by the decrease in the oil supply. Even though their cost was high, hydrogen storage electric peaking plants were used in place of gas turbines in minimizing the cost of operating the system. If additional oil were available the use of gas turbines would be favored and the cost lower. When oil is limited any alternative storage device not dependent on oil would be used if the capital cost were low enough. For example, pumped storage would be installed at higher levels than were allowed if sites were available and could be developed in time.

Case 3–All Hydrogen Production for Electric Peaking: In this case none of the hydrogen produced is allowed to satisfy any process heat demands. The total amount of hydrogen produced decreased to 0.631 x 10^{15} Btu_t from the level 1.2 x 10^{15} Btu_t in Case 2. Available oil is constrained to the same level as in Case 2, 46.1 x 10^{15} Btu_t. Some of the available off-peak capacity goes unused, but the LWR is still base loaded as in the previous two cases. The total cost of operating the system increases slightly to \$188.1 x 10^9 from the level of \$187.6 x 10^9 in Case 2.

Installed electric capacity increases slightly and the amount of electricity delivered from hydrogen storage peaking plants increases also to 0.347 x 10^{15} Btu_e compared to the level of 0.324 x 10^{15} Btu_e in Case 2. Thus, the use of part of the hydrogen output generated from off-peak electricity to satisfy thermal demands, such as process heat, reduces the annual cost of operating the energy system as evidenced by the increased cost when this option is not allowed. Pumped storage remains at the maximum allowed level as in the previous cases and would increase if the capacity constraint could be relaxed.

Case 4–Coal Derived Gas for Electric Peaking: This case is similar to Case 3; however, natural gas or hydrogen derived from coal is allowed to be used in fuel cells as in Cases 2 and 3 where hydrogen from off-peak electricity is used for electric peaking. Given this choice the system does not choose this technological option, but instead continues to use pumped storage and hydrogen storage electric peaking plants to meet peak demands to achieve a minimum total cost. The resource allocations and total annual cost obtained in this run were identical to Case 3.

Gas derived from coal was not used in the minimum cost system for peaking, all the gas projected to be produced in 1985, 2.1 x 10^{15} Btu_t, being used to satisfy other demands, such as petrochemical and process heat, because of the great flexibility in the utilization of gaseous fossil fuels. If larger amounts of synthetic gas were available, that is, quantities comparable to the decrease in the oil supply, the optimal solution would change, the annual cost would decrease, and some of the synthetic gas supply might be used to meet peak demands.

Conclusions

[1] As long as a cheap abundant oil supply exists, hydrogen storage electric peaking plants or any other storage device more expensive than pumped storage are not competitive with the gas turbine for electric peaking applications.

[2] When the oil supply is constrained hydrogen storage peaking plants are an economic alternative for peaking applications and minimize the total cost of operating the energy systems.

[3] Under the above condition [2], alternative storage devices not dependent on liquid or gaseous fossil fuels are also attractive options which minimize cost even at capital costs in excess of $300/kw.

[4] The use of hydrogen generated from off-peak electricity is attractive for use in fuel cells to meet peak demands or to supply industrial process heat. In a sense, the hydrogen in this case is being used to supplement the natural gas supply.

[5] In general, there are some significant benefits which accrue when hydrogen derived from off-peak electricity is used in the United States energy supply system. This study has quantitatively investigated some system options; others remain to be explored.

ENERGY STORAGE FOR UTILITIES–PSE&G MAPLEWOOD TEST SYSTEM

Description of System

To prove the technical feasibility of hydrogen based energy storage concepts and evaluate the concept in a utility context, Public Service Electric and Gas Company (PSE&G) established what is believed to be the first complete hydrogen energy storage system at its Maplewood, New Jersey Testing Laboratory. In addition to energy storage experience, the facility provides fuel cell experience beneficial to PSE&G. Also the experience with electrolysis and storage equipment is potentially applicable to hydrogen used for generator cooling systems.

Supply of hydrogen at generating stations is now accomplished with truck delivery and storage in high pressure (2,000 psi) cylinders. With the advent of floating nuclear power plants where delivery schedules may be uncertain and space is a more critical consideration than previously, on-site production coupled with the compactness of storage offered by metal hydrides may be of substantial value.

The Maplewood system consists of available components designed for other purposes. Consequently, while the system is functionally complete, performance is not representative of that possible with advanced technology components specifically designed for energy storage applications.

In the Maplewood facility, a rectifier converts to DC power for use in a Teledyne Isotopes 125 SLM hydrogen generator which produces hydrogen at approximately 70 psig. In initial operation of the system, hydrogen was stored in a tube trailer of the type used routinely to deliver hydrogen at 2,400 psi. To generate power, hydrogen is delivered through a regulator at atmospheric pressure to a 12.5 kw hydrogen-air fuel cell furnished by Pratt & Whitney Aircraft.

The experimental fuel cell is a TARGET installation program PC11 power plant with the fuel processing section (steam reformer) removed. DC power from the fuel cell is fed to a single phase inverter. This equipment and a control/display panel are located in a 500 square foot room. The hydrogen storage unit is located outside the building. Operation of the system began in April 1973 and continued throughout 1973 with storage of hydrogen as a pressurized gas in the tube trailer. Except for the normal problems associated with initial startup of a new experimental system, system operation was very satisfactory.

Major conclusions of testing conducted with gaseous hydrogen storage are:

[1] System performance can be predicted accurately using manufacturer's data for individual components.

[2] Although megawatt scale systems would permit substantially improved characteristics, storage systems based on hydrogen can be operated at power levels as low as 10 kw. This suggests the feasibility of a modular approach whereby total system capacity can be increased readily in phase with demand growth.

[3] Installation, operation and maintenance of hydrogen storage systems can be carried out with the personnel, skills and equipment available in a utility operation.

[4] The system is quiet and safe; it produces essentially zero pollution. These features plus the modular approach enhance siting feasibility.

[5] Charging and discharging rates can be varied continuously over the range from zero to maximum capacity. The state of charge of the storage system has no effect on system output capability. This permits storage system output to be varied to achieve the most economic system operation. Once at operating temperature, the fuel cell power plant responds within a cycle ($\frac{1}{60}$ second) to system load changes. These features permit a flexible response to the system load demand.

[6] Operation of system components can be decoupled with charging, storage and discharging functions optimized in design and use.

A scaling up of the experimental system at the PSE&G Maplewood Laboratory indicates that the hydrogen storage portion will probably be the most serious limitation to siting flexibility. Although the technical feasibility of gaseous and cryogenic storage is certain, these approaches do not have all the attributes of hydride storage from the standpoint of compactness, quietness, and safety.

These limitations may be overcome by using metal hydride storage. The intermetallic compound iron-titanium (50:50 atom percent) is of particular interest for hydrogen storage applications because of a favorable hydrogen partial pressure temperature relationship for the known hydride phases. The general chemical reaction forming these ternary hydrides is:

$$FeTi + \frac{x}{2}H_2 \rightleftharpoons FeTiH_x$$

The series of pressure-composition isotherms shown in Figure 5.9 indicates that hydrogen at low pressures can be added to or removed from the solid at temperatures within a few degrees of the ambient. The formation reaction is exothermic and produces approximately 7,000 Btu/lb of hydrogen, which is only a small fraction of the heat of combustion, 61,000 Btu/lb of hydrogen. When discharging, the same quantity of heat must be added to release one pound of hydrogen. Enhanced safety of storage arises because the hydrogen is contained as a chemical compound with a nonreactive metal. The probability of rupture of a high pressure vessel is reduced and, further, the release of stored energy in the form of hydrogen is inherently self-limiting. This arises because the release rate is reduced by cooling associated with discharge.

FIGURE 5.9: PRESSURE-COMPOSITION ISOTHERMS FOR THE $FeTiH_x$ SYSTEM

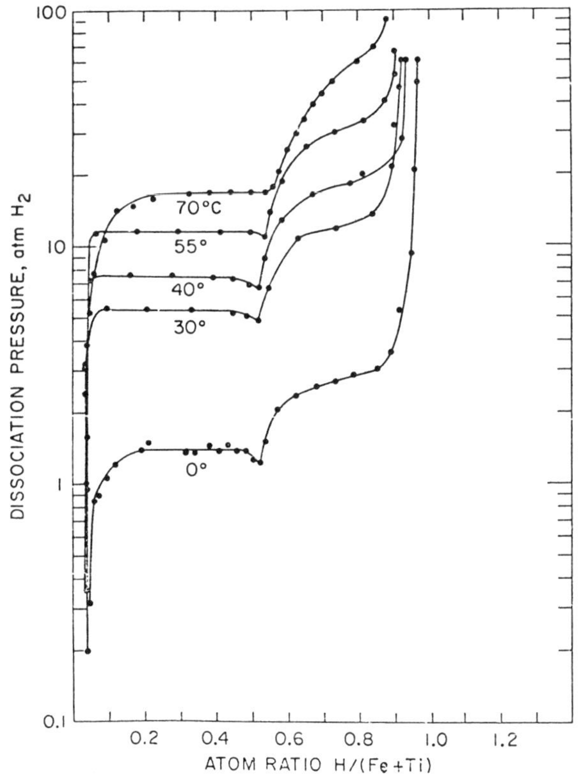

Source: BNL-19249

In view of these potential advantages, PSE&G in cooperation with Brookhaven National Laboratory (BNL) has installed an iron-titanium bed in its hydrogen energy storage experiment. This is the first single bed of such size, and can store about as much hydrogen (12 pounds available) as in previous tests.

Briefly described, the outside container is 6½ feet long by 1 foot in diameter and the entire storage unit weighs 1,240 pounds. The 880 pound iron-titanium bed which fills the container shown in Figure 5.10 consists of granular particles ranging in size from 4 to 100 mesh. The density of hydrogen within the bed when fully charged is about ¾ liquid hydrogen density. Hydrogen is introduced and withdrawn through six, dead-ended porous metal tubes attached to a common header. Heat is removed on charging and supplied on discharging by passing cold

FIGURE 5.10: DIAGRAM SHOWING INTERNAL COMPONENTS OF RESERVOIR

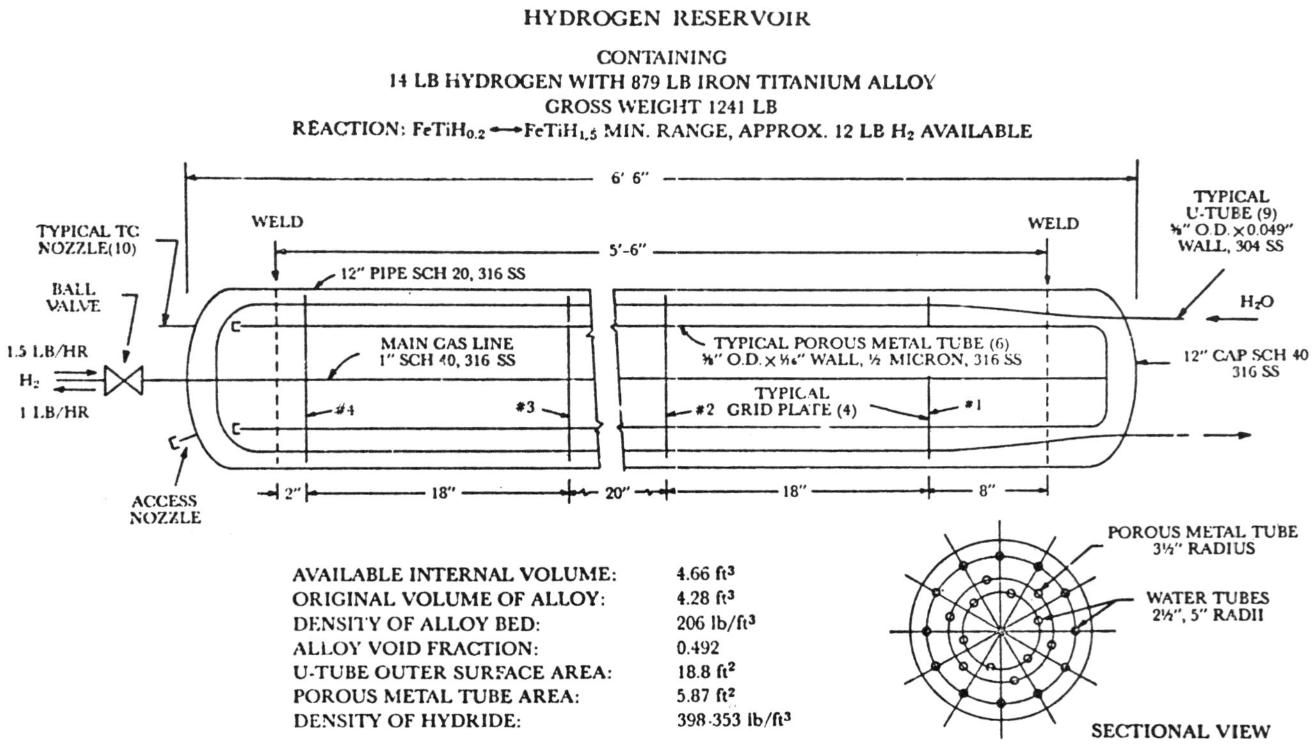

Source: BNL-19249

and hot tap water through nine U-tubes within the bed at a design flow rate of 6 gallons per minute. The design flow rate for hydrogen delivery is 1.5 pounds per hour for 8 hours but the bed has already been operated at twice this flow rate for a period of several hours. A more complete description is contained in Reference (61). The unit is now fully operational at the PSE&G Maplewood Laboratory.

This bed is of a sufficiently large size so as to provide meaningful engineering and performance data for the bulk storage of hydrogen. In particular the effect of repeated cycling of the bed on storage capacity, charging rate and physical integrity of the particles is being examined. Degradation as a result of contamination introduced from other system components is being studied.

Data on operating characteristics will be used as an input for modeling studies which are directed toward the design and optimization of larger systems. Modelling studies are now in progress at BNL (see description of BNL modelling studies in previous chapter).

Conceptual Design of Metal Hydride Bed

A design consideration of a hydrogen storage system based on the use of iron-titanium is heat removal during charging and heat addition during discharge of the hydrogen. Three modes of heat transfer can be considered.

Fluid Bed Processing: A system can be envisioned in which hydrogen would be the fluidizing gas used to move granular iron-titanium from large storage tanks to small external hydriding or dehydriding units. Hydriding and dehydriding could be accomplished in a single piece of process equipment. The capacity of the system would be determined by the size of the storage tank, and the hydrogen discharge rate determined by the external process capacity and rate of heat addition.

This concept is shown in Figure 5.11. The metallic hydride is fluidized in a gaseous hydrogen stream and flows from a storage hopper through the reactor to another hopper. Among the advantages of this scheme are flexibility in the heat exchange design. The cooling-heating coils can be suspended in the fluidized bed where the high turbulence of the bed will greatly assist the heat transfer. To simplify the reactor design, the heat exchange can be done elsewhere in the gaseous hydrogen circuit in conventional apparatus.

Another advantage of the fluidized bed concept is the decoupling of the hydriding/dehydriding function from the storage function. On the other hand, the relatively high particle density of $\approx$350 lb/ft^3 coupled with the low density and low viscosity of the fluidizing gas requires high-pressure, high-volume gas handling equipment.

Additional power consuming apparatus is required to transport the heavy powder from the storage hoppers to the reactor and back to the storage hoppers. However, the anticipated attrition resulting from the constant movement of the metallic hydride and the continuous reduction in particle size limit the application of the fluidized bed concept. After preliminary design work, this concept was found unsuitable for utility applications. Fluid bed systems are energy intensive and require significant quantities of electricity (similar to liquefaction storage

FIGURE 5.11: FLUID BED PROCESSING SCHEME

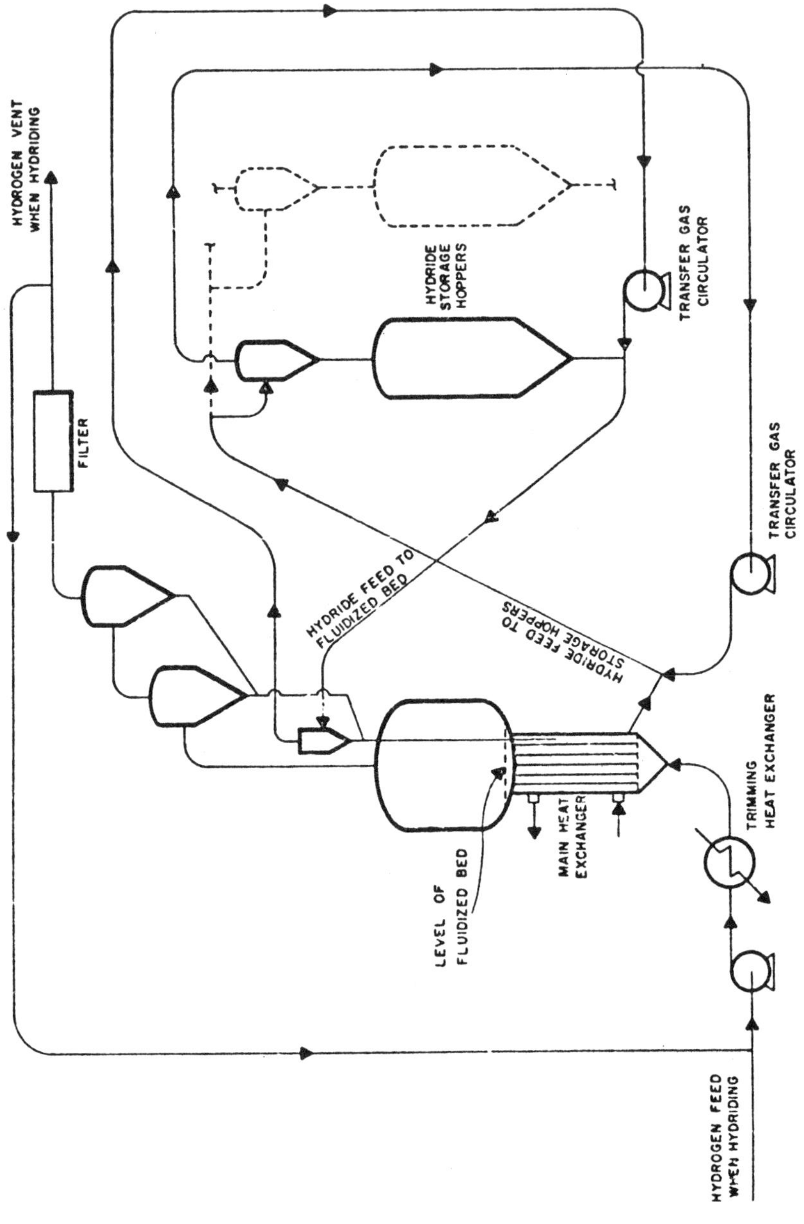

Source: BNL-19249

systems) for fluidizing and circulating the gas. Aside from the energy costs, the capital cost of the circulators is significant, and the effect of erosion and particle size degradation on the cost and operation of the system over a long period of time is uncertain.

Fixed Bed-Indirect Heat Exchange: In one approach to fixed bed systems, the basic storage module is a shell and tube heat exchanger (Figure 5.12). Granular iron-titanium is contained on the shell side and heat is added and removed through fluids flowing in tubes spaced throughout the hydride bed. Hydrogen is added and removed through porous metal tubes also dispersed in the granular metal hydride bed.

FIGURE 5.12: FIXED HYDRIDE BED WITH AN INTERNAL HEAT EXCHANGER

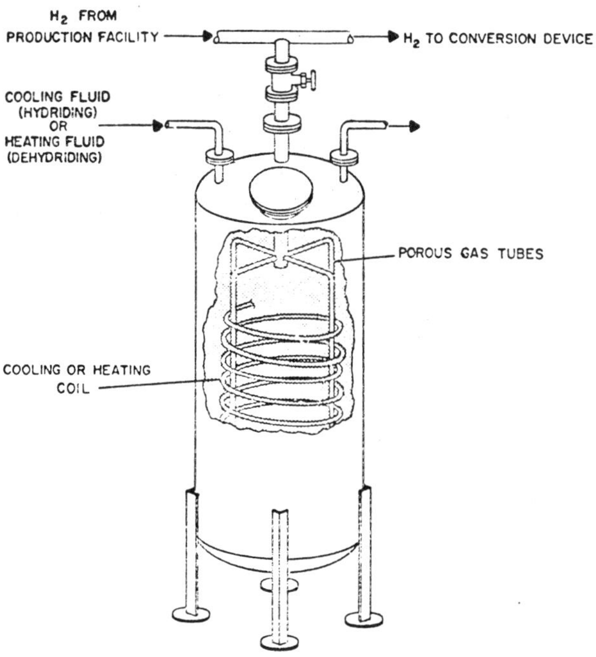

Source: BNL-19249

Hydriding/dehydriding rates are determined by the heat transfer rate out of and into the bed, respectively; thus, the two important design factors are the amount of heat transfer surface and the effective thermal conductivity of the bed. The 12 pound storage reservoir made for PSE&G is an example of this concept (62). This is a workable concept which can be applied to larger scale systems; however, it does involve the complexity and cost of building a shell and tube heat exchanger.

Fixed Bed-Direct Heat Exchange: Another arrangement involving a fixed bed is possible. Hydrogen is circulated through the bed and through an external heat exchanger (Figure 5.13). During a charging cycle a small fraction of the circulating hydrogen would be captured and stored in the bed. Heat produced by the reaction would be carried away from the fixed bed by the circulating process hydrogen stream and rejected in the external heat exchanger.

The rate of addition would be controlled by the pressure and temperature of the circulating gas. The system would very likely be designed to operate at a constant pressure, and the charging rate would be limited by heat transfer within the bed. In discharging hydrogen from the system, the process is reversed.

FIGURE 5.13: FIXED BED-DIRECT HEAT EXCHANGE CONCEPT

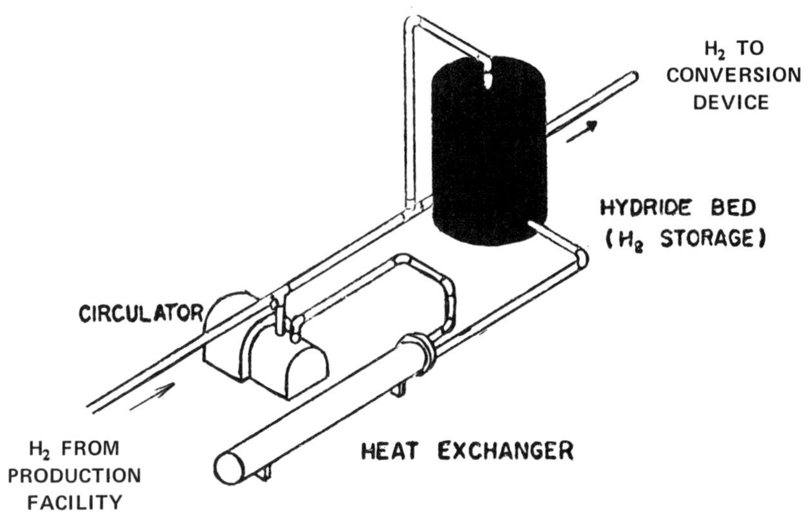

Source: BNL-19266

Preliminary design and modeling work has shown this to be a viable system. However, particle size degradation and the resultant pressure drop increase must be examined. Work is presently in progress to demonstrate the feasibility of this concept in a device similar in size to the reservoir built for PSE&G.

Overall System Design Considerations

In defining the overall design of hydrogen energy storage systems, the interfaces between system components, systems integration, and the trade-offs within the system and with external components must be considered. The interfaces between components include: power conditioning-electrolyzer; electrolyzer-storage system; storage system-power generator; and power generator-power conditioning. The primary interface consideration involving the power conditioning system involves voltage levels and control when fuel cells are used as the power generator.

If one unit is to be used as a combination rectifier-inverter the input voltage to the electrolyzer and the output voltage from the fuel cell must be nearly the same. The interface consideration between the electrolyzer and the storage system is the pressure level. The pressure required for charging iron-titanium will be in the range of 20 to 30 atmospheres. Pressures in excess of this range are achieved in commercial water electrolysis systems and would avoid the need for a compressor between the electrolyzer and storage unit.

The operating pressure of a hydrogen storage system involving circulating hydrogen gas is an important variable which affects the cost of the circulating compressor and the size and cost of the storage tanks. Hydrogen from the electrolyzer would have to be dried before being charged to the storage system; however, this process would be based on available technology.

Waste heat utilization is a major consideration involving the storage-power generation interface. The energy involved in discharging a hydride bed can be supplied by waste heat from the power generation equipment. The mode of disposal of waste heat from the conversion device must be considered in determining the design of the heat exchanger in the iron-titanium storage system.

The final configuration of the overall system will depend on information on the trade-off between cost and energy requirement for the fixed bed-direct heat exchange and fixed bed-indirect heat exchange concepts. If after many cycles the particle size of the iron-titanium decreases, the fixed bed-indirect exchange concept would be favored.

Another important question is the value of oxygen produced in the electrolysis system. Hydrogen-oxygen fuel cells and turbines are more efficient conversion devices than the hydrogen-air counterparts. Oxygen could be saved and stored; however, it may be more economic to produce oxygen as required in an air separation plant for a hydrogen-oxygen fuel cell.

The size relation between the electrolyzer, hydrogen storage system, and the fuel cell depends upon the load characteristics of the utility system. If off-peak electrical capacity is insufficient, it would be possible to supplement the hydrogen supply during certain times of the year by converting fossil fuel to hydrogen in a reformer system. If at times excess off-peak capacity were available, hydrogen could be dumped into the natural gas supply as a supplement. Up to 8% hydrogen could be added without change in the distribution system or end use devices.

Economic Perspective

It is not clear at this time when and at what cost level hydrogen energy storage systems will become practical for electrical utilities. At 1974 costs and given the span of reasonably projected trends in fuel availability, an allowable cost of $350/kw electric for an energy storage system with 8 hours of capacity and an efficiency in the 40 to 50% range appears possible.

The cost of advanced high efficiency electrolysis systems has been estimated to eventually be below $50/kw thermal hydrogen output and a future fuel cell and power conditioning system may cost $150/kw electric with an efficiency of 50% or more.

If these estimates can be met, a cost of about \$6/kwh thermal for the hydrogen storage subsystem is implied. Fuel cell power plant efficiencies in excess of 50% would raise this allowable cost. Since the cost of iron-titanium in bulk quantities has been projected to be as low as \$0.50/lb corresponding to \$2/kwh thermal for the BNL-PSE&G unit, \$6/kwh thermal appears to be a reasonable goal for hydrogen storage subsystems.

FUEL FOR TRANSPORTATION SYSTEMS

Ground (Surface) Transportation

Private Rapid Transit (People-Mover) and Automated Material Transfer Systems: In urban and other congested areas, a large number of alternative approaches to the automobile and light utility vehicle (delivery van, etc.) are under investigation. For the transport of people, automated centrally powered units, typically moving on fixed guideways, are being presently demonstrated. Those devices have been referred to as "people-movers," or as "Personal Rapid Transit (PRT)" systems. Because of the obvious analogy, the function of the PRT is noted to be that of a horizontal elevator, a useful notion.

For the movement of commercial freight and other material items for relatively short distances within warehouses, wholesale and retail outlets, and in similar materials handling activity, advanced automated conveyor-like systems are also under constant development.

PRT systems of various types were prominently displayed in the form of working units at the TRANSPO 72 exhibition in the spring of 1972. A full-scale demonstration system is being installed at Morganton, West Virginia. A full-scale city-wide system has been announced for incremental deployment in Denver, Colorado. Similar innovative advances are being progressively introduced in the materials handling field for the moving of materials and goods.

Since, by and large, these transportation systems employ electrical drive and control, the relevance of hydrogen would normally be in the application of this fuel to provide electrical energy for them. Power can be taken from the main electrical net in an urban area, or be specifically generated for the system.

Private Automobiles and Taxis: Dr. A.L. Austin of the Lawrence Livermore Laboratory has dealt specifically with hydrogen as an automotive vehicle fuel. His work is discussed in detail in a later section entitled Metal Hydride Storage System for Vehicles. Work to date has made it clear that conventional automotive engines can be operated on hydrogen fuel. A number of modification approaches have been successfully demonstrated in this connection including direct fuel-injection, water-injection, exhaust gas recirculation, and special valving arrangements.

Both reciprocating and rotary (that is, Wankel) engines have been so modified. Also, gas turbine engines have been operated on hydrogen with excellent results (see discussion of aircraft applications of hydrogen following). As reported by Weil (3) and others who have reviewed the history of the hydrogen internal combustion engine, work began as early as the twenties and reached a level of successful development by the late thirties. Maybach and Daimler-Benz engines used in the Zeppelin lighter-than-air craft demonstrated the range-extending advantage of hydrogen-conversion. Submarine power systems were developed in which water electrolysis provided hydrogen and oxygen to operate converted diesel power plants for underwater propulsion. Hydrogen by itself, and as an additive was thoroughly explored in research laboratories in the United States, Britain, and in other countries. A number of classical automotive engineering texts note this achievement.

Today, in the United States alone, a significant number of hydrogen automotive projects are underway. Several of these reported on their work at the 7th IECEC (63). Others, such as UCLA and the Brigham Young University entered the recent Urban Vehicle Design Competition with notable results (64) using hydrogen.

The principal and obvious technical difficulty with the hydrogen-fueled automobile, as evidenced in all of the projects carried out so far, is the problem of onboard fuel supply. Also, assuming an acceptable means of tanking hydrogen can be developed, there is the general problem of servicing vehicles with hydrogen. For hydrogen, unfortunately, is by no means as convenient to handle as gasoline.

Hydrogen's low density as a gas and as a cryogenic liquid is a marked disadvantage in achieving compact storage aboard a vehicle. If high-pressure gas is selected, two problems develop: [1] a thick-wall, hence heavy tank is required; and [2] pneumatic energy of the compressed fuel makes for hazards unrelated to combustion.

On the other hand, as a cryogenic liquid hydrogen requires a sophisticated, hence expensive container of the Dewar type. Even then, a certain rate of loss of the fuel by its boiling off will be encountered. Certainly far more complex servicing equipment will be needed than is the case with gasoline or diesel fuel.

Alternatives to gaseous and liquid hydrogen storage have been identified, and at least limited research undertaken in: [1] storage in the form of metal hydrides; and [2] production aboard the vehicle from compounds containing hydrogen such as ammonia and hydrocarbons.

Work by Wiswall and Reilly of Brookhaven National Laboratory, and Sorensen of International Materials Corporation, respectively, deal with these two alternatives (41)(49). Further, Austin (see next section) considers the hydrogen vehicular storage problem in some depth.

Other synthetic fuels which generally derive from hydrogen, or would do so in a fossil fuel-short era of the future, represent, in one sense, an extension of the second alternative above. Alcohols, ammonia, and hydrazine can thus be viewed as hydrogen carriers. Though more compact and convenient to carry as an onboard fuel supply than hydrogen itself, all suffer from a much lower gravimetric heating value. The Synthetic Fuels Panel included consideration of such hydrogen-based fuels (65).

Noting the premise of the metal hydride storage approach (and the need for amplified applied research), it would appear that the liquid hydrogen, or cryohydrogen tank is the most acceptable developed means of vehicular storage. Certainly, further development will be necessary before overall feasibility can be stated for even it, however. Here, the space program, particularly the Apollo Program, has accomplished rapid and impressive developments. Liquid hydrogen has proven to be the staple high energy rocket fuel, superseding kerosene-base fuels used previously. Combusted with liquid oxygen, hydrogen provided a major portion of the energy required to transport astronauts to the moon.

This usage developed relatively large-scale production, transportation and storage means for handling cryohydrogen fuel. Concomitant advances in the technology of liquid hydrogen took place. Today, instrumentation, containers, valving, controls, and other items for cryohydrogen service are readily procurable at competitive costs. This trend will continue as widespread usage of the fuel progresses.

The fact that a cryogenic fuel can be used in automotive applications is demonstrated by the rather widespread operation of liquiefied natural gas (LNG) powered automobiles. LNG is considered a mild cryogenic at –260°F, but requires a vacuum-jacketed storage tank to minimize boil-off, as does hydrogen.

A system, developed by Ametek-Straza of El Cajon, California, permits the driver to switch from gasoline fuel to LNG merely through the operation of a dashboard-mounted switch. The incentive here, of course, is to minimize automotive emissions in the urban areas by operation on the cleaner-burning LNG.

Assuming a point of supply for liquid hydrogen, or gaseous hydrogen which can be consequently liquefied on a local basis, there appears to be little question that hydrogen can be loaded into small, individual storage tanks of vacuum-jacketed construction. Many research laboratories, and a number of industrial facilities are presently equipped with similar cryogenic hydrogen local distribution means. From these, hydrogen is off-loaded into smaller mobile vessels, and thence into laboratory Dewars or equivalent units. Proper precautions are, of course, observed; but the operations are characterized as routine.

In this connection, a number of workers active in liquid hydrogen handling have been queried as to their opinion about the practicability of the Cryohydrogen Service Station concept as an ultimate development. The consensus is that it can be done if properly engineered. One specific device which must be developed is the cryohydrogen equivalent of the gasoline pump fill nozzle. No doubt, this will be considerably more complex than a conventional unit with positive sealing, safe venting, etc.

Several of the entries and would-be entries in the 1972 Urban Vehicle Design Competition (64) had been scheduled for liquid hydrogen. The UCLA entry changed over to a 6,000 psi gaseous hydrogen storage system when it became apparent that the cryohydrogen installation could not be installed within the constraints of their effort. Accepting the overall prize in the internal combustion engine class, the team said that they hoped their entry might presage a long, hard look at hydrogen as the fuel of the future. However, liquid hydrogen was not used.

One project to demonstrate the use of cryogenic hydrogen (and cryogenic oxygen) is that of the Perris Smogless Automobile Association tested in 1970. A technical report on the system was presented at the 6th IECEC in Boston in August 1971 (48).

In summary with regard to the hydrogen, or bireactant hydrogen-oxygen fueled private automobile and taxis, the following statements appear supportable. Given the appropriate development effort the prospect for a feasible hydrogen powered automobile in the not-too-distant future appears rather promising. Specifically, it appears that:

[1] Hydrogen can be used by most conventional automotive engines, with some modifications.

[2] Unconventional power cycles, such as that employing the stoichiometric mixture of hydrogen and oxygen also appear technically feasible and may offer significant advantages (44).

[3] The principal problem area is that of onboard storage of hydrogen; hydrogen's extremely low density and availability as a liquid at cryogenic temperatures pose difficult technical problems.

[4] Identified methods of storing hydrogen on board automotive vehicles are: as a gas under pressure, as cryohydrogen, in a metal hydride, and in a nitrogen or carbon compound which can be decomposed into hydrogen (and other constituents) prior to use in the engine. Sporadic, generally undirected research in each of these has been undertaken; focus is lacking.

[5] The concept of a hydrogen service station appears to be feasible, given proper engineering-design attention.

Trucks, Buses and Utility Vehicles: Much of the foregoing discussion of the applicability of hydrogen fuel to the automobile bears also on trucks and general utility vehicle usage. Further, there are at least two advantages of the latter application for hydrogen:

[1] Servicing of fuel for trucks, particularly in fleet usage, is often handled by special fueling stations employing personnel of upgraded skill (vis-a-vis the typical service station attendant).

[2] The size of the fuel load is considerably larger than that of an automobile, a fact which would tend to minimize boil-off and loading losses (more favorable surface/volume ratio).

Perhaps an ideal transportation system situation for introduction of hydrogen fuel would be a point-to-point fixed route trucking operation. With hydrogen servicing facilities required at each end of the route only (with the possible need for a few at intermediate locations), considerable experience would be gained at a relatively low cost in facility construction or modification. Another example is city bus operation. Depending on the size of the urban area serviced by the bus fleet, a single central depot or a number of dispersed depots, are normally employed by a transit operator.

Typically, at the end of the day's operation the buses are returned to these depots for servicing and overnight storage. The buses are cleaned, maintenance is performed as required, and they are fueled for the next day's operation. It is noted that buses are not normally refueled during an operating day, operating off the single tanking mentioned. The following morning the buses are dispatched according to established revenue traffic patterns and enter service fully prepared for the day's operation.

Such an operational mode is compatible with cryogenic hydrogen servicing techniques. Low or zero boil-off losses would be anticipated during the overnight wait using conventional vacuum-jacketed tanks similar to those in use now for LNG service as already described.

The fact that fuel of considerably different characteristics than diesel fuel (used predominantly in city buses) can be handled routinely in a bus system is evidenced by the experience of the Chicago Transit Authority with propane fuel. Propane, a high vapor-pressure liquid at ambient temperatures has been used in a large fraction of CTA's bus fleet for many years.

The conversion of urban delivery truck service to a hydrogen fuel base would profit the city significantly in terms of reduced air pollution levels as caused by hydrocarbon-fueled vehicles operating in the city center. This would be especially the case if trends to discourage the operation of private automobiles within the city continue toward the alleviation of traffic congestion and land-use for downtown parking space.

A special application area of potential interest in the utility vehicle area is that of vehicles such as forklifts, small tractors, and similar kinds of vehicles for indoor operation. Hydrogen engines could operate in this environment with little if any modifications (perhaps a device to minimize the production of NO_x), whereas gasoline-fueled units require extensive exhaust-treatment systems to eliminate carbon monoxide (by complete conversion to carbon dioxide through catalytic devices). The advantages of centralized fueling for hydrogen would normally apply in this case.

Noise from trucks and utility vehicle power plants has been a notable problem in many instances. Diesels used in cross-country trucks and in off-road machinery are particularly well known offenders. City buses with diesel engines are frequently cited as noise sources by the public. Conventional hydrogen-fueled engines, that is, piston type, can be operated at gasoline engine noise levels using conventional silencing equipment. Practical usage of hydrogen in compression ignition engines (for example, diesel) has not been demonstrated, but remains a possibility. High-pressure fuel injection of hydrogen is feasible as shown by work at Oklahoma State University (26).

Employing the bireactant hydrogen-oxygen system in at least one configuration, the engine exhaust is eliminated as a noise source, as is the intake. Hence, the engine could be virtually noiseless at no sacrifice to operating efficiency (as occurs in muffling completely the diesel).

In summary, regarding hydrogen for trucks and utility vehicles, it would seem that there are no significant problems to preclude its introduction. Demonstration systems in the truck and city bus field can be recommended to initiate this transition. In many respects, especially those of hydrogen logistics and servicing, it would seem to be relatively simpler to initiate hydrogen in this sector, than in that of the private automobile.

Rail and Guideway-Supported Vehicles: Although the railroads, long operating on diesel-electric power, utilize less hydrocarbon fuel than trucking nationally (and both together use much less than the automobile), the greater share of the nation's bulk cargo transported is moved over the rails.

A breakdown of the ton-mileage apportionment follows (66): rail (36.7%), water (24.7%), pipeline (19.4%), highway (19.1%), and air (0.1%). Trains, then, represent an important fuel-consuming sector of transportation. As such, rail systems are candidates for ultimate conversion to hydrogen.

On first examination, a number of approaches are possible in making such a hydrogen conversion inasmuch as both power plant modification and new power plant development should be considered. Following are the options:

Conversion of present diesel-electric systems to hydrogen fuel

Employment of hydrogen-fired steam systems, probably turboelectric

Adoption of hydrogen gas turbines

Development of hydrogen-oxygen/steam power plants

It should also be noted that electrified systems (third rail, overhead pantograph pickup) are operated in this country for commuter service and on routes such as that of the Penn Central New York to Washington route (operated for the Amtrak system). As will be discussed under electrical generation, hydrogen can be used to supply the electric trains' motive power in this mode also, that is, as local generation.

Just as over-the-road liquid hydrogen (and liquid oxygen) truck trailers are in use on the highways, rail transporters for these cryogenic fluids are well developed. Thus, cryohydrogen has been already demonstrated as being compatible with the railroad environment to the extent of being a rail-shipped product.

The concept of a cryohydrogen tender for fueling a rail prime mover is easily imaginable considering that the scale is large (vis-a-vis the automobile), and trains are not volume-limited in the sense highway vehicles are. One can consider adding another, or longer, car for fuel. A practical technological basis for configuring such a tender clearly exists.

Conversion of diesel engines to hydrogen may be possible, there being a number of historical precedents (3). In more recent times, the work of Schoeppel, Murray, et al at Oklahoma State University on hydrogen direct-cylinder injection may be applicable to advancing the state of the diesel-conversion task (26). This is an area where exploratory activities can be recommended immediately.

Gas turbines operating on aircraft-type turbine fuel (JP-4, etc.) have been used on trains for demonstration programs in the past. The best known system in present use is the United Aircraft Turbo Train which has been demonstrated in New England in recent years under Department of Transportation sponsorship. These gas turbine powered units have been placed into normal revenue service in Canada to a limited extent.

Hydrogen-fueled gas turbines are eminently feasible (see discussion under Air Transportation) and offer unique advantages over their hydrocarbon-fueled counterparts. Therefore, the concept of a hydrogen gas turbine rail prime mover can be readily demonstrated as being feasible technically. Another possibility is the water-based Rankine cycle system, perhaps in the form of a steam-turbo-electric drive. Hydrogen-fired boilers are certainly feasible. Such a return to steam in rail systems is worth studying to determine if increased energy conversion efficiency can be achieved (over that of the other options noted) in reasonably compact hardware. The design of acceptable air-cooled condensers may be one of the more difficult tasks at hand, particularly so if vacuum-condensing is required to extract energy from the steam working-fluid.

A final candidate for rail power plants is the hydrogen-oxygen/steam cycle described by a number of researchers (48)(44). Although this would require the carrying of liquid oxygen (as in the case of hydrogen, cryo-oxygen rail tankers are in use), in addition to hydrogen, possible gains in conversion efficiency (44) may make this system attractive. Further, the elimination of all emissions and the capability to completely silence the power plant may be further inducements. Although the hydrogen-oxygen system resembles the standard steam cycle already mentioned, it offers some salient advantages for increased efficiency and decreased size and weight:

[1] The turbine inlet temperature (throttle temperature) can be pushed well beyond the normal 1000° to 1200°F boiler limit, given a cooled expander, with significant gains in efficiency and a reduction in size.

[2] If the water exhaust is prechilled (say, in heat exchange with the cryogens), it may be used on the external surface of the condenser to enhance heat transfer (reduction in size and air throughput).

Another possibility which would be compatible with the hydrogen-oxygen/steam system for rail operation is the notion of regenerative-braking powered electrolyzers. On downhill grades, were the electric wheel-drives reversed such that they act as generators (thus providing braking torque), the electricity could be used to regenerate water (exhaust) into hydrogen and oxygen. This would require the addition of electrolysis equipment, but this is industrially available, and a means of storing the produced gases. On-board liquefaction can be considered, in which case the fuel produced in braking could be reentered into the main tanks.

An extension of conventional rail transportation systems into higher speeds, specifically for passenger service, is the category High Speed Ground Transportation or HSGT. A special office has been set up in DOT's Railroads Administration to conduct studies and demonstration programs leading to future developments with speeds as high as 300 mph and even higher. A testing laboratory has been established near Pueblo, Colorado capable of full-scale system evaluation and engineering development of such HSGT systems.

Emphasis has been given to the concept of off-board power, that is, guideway electrical power pickup (in analogy to third rail/pantograph systems in use), as opposed to on-board power. The principal incentives have seemed to be the opportunity to eliminate noise and emissions from the HSGT system's prime mover.

However, two principal factors mitigate against the wholesale adoption of the guideway power pickup approach, and have caused attention to return to conventional and unconventional on-board power systems (a recently completed study by NASA for DOT went into this subject). These factors are:

[1] Electrification of long-distance routes is extremely expensive in

terms of initial and operating costs in comparison with a passive nonelectrified guideway.

[2] The technology of high-speed, high-power electrical transfer through moving contacts, say above 200 mph, does not yet exist; research results obtained so far are not particularly encouraging.

In examining the applicability of hydrogen for powering HSGT systems, it is once again noted that hydrogen systems for conventional electrical power networks are feasible. These systems might well be applicable for the special requirement of guideway electrification if this approach is pursued.

Hydrogen as an on-board power plant fuel, as discussed for conventional rail rolling stock application, is of interest. Particularly noting that the incentive to remove the on-board power system for the HSGT application (that is, to go to electrified guideways) is that of air pollution/noise, hydrogen may well be the avenue to a desirable on-board power plant.

Much more so than is the case with low-speed freight equipment, the HSGT power plant must be lightweight and relatively compact for vehicle overall weight and aerodynamic drag must be minimized in advanced HSGT systems, as for aircraft. For one thing, suspension systems based on the air-cushion and magnetic system approaches (rail/wheel support and drive are apparently no longer feasible above approximately 200 mph) must be sized according to gross vehicle weight, and the power consumed varies likewise.

For this reason the on-board power plant must be compact and of minimum weight. Hydrogen gas turbines would be superior to converted diesels and conventional steam turboelectric units. But the hydrogen-oxygen/steam power unit may also be a contender. For it attacks the two reasons for seeking off-board power initially in the case of the HSGT: noise and air polluting emissions. Whereas the hydrogen gas turbine may still have a NO_x problem, as well as the need for extensive muffling of intake and exhaust to limit noise, the semiclosed Rankine hydrogen-oxygen power cycle has zero emissions and can be silenced at no loss in efficiency or significant increase in weight or size. Clearly, comparison studies are in order in this area.

In summary, although the railroads are not the predominant consumers of petroleum products (it is automotive, and, the fastest-growing consumer, aircraft which dominate), they are critical to the national transport system. Consideration of the impact of escalating fuel costs and potential interruptibility problems based on foreign-source dependence on rail transportation justifies consideration of hydrogen fuel.

The compatibility of liquid hydrogen with the rail environment has been demonstrated. The approach in selecting the specific hydrogen power system, though not yet clear, can be made flexibly from a number of alternatives as discussed. These range from direct conversion of diesel-electric systems to the more exotic hydrogen-oxygen system. The special area of High Speed Ground Transportation systems may be the leader, because of its special set of incentives, in the development of an advanced hydrogen-fueled prime mover for guideway/rail service in the future.

Water Transportation

Applicability of Hydrogen Fuel to Ship Propulsion: The fueling of shipboard power systems with hydrogen, in generally the same vein as discussed for rail transportation, appears quite feasible. The precedent of using a cryogenic fuel at sea is already well-established in that liquefied natural gas (LNG) carriers use the vaporized methane to fuel their boilers. With appropriate burner modifications, there are no basic problems in adopting hydrogen in the same way. (Nevertheless, applied research is needed).

Indeed, if hydrogen becomes a world energy commodity at some point in the future, very large "cryotankers" transporting liquid hydrogen over the seas may become common. No doubt these ships will use hydrogen to fuel their prime movers.

Alternatives to the hydrogen-fueled conventional steam system should be explored however. Specialized applications, such as smaller than tanker class vessels capable of higher speeds, etc. may profitably select hydrogen gas turbines, or the hydrogen-oxygen/steam systems already described.

Special Applicability to Nuclear Powered Systems: It is for ship propulsion, particularly for naval applications, where nuclear power plants have had widespread application. This fact lends itself for a special consideration of hydrogen-energy. Specifically, a combination nuclear-chemical propulsion system in which the nuclear power source provides ordinary cruising speed capability, or light-load service, wherein hydrogen augmentation can be called upon for special high-power mode operation, may be of interest. The hydrogen (and oxygen) would be generated from seawater via electrolysis or possibly by thermochemical processes using nuclear heat directly.

Such a vessel would have special performance characteristics, with hydrogen serving as an energy storage reservoir. An analogy with the electrical utility's system of baseload and peaking operation is obvious. Storage of hydrogen (and oxygen, if used) would likely be in the cryogenic form, although other options, such as metal hydride storage, should be studied.

The ability of a ship to develop very large output power such as required for high-speed dash capability, or for heavy loads (for example, towing), for limited periods could be of considerable operational value. Hydrogen-oxygen prime movers, for example, as supplements to the direct nuclear power system, would be relatively lightweight and compact. The availability of water for condenser cooling is a distinct advantage (not shared by the rail system, for example).

Hydrogen-Oxygen for Submersibles: Much of the published work describing and evaluating the hydrogen-oxygen/steam power cycle [e.g., Reference (67)] has been in view of its applicability for powering submersibles. With the normal air supply unavailable in the depths, a stored-energy system must be used. For smaller submersibles, such as those used in oceanographic research, nuclear power may not be applicable, in which case chemical energy is used. Batteries, fuel cells, and dynamic chemical reactant approaches have all been extensively looked at. The reactant combination of hydrogen-oxygen offers the following uniquely interesting advantages for submersible power applications.

[1] It provides for a highly concentrated energy storage form on a mass basis (less so volumetrically).

[2] It is adaptable to powering a number of prime mover types, for example, turbines and reciprocating engines, and can be used in fuel cells.

[3] It produces only water exhaust, a completely condensible liquid, obviating the high back-pressure exhaust required with gaseous effluents under depth conditions.

[4] The water exhaust is compatible with the life-support/environmental control system, as is the oxygen.

[5] If water exhaust is retained aboard, the overall power system is a constant-mass one, thus simplifying ballast control for effecting depth settings.

Added to these advantages is that of fuel extraction from seawater through the direct employment of an energy source such as nuclear, solar, etc.

Hydrogen—Unique Compatibility with the Sea: Presently, oil is the basis for ship propulsion; it is also shipped in large supertankers as an important commodity on the sea in increasing amounts. However, oil in the sea environment has its special problems, oil spills and off-shore oil well leakage being a prominent one. The adverse effect on marine and coastal life-forms, and the degradation of natural scenic areas along the coast, as well as direct property damage, is well known from circumstances like the recent Santa Barbara incident.

Viewing a future scenario in which hydrogen cryotankers ply the seas in large scale, what about an inadvertent gross cryohydrogen spill? Although not demonstrated experimentally, the hydrogen will vaporize more or less immediately when it contacts the relatively hot water surrounding the spill. The water so-chilled will tend to sink; perhaps ice will be formed. However, there should be no lasting effect which will grossly affect the environment, certainly nothing like oil-damage to the coast from a major spill.

Thus, the shipment of hydrogen in very large quantities, and its general use in shipboard power systems will bring to the scene a much more benign environmental interaction than that associated with hydrocarbon fuel. If hydrogen is eventually adopted as a prominent energy form, surely it will see service at sea. It has even been suggested that it may be produced at sea from solar energy [The Helios-Poseidon concept (68)].

Air (and Space) Transportation

It is the field of aeronautical propulsion where hydrogen's unexcelled heating value (gravimetric) is of ultimate importance. The range/payload performance of aircraft is directly affected by the heating value of the fuel, other factors such as propulsion and aerodynamic efficiency being the same. Hydrogen's superiority as a fuel for both subsonic and supersonic aircraft was documented as early as 1955 (69). Much earlier in 1923, J.B.S. Haldane remarked:

> "Liquid hydrogen is weight for weight the most efficient known method of storing energy, as it gives about three times as much

> heat per pound as petrol. On the other hand it is very light, and bulk for bulk has only one-third the efficiency of petrol. This will not, however, detract from its use in aeroplanes, where weight is more important than bulk." (70)

The principal drawback for the application of hydrogen to aircraft, outside of its high cost (relative to kerosene-based fuel) which remains today, is the technology for the containment of cryohydrogen in an aircraft-suitable tank, and general problems of logistics and servicing.

The wholesale employment of liquid hydrogen in the Apollo program, itself stimulated by the successes of the earlier Centaur project, has done much to advance the liquid hydrogen technologies to the point that aircraft fueling is judged to be feasible. The safe-handling of hydrogen has been demonstrated convincingly by this same space program effort.

Adaptation of hydrogen fuel to the aircraft itself remains to be demonstrated in a practical application. Hydrogen-fueled aircraft engines appear to be no great problem, as will be noted below. Studies backed up by limited experimental work in the late 1960s on the use of LNG (essentially liquid methane, a cryogenic at −260°F) have provided engineering insight for integrating cryogenic fuel into aircraft. However, in-depth analysis of the liquid hydrogen case remains to be done (efforts within NASA are underway).

One salient experiment with a liquid hydrogen-fueled aircraft has been recorded, however. The Lewis Flight Propulsion Laboratory of the NACA (now NASA Lewis Research Center) successfully flight-tested a modified B-57 aircraft in which one engine (Curtiss-Wright J-65 turbojet) was converted to hydrogen fuel in flight. The liquid hydrogen was pressurized with helium, vaporized in an air heat exchanger, and fed to the engine as a gas. Predicted performance was achieved in this early program, with specific fuel consumption (SFC) about one-third of that operating with JP-fuel.

At roughly the same time, an Air Force sponsored effort at Pratt & Whitney Aircraft successfully demonstrated the conversion of a J-57 turbojet to hydrogen operation (this engine and its fan-version power the well-known 707 and DC-8 commercial transports). Having demonstrated the feasibility of hydrogen as an aircraft fuel from the engine standpoint, the P&WA program continued toward the development of a unique engine which capitalized on hydrogen's characteristics more optimally (5). According to the Air Force Aero Propulsion Laboratory at Wright-Patterson AFB, other engines have been successfully converted to hydrogen operation.

Hydrogen-Fueled Subsonic Aircraft: As noted by Austin (Report UCRL-51228) by 1990 aircraft fuel usage will rival that projected for automobiles at approximately 70 billion gallons of petroleum consumed annually. This figure is for United States consumption assuming no SST. This is clearly a very large energy market to consider for conversion to hydrogen fuel, and a fruitful one to attack in our attempts to get off the hydrocarbon standard (balance of payments, national energy security, etc.).

As already noted, the development of power plants for hydrogen operation does not appear to be a technological problem, but one where straightforward engineering can be immediately undertaken. With respect to its hydrocarbon antecedent, the hydrogen turbofan engine (for instance) will have some interesting features:

[1] Higher efficiency due to higher turbine inlet temperature capability because of hydrogen's excellent combustion and cooling characteristics (direct and indirect blade cooling opportunity).

[2] Lighter weight, more compact design due to [1], and reduction in combustor length.

[3] Longer time between overhaul and less maintenance, due to the cleanliness of hydrogen (no deposits, tars, particulates).

The hydrogen engine will provide low emissions and zero smoke. This will be significant to airport-locale operations, and will reduce the annual cost of cleaning buildings, facilities, and the aircraft themselves due to exhaust products of hydrocarbon fuel. The oxides of nitrogen remain as an intrinsic emission, and will tend to increase if turbine inlet temperatures are further raised. However, as carbon monoxide and hydrocarbon emissions, as well as smoke, are not possible with hydrogen, cleanup measures undertaken can be more single-mindedly directed to NO_x as the one remaining pollutant.

In this connection, selective use of water injection as developed by the Ginter Corporation, may be effective in limiting NO_x emission in the vicinity of airport operations. Flat temperature profile combustors of innovative design may also prove to be effective for this purpose.

However, it may be that the superior performance of hydrogen-fueled subsonic aircraft will offer the motivating economic inducement for its development as a successor to the hydrocarbon-fueled present generation aircraft. Although a definitive study of a modern-technology liquid hydrogen fueled commercial aircraft has not been noted, preliminary indications from two limited-scope studies within the aircraft industry have been examined with interest. Both yielded very promising results favoring the hydrogen system.

Stimulated by the Synthetic Fuel Panel's efforts, the Lockheed-California Company investigated hydrogen-converting the present-day medium-range wide-bodied transports (L-1011, DC-10 class aircraft).

Basically, holding payload and range fixed, a reduction in take-off gross weight (TOGW) was sought to reflect the benefit of hydrogen's high heating value (or engine SFC reduction). In lieu of an in-depth structural analysis and optimization, the required hydrogen fuel was carried in large (11 foot diameter) wing tip tanks (not in integral wing tanks as with the Jet A fuel (that is, kerosene).

The net finding here was that the TOGW could be reduced for the same mission (range, payload) as much as by 30%. This is a significant gain which could also be reflected as an increase in payload and/or range at the initial TOGW.

Workers at NASA's Langley Research Center have unofficially confirmed the trend of the Lockheed finding (71) and have had carried out a preliminary assessment of a very large hydrogen-fueled cargo plane (roughly two times the size of the 747 transport). Boeing presented a hydrogen-converted 747 concept with a 25% reduction in TOGW. Range and payload were held at the standard values. A general conclusion from this report (71) is worth noting here:

> "Hydrogen fueled aircraft have the distinct advantage of performing the same mission with lower gross weight and greatly reduced fuel consumption. Since hydrogen is not a fossil fuel, an operating fleet of hydrogen fueled aircraft would represent a tremendous conservation of fossil fuels. Additionally, the hydrogen combustion process would solve most of the emission problems inherent with fossil fuels."

These two preliminary assessments, combined with older studies such as (69), clearly indicate the ultimate superiority of hydrogen for large subsonic aircraft of the future.

Hydrogen-Fueled Supersonic Aircraft: What has been stated generally about the potential of the hydrogen-fueled subsonic aircraft is also valid for supersonic aircraft, such as a projected liquid hydrogen-fueled Advanced SST. The sensitivity to hydrogen's bulkiness because of aerodynamic and/or heating constraints may somewhat limit any overall gains in performance. Detailed design studies will be required to pin this down; such are understood to be underway within NASA and at several airframe companies.

In an unpublished paper by Kirkham, NASA Langley (72), it was shown that a Mach 3 SST system designed for hydrogen would show the following gains over a JP-fuel version. The two comparison aircraft were of differing design, but once again the study was not in great depth. The advantage of hydrogen was revealed in these alternative gains: a 32% improvement in range; or a 45% improvement in payload; or a 25% reduction in TOGW.

If the advantage of hydrogen is taken as a reduction in aircraft weight, it has been pointed out that this could lead to a reduction in sonic boom intensity. This may be a significant point in hydrogen's favor. Certainly the reduction in pollution of the air from the SST's high-output engines would be welcomed; both the Concorde and the TU-144, present-day SST contenders have been criticized on this point.

Unfortunately, and this applies to the subsonic aircraft application also, it is not clear that hydrogen will open any striking avenues for the reduction of aircraft engine noise; as noted by Kolk of American Airlines, recently (73) in the 1972 Wright Brothers Lecture, it may be the solving of the engine noise problem which will limit our technical advances. Perhaps an innovative engine design, capitalizing on hydrogen's characteristics in some way, will be derived which will provide suitably quiet operation for future aircraft.

Hypersonic Aircraft and Space Vehicles: Interesting to note, with regard to studies and experiments directed toward hydrogen usage in aircraft, it is the hypersonic field where most of the extant work has been accomplished. Hypersonic (that is, flight speeds greater than about Mach 4) aircraft and propulsion research in the United States extends considerably back in time, but essentially from the beginning it has been based on the concept of liquid hydrogen fuel. A recent paper by Witcofski (74) points out the economic trends of hydrogen vis-a-vis hydrocarbon fuels as they ramify into direct operating costs. He predicts the same cost/unit-energy for the two fuels by the 1990-2000 time period, about as early as a hypersonic aircraft might be expected to appear. However, the reason hydrogen is essentially exclusively treated for hypersonic applications is due to its excellent characteristic as a coolant for propulsion systems and, ultimately, for airframes. Aerodynamic heating constrains the entire aircraft design at these Mach numbers. The use of hydrogen directly and indirectly, prior to its consumption by the engines, for cooling is of critical importance. Conventional hydrocarbon fuels have neither the heat-sink capability nor the chemical stability to meet this requirement at flight speeds much in excess of Mach 4 or so.

Representative of a number of government sponsored research efforts on hydrogen-fueled hypersonic engines (Air Force and NASA) is the Hypersonic Research Engine (HRE) being evaluated by NASA under the leadership of Dr. Kennedy Rubert of the Langley Center (75). The AiResearch Division of the Garrett Corp. is the contractor for this program presently underway. The engine is a hydrogen-fueled ramjet capable of operating in both the subsonic- and supersonic-combustion mode for operation at (simulated) flight speeds of Mach 3 to 8.

Finally, in the space transportation sector, hydrogen fuel with liquid oxygen to support its combustion at elevated pressures and at nearly the stoichiometric ratio, is undoubtedly the staple rocket propellant combination for the forseeable future. As noted previously, it was the technology and logistical requirement involving hydrogen which established the potential base for its domestic application deriving from the Apollo effort.

In a rocket engine hydrogen-oxygen offers the highest specific impulse (thrust produced for a unit mass flow-rate) of all the practical chemical propellants worked with to date. Short of nuclear rocket propulsion (which uses liquid hydrogen alone as propellant after heating it in the reactor) it is not apparent that hydrogen-oxygen will be supplanted in the future for rocket propulsion in space transportation systems. The Space Shuttle orbiter system is, for example, predicated on high-pressure rocket engines based on hydrogen-oxygen.

Beyond the Space Shuttle is the possibility of high-performance air-breathing engines capable of booster stage propulsion. Combinations of air-breathing and rocket engines may be used; or a new class of composite rocket/air-breathing engines may be developed (76). However, it appears that hydrogen, and hydrogen-oxygen will be the preferred fuel for all of these possibilities because of its overall excellence in performance and cooling characteristics. As noted by Witcofski (74), its cost is anticipated to be reasonable in the future also.

LIQUID HYDROGEN SYSTEM FOR AUTOMOTIVE TRANSPORTATION

Vehicle Fuel Storage

Hydrogen may be stored on board a vehicle in several forms: as a compressed gas, as a cryogenic liquid, as a metallic hydride, and as a chemical compound. The compressed gas and cryogenic liquid forms are better developed and appear to be the most straightforward. The compressed gas form involves considerable weight and volume for tankage. The cryogenic liquid form requires less weight and volume for tankage than the compressed gas form but involves handling a cryogenic fluid with its associated losses from boil-off, chill-down, flashing, etc., and requires additional energy and equipment to liquefy and handle.

Metallic hydrides definitely offer several desirable features such as: H_2 is generated only as needed, energy from the engine exhaust may be used for hydride dissociation, and the hydride may be recharged with H_2 gas. Storage as a cryogenic liquid on board the vehicle was selected for further study to show that this method is not only possible but may be a practical alternative. A H_2 production and distribution system will be required no matter which form turns out to be the most favorable.

A vehicle with a 20 gallon (76 liter) gasoline tank would require a 50 gallon (189 liter) capacity LH_2 Dewar for an equivalent driving range. This equivalence is based on the data given by the Perris Smogless Automobile Association that their vehicle obtained 10 miles per pound (35.5 km/kg) of H_2 or 2.5 miles per pound (8.9 km/kg) of gasoline; that is, 2.5 gallons (9.5 liters) of H_2 are required to replace 1 gallon (3.8 liters) of gasoline.

A rugged 50 gallon (189 liter) LH_2 Dewar [equivalent to 20 gallons (76 liters) of gasoline and suitable for use in an automobile] could be built to withstand most collisions, and would cost approximately $2,500 in 1973. However, it is estimated that mass production could lower the cost to less than $750. A 200 liter (52.8 gallon) Dewar with 10% ullage, 50 pounds per square inch gauge (3.5 kg/cm^2 gauge) working pressure and 1% per day boil-off rate would require one week to reach working pressure; that is, a vehicle equipped with such a Dewar and operated at least once a week would not lose any H_2 through venting.

A 175 liter (46 gallon) LH_2 Dewar with a 1.5% per day boil-off rate and capable of withstanding up to 16 gram acceleration loads is already available (Cryenco). The Beech Aircraft Corporation has announced that it intends to develop and build a prototype container system to demonstrate the feasibility of using liquefied gas as a fuel for automobiles and other surface vehicles. This will probably lead to the development of a LH_2 container system for automobiles. A cylindrical 200 liter Dewar would be of approximately 24 inch (61 cm) outside diameter, 39 inches (99 cm) long, and weigh about 400 pounds (181 kg). A spherical Dewar of equal capacity would be of about 30 inch (76 cm) outside diameter.

A one-half inch (1.3 cm) inside diameter fill line and a 2 inch (5.1 cm) inside diameter vent line appear adequate to fill the car Dewar at a rate of 15 gallons LH_2 per minute (0.95 l/sec) to permit a refill in approximately three minutes without excessive pressure losses.

The typical United States automobile gasoline mileage of 13 miles per gallon (5.5 km/l) corresponds to a 5.2 miles per gallon (2.2 km/l) LH_2. An automobile equipped with a 50 gallon (189 liter) capacity LH_2 Dewar having a boil-off rate of 1% per day could only be driven at a speed of 0.1 mile per hour (0.04 m/sec) on the storage Dewar boil-off gas. Therefore a heat source is required to increase the boil-off rate during vehicle operation.

A 650 watt heat source would enable the vehicle to travel at a speed of 100 miles per hour (44.7 m/sec). The heat source could be operated by a pressure switch set to turn on at a tank pressure of 16 pounds per square inch absolute (1.1 kg/cm^2 absolute) and off at 20 pounds per square inch absolute (1.4 kg/cm^2 absolute). The vehicle could be driven only 0.7 mile (1.1 km) on the 50 pounds per square inch gauge (3.5 kg/cm^2) GH_2 remaining in the Dewar if the Dewar were warmed up. If the vehicle were allowed to stand until maximum pressure was reached in the Dewar, before refilling, it would be necessary to relieve the pressure by either venting or driving the vehicle. The maximum distance that the vehicle would have to be driven to reduce the pressure to 16 pounds per square inch absolute (1.1 kg/cm^2 absolute) is 18 miles (29 km).

The vehicle LH_2 Dewar may be refilled or replaced at a service station whenever resupply is required. The number of connections for refilling the Dewar could be as many as five (electrical, LH_2 supply, tank vent, vent line purge, and supply line purge) or as few as two (electrical and gas supply to engine) for replaceable Dewars. The number of connections should be minimized and reliable quick-disconnect type fittings should be developed to expedite refueling. A sketch of a car Dewar is given in Figure 5.14. A periodic (annual) warmup to remove impurities, inspection for leaks and damage, vacuum testing, etc. is anticipated.

The only known vehicle operated to date on LH_2-liquid oxygen (LOX) carried both cryogens in separate Dewars (48) but the refrigeration in the LH_2 may be used to separate air to provide oxygen or O_2 enriched air, on board the vehicle thus eliminating the need for the LOX Dewar. A United States patent was granted in 1960 for a system for the production of LOX from atmospheric air on board a vehicle such as an aircraft (77). The use of such a system on an automobile is feasible but not known to have been accomplished.

During periods when the vehicle is not being operated, the boil-off H_2 could be utilized in numerous ways such as: to power a fuel cell for battery recharging; to store energy (in a battery, etc.) for operation of accessory equipment (power brakes, power steering, etc.); and to react in a catalytic burner to keep the engine and vehicle warm during cold weather or in a refrigerator to keep the engine and vehicle cool during hot weather.

A comparison of the weight and volume for fuel and tankage for the different methods of H_2 storage on board a vehicle is given in Table 5.10. The following observations stem from comparisons of the data presented in Table 5.10. The most compact tank is that for gasoline. The lightest fuel storage system weight is that for gasoline with its simple, single-wall, sheet-metal tank. Less H_2 by mass is required than gasoline.

The H_2 fueled conventional engine will be the heaviest system of those compared here. It shares the disadvantage of a heavy engine (an unchanged conventional engine) and the heaviest tankage. Some improvement in specific power output and thermal efficiency may be intrinsic with the shift to H_2 in a conventional engine; however, major gains can be obtained from the development of basic cycle improvements and/or new operating cycles that capitalize on the outstanding heating value, combustion characteristics, and the excellent cooling performance of H_2.

Such improvements in efficiency would help to counter the otherwise large volume of H_2 that must be carried for the equivalent range achieved by gasoline powered vehicles. A H_2-O_2 fueled engine should be both smaller and lighter than a conventional engine, weighing about half as much. This will help offset the heavier fuel load required for a H_2-O_2 powered vehicle (44). The total weight of the metallic hydride fuel and tank is slightly greater than for a cryogenic liquid but the volume requirements of the metallic hydride appear to be slightly less.

FIGURE 5.14: AUTOMOBILE DEWAR SYSTEM

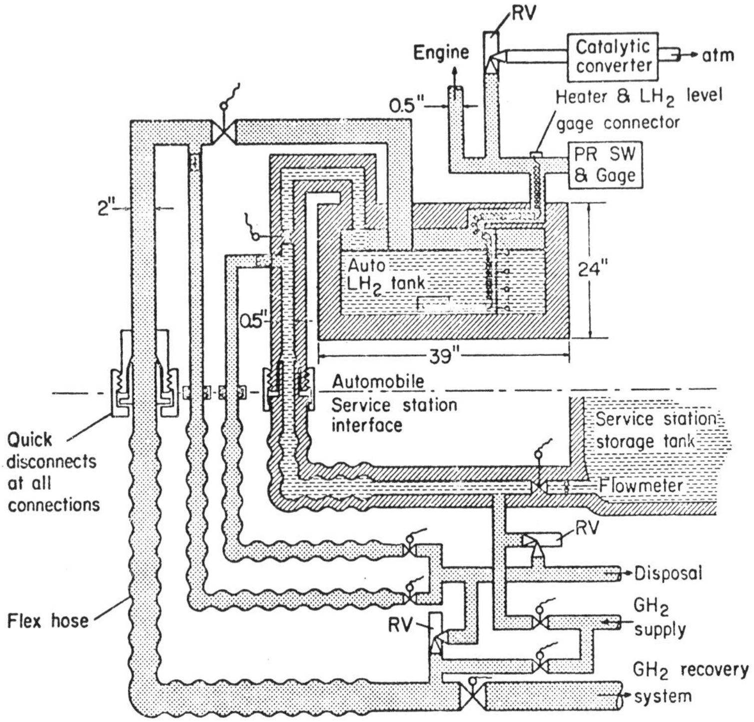

TABLE 5.10: COMPARISON OF FUEL STORAGE SYSTEMS FOR A VEHICLE RANGE OF 260 MILES (418 KM)

	Gasoline	Cryogenic LH_2	Compressed GH_2	Metallic Hydride
Fuel:				
weight, lb (kg)	118 (53.5)	29.5 (13.4)	29.5 (13.4)	400 (181) (Mg H_2)
volume, ft^3 (m^3)	2.6 (0.07) (20 gal)	6.7 (0.19) (20 gal)	35 (1.0)	8 (0.23)
Tankage:				
weight, lb (kg)	30 (13.6)	400 (181)	3000 (1361)	100 (45.4)
volume, ft^3 (m^3)	3 (0.08)	10 (0.28)	54 (1.53)	9 (0.25)
Total:				
weight, lb (kg)	148 (67)	430 (195)	3030 (1374)	500 (227)

Source: LAUR-73715

The weight and volume requirements for storage as a compressed gas are excessive. However, some improvements may be possible by using high-strength, low-density materials for pressure vessel fabrication.

The volume and weight advantages of the cryogenic liquid form over the compressed gas form of storage may be illustrated as follows: a standard cylinder containing approximately 220 ft^3 (6.2 m^3) of compressed H_2 weighs approximately 125 pounds (56.7 kg) and contains only 1.2 pounds (0.5 kg) of H_2 which is equivalent to 11 miles (17.7 km) range at 13 miles per gallon (5.5 km/l) of gasoline. Twenty-four cylinders with a weight of 3,000 pounds (1,361 kg) would be required for a range of 260 miles (418 km). At a speed of 60 miles per hour (26.8 m/sec), about 6 cylinders per hour would be used.

Hydrogen Production and Distribution System

Consideration of the amount of H_2 that must be annually produced and liquefied to replace the annual consumption of gasoline in automobiles in the United States must include boil-off, cool-down, and transfer losses from the production plants, transport trailers, service stations, and automobiles. These losses are a small percentage of the total H_2 required and are not necessarily total losses since the H_2 may be recovered as a gas and be reliquefied or used as a gas.

Estimates of the number and cost of the production plants required to produce this amount of LH_2, the number and cost of the service stations required, and the trucks required to transport LH_2 from the production plants to the service stations indicate that the total cost is not much greater than the capital equipment expenditures forecast by the petroleum industry over the next thirteen years.

Assuming an annual consumption of 10^{11} gallons (3.8×10^{11} liters) of gasoline in motor vehicles in 1972 [8.9×10^{10} gallons were consumed in 1970 (78)], 2.5×10^{11} gallons (9.5×10^{11} liters) LH_2 per year would be required to replace gasoline. Losses from boil-off, transfer, cool-down, etc. must be added to this to determine the required annual production of LH_2.

The typical gasoline service station assumed for this study has a storage capacity of 15,000 gallons (56,775 liters) gasoline, sells 3.6×10^5 gallons per year (1.4 times 10^6 liters per year) and is resupplied by 41 truck deliveries per year (8,650 gallon capacity/truck). The total number of gasoline service stations is estimated to be 300,000.

The typical H_2 service station would have a storage of 38,000 gallons (143,830 liters) of LH_2. The typical LH_2 transport trailer carries 13,000 gallons (49,205 liters) of LH_2 but a few 22,000 gallon (83,270 liter) capacity trailers are in operation. A comparison of the H_2 and gasoline systems is presented in Table 5.11.

A possible H_2 service station would have the LH_2 Dewars 25 feet (7.6 m) from a roadway and 75 feet (22.9 m) from a building in accordance with existing safety standards. Figure 5.14 shows a possible car Dewar service station interface. Standardized fittings (and their location) should be used on all cars and stations. The predicted boil-off, transfer, and cool-down losses of LH_2 for the total system are summarized in Table 5.12.

TABLE 5.11: COMPARISON OF GASOLINE AND LIQUID HYDROGEN SYSTEMS

	Gasoline	LH_2
Total usage, billions of gal per yr (10^9 ℓ/yr)	100 (379)	250 (946)
Station storage, thousands of gal (10^3 ℓ)	15 (57)	38 (144)
Distribution per station, millions of gal/yr (10^6 ℓ/yr)	0.36 (1.4)	0.90 (3.4)
Delivery vehicle capacity, gal (ℓ)	8650 (32,740)	13,000 (49,205)
Average number of deliveries/yr	41	72 or 42*
Car tank size, gal (ℓ)	20 (76)	50 (189)

*If 22,000 gal (83,270 ℓ) deliveries were made.

TABLE 5.12: SUMMARY OF LIQUID HYDROGEN REQUIREMENTS

Item	Liquid Hydrogen, Billions of Gal per yr (10^9 ℓ/yr)	
Consumed in automobile engines		250 (946)
Losses:		
Producer to Transport	1.094† (4.141)	
Transport Boiloff	0.234* (0.889)	
Transport to Service Station	2.105† (7.967)	
Station Boiloff	8.322 (31.499)	
Station to Automobile	11.695 (44.266)	
Automobile Dewar Boiloff	18.25* (69.076)	
Automobile Dewar Chilldown	0.31 (1.173)	
Total losses		42 (159)
Total production required		292 (1105)

†Includes line chilldown, line boiloff, left in line, and flashing losses.

*Avoidable losses

Source: LAUR-73715

Some of the assumptions made in these estimates are as follows:

[1] A boil-off rate of 1% per day for the automobile Dewar.

[2] A total of 10^8 automobiles.

[3] The number of new automobiles requiring Dewar cool-down per year is 10^7. (Automobiles are replaced every ten years).

[4] An average of 50 refills per year for each automobile.

[5] A boil-off rate of 0.2% per day for the service station Dewars.

[6] A boil-off rate of 0.25% per day for the transport trailers.

[7] The transfer hoses from the production plant to transport trailer and from the transport trailer to service station are 15 feet (4.6 m) long, 2 inches (5.1 cm) inside diameter, insulated with superinsulation, and have an inner line wall thickness of 0.020 inches (0.51 mm).

[8] The pressure drop in the supply line is 1.5 psi (0.1 kg/cm^2) and the pressure drop in the vent line is 1 psi (0.07 kg/cm^2). Thus the service station tank operating pressure should be 17.5 psia (1.23 kg/cm^2 absolute) if the vehicle Dewar pressure is 16 psia (1.12 kg/cm^2 absolute) at the time of refilling. The transport trailer pressure must be 19 psia (1.34 kg/cm^2 absolute) and the producer's storage tank pressure must be 20.5 psia (1.44 kilograms per square centimeter absolute).

The initial cool-down of the 3×10^5 stations will require approximately 2×10^9 gallons (7.57×10^9 liters) of LH_2.

The total annual production of LH_2 to replace gasoline (based on 1972 usage) is 2.92×10^{11} gallons (1.11×10^{12} liters), of this amount 2.5×10^{11} gallons (9.46×10^{11} liters) are consumed in automobile engines and 0.42×10^{11} gallons (1.59×10^{11} liters) are losses. The losses may be recovered and reliquefied or compressed and used for purging, heating, etc. The losses amount to 14% of the total production, but some of these losses are avoidable (transport truck and car boil-off) and in this case the losses are reduced to 8.6%. A more comprehensive design can reduce the losses even further.

The largest LH_2 production capability that once existed in the United States was 1.8×10^8 gallons per year (6.81×10^8 liters per year). The production of LH_2 would need to be increased by a factor of 1,622 to produce the LH_2 required for automobiles. A total of 800 LH_2 production plants each with a production rate of 300 tons per day (2.72×10^5 kilograms per day) are required. The estimated cost of these plants is \$60 million each for a total of \$48 billion. If these plants were distributed equally across the United States, each would service an area approximately 67 miles (107.8 km) by 67 miles (107.8 km).

The total United States production of GH_2 in 1968 was 2.28×10^{12} Ncf (6.46 times 10^{10} cubic meters) which is equivalent to 2×10^{10} gallons (7.6×10^{10} liters) of LH_2; therefore, GH_2 production would need to be increased by a factor of 14.6. A total of 135 electrolysis plants each with a production rate of 73 tons per hour (66,226 kg/hr) would be required to produce the required amount of GH_2. The estimated cost of these plants is \$170 million each for a total of \$23 billion.

The estimated cost of the 300,000 service stations is $200,000 each for a total of $60 billion. This estimate is based on a cost of $100,000 for Dewars and $100,000 for buildings and auxiliary equipment.

A total of 2,000 transport trailers with a capacity of 13,000 gallons (49,205 liters) each will be required assuming that each makes three deliveries per day and that each station receives 72 deliveries per year. The estimated cost of these trailers is $100,000 each for a total of $2 billion.

The total H_2 system cost, as summarized in Table 5.13 is estimated to be $133 billion. While this cost is great it should be compared to a National Petroleum Council forecast of $110 billion expenditure over thirteen years for exploration, capital equipment, etc. Because the need for a portable fuel is a problem that must be solved, whatever capital investments are required will be made.

TABLE 5.13: LIQUID HYDROGEN PRODUCTION AND DISTRIBUTION SYSTEM COSTS

	Billions of Dollars
GH_2 Electrolysis facilities	23
135 plants of $170 million each	
73 ton/h (66,226 kg/h) capacity each	
Liquefaction facilities	48
800 plants at $60 million each	
300 ton/day (272,160 kg/day) capacity each	
Service stations	60
300,000 stations at $200,000 each (includes $100,000 for LH_2 tanks)	
LH_2 Transport trailers	2
20,000 trailers at $100,000 each	
13,000 gal (49,205 ℓ) capacity for each	
Total cost	133

Source: LAUR-73715

At present the cost of LH_2 is estimated to be 15¢ per gallon (10¢ per gallon to produce the GH_2 plus 5¢ per gallon to liquefy). The costs of the distributor and the seller (service station) were estimated and added to this for a total cost of 20¢ per gallon (5.3¢ per liter) of LH_2 which is equivalent to a cost of 50¢ per gallon (13.2¢ per liter) for gasoline. For a post-fossil fuel era the cost of LH_2 equivalent to one gallon of gasoline is estimated to be $1.15 (30.4¢ per liter). These estimates are summarized in Table 5.14.

TABLE 5.14: OPERATING COSTS FOR LIQUID HYDROGEN AND GASOLINE FUELS

Time	Gasoline, ¢/mile (¢/km)	Liquid Hydrogen, ¢/mile (¢/km)
Present	3 (1.9)	4 (2.5)
Post-Fossil	-----	9 (5.6)

Source: LAUR-73715

In this case LH_2 is assumed to be produced by electrolysis of water and liquefaction. The cost of production of H_2 by electrolysis was taken from an estimate by Gregory (79).

A rough estimate of a schedule for the installation of the H_2 production and distribution facilities is given in Figure 5.15. Obviously many problems exist in making a change-over from gasoline to H_2. A dual system of H_2 and gasoline will probably be required for several years with a gradual decline in the use of gasoline accompanied by a gradual increase in the use of H_2. Automobiles will probably need a dual fuel capability during this period.

Initially H_2 production and distribution facilities could be built across the country so that each would serve equally sized areas. Each area could then be subdivided into smaller and smaller areas with the construction of new facilities. Alternatively, the switch-over could be initiated in the more populous regions of the country and progress to the less densely populated areas, thus attacking the areas of greater pollution first.

Another alternative method would be to divide the country into ten regions and complete the conversion in one region each year. If the total system were constructed over a ten year period, the number of facilities and the costs involved annually would be one-tenth of that given in Table 5.13.

Other facilities, such as power plants to operate the electrolysis and liquefaction equipment, must also be built on a parallel schedule. Problems of financing, materials availability, site or land availability, etc. may greatly affect this schedule.

An orderly conversion from gasoline to H_2 fuel will probably require a national commitment on the order of the Apollo program. An educational program will be required for training production plant and service station personnel and for indoctrination of the general public. Safety and logistics problems are real but not insurmountable.

Safety Considerations

Safety is a usual objection made to the use of liquid hydrogen (LH_2) as a fuel, but in many respects it is no more hazardous than gasoline. Hydrogen has been safely produced, transported and handled for several years in the space program and in ever increasing industrial applications. Federal regulations already exist

FIGURE 5.15: ESTIMATED SCHEDULE FOR INSTALLATION OF HYDROGEN PRODUCTION AND DISTRIBUTION FACILITIES

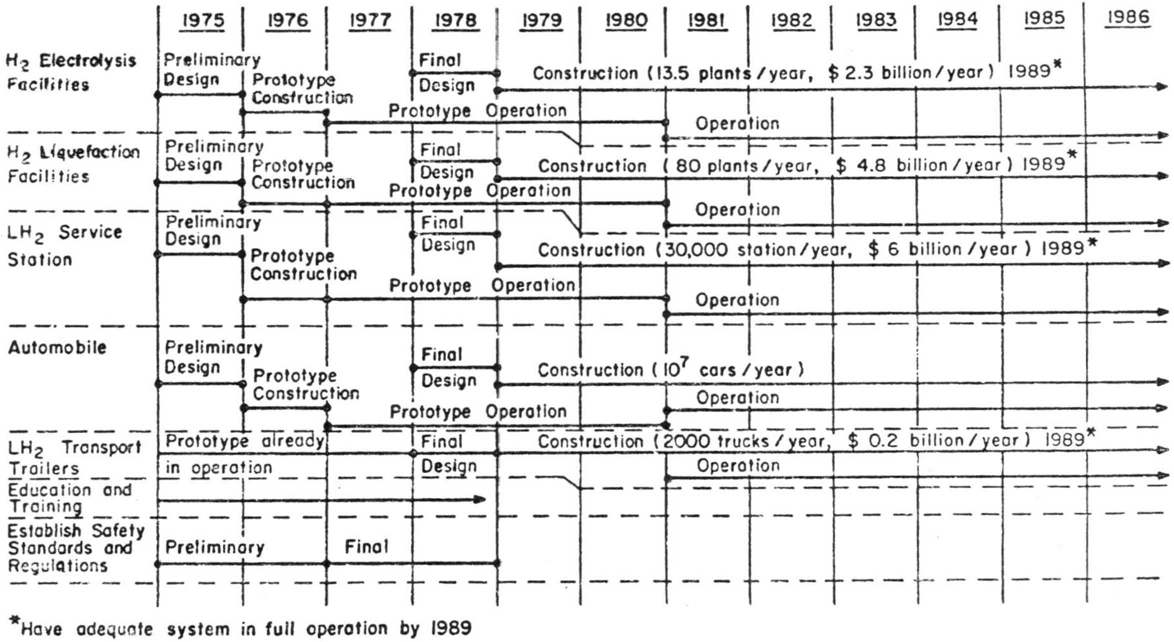

Source: LAUR-73715

for its transportation and storage. These regulations and other guide lines are summarized in Table 5.15. Personnel involved in the production, transportation, and transfer of hydrogen should receive some training in its handling but highly trained or skilled personnel will not be required for all operations since most operations will be automated to minimize human error and safety problems.

TABLE 5.15: REGULATORY GUIDELINES FOR DISTRIBUTION OF HYDROGEN

Distribution Method	Equipment Specifications	Shipping Regulations	Installation Standards
Liquid Cylinders	Title 49 [20] CFR 178.57	Title 49 CFR 173.316	NFPA 50B [22]
Liquid Trailer	ASME/ (Ref. CGA 341)	Special Permit	-----
Liquid Tank Car	Title 49 CFR 179.400	Title 49 CFR 173.316	-----
Liquid Customer Station	ASME [23]	-----	NFPA 50B
Gas Cylinders	ASME/Title 49 CFR 178.36-.37	Title 49 CFR 173.302	NFPA 50A [24]
Gas Cylinder Trailer	ASME/Title 49 CFR 178.36-.37	Title 49 CFR 173.302	NFPA 50A
Gas Pipe Line	ANSI B31.8 [25]	-----	Title 49 CRF Part 192

Source: LAUR-73715

Unfortunately, in considering the automotive (or any other) use of hydrogen, one has to solve not only the safety problems (which are real) but one must also consider the inordinate fear of hydrogen which has been termed the "Hindenberg Syndrome." Actually, two-thirds of the people in the Hindenberg did survive and compared to more modern disasters the total of 38 deaths, though tragic, is not quite so overwhelming.

Clearly a more studied approach to safety problems is required. Obviously there are safety problems; however, not all properties of hydrogen are bad from a safety standpoint, and the safety problems are not insurmountable. Several properties which have significance in safety consideration are listed in Table 5.16.

The wide range of flammability of hydrogen is the property that is usually cited as making hydrogen more hazardous than gasoline. However, it is doubtful that one would ever depend upon a safe condition obtained by increasing the concentration of a combustible above the upper flammable limit. Therefore, the lower limit of flammability becomes more important to safety than the flammability range and on this basis hydrogen should be considered safer than gasoline. On the other hand, the ignition energy for hydrogen is an order of magnitude lower than that for gasoline and must make hydrogen less safe. The higher flame velocity and lower quenching distance of hydrogen also make gasoline easier to handle safely.

TABLE 5.16: COMPARISON OF SAFETY PARAMETERS FOR HYDROGEN AND GASOLINE

	Property	H_2	Gasoline	H_2 Safety Comparison
1.	Ignition energy (mj)	0.02	(0.25)*	-
2.	Quenching distance (cm)	0.06	(> 0.25)	-
3.	Ignition temp. (°K)	858	530	+
4.	Combustion range (%)	4-75	1.5-7.6	+
5.	Diffusivity (cm^2/sec)	0.6	(0.08)	=
6.	Flame velocity (cm/sec)	270	(< 30)	-
7.	Flame emissivity	0.1	(> 0.1)	=

*Properties enclosed in parenthesis are estimated.

Source: LAUR-73715

The higher ignition temperature of hydrogen is a safety plus for hydrogen. The ignition temperature is such that a hydrogen-air mixture would not be ignited by a lit cigarette, but could be ignited by an open flame from a match or cigarette lighter. Thus, personnel involved in hydrogen operations will not be permitted to smoke just as they are not permitted to smoke during gasoline handling.

The relative values of diffusivity and flame emissivity are more difficult to assess. The low radiant energy from a hydrogen-air flame is certainly responsible for a smaller amount of radiant energy reaching objects in view of the flame. However, a hydrogen flame is nonluminous (in the absence of impurities) making it easier for a person to contact such a flame inadvertently.

The low molecular weight and small molecular size of hydrogen will allow it to leak easier and spread faster than other gases, but its high diffusivity will allow it to dissipate faster. A gasoline spill would present a hazard for a much longer period of time since the gasoline would flow into low areas and accumulate whereas LH_2 would rapidly vaporize and disperse or be carried away by the wind. For example, a spill of 500 gallons (1,893 liters) of LH_2 in open air will disperse to where all the hydrogen is below the explosive limit after approximately one minute. Thus, for the properties of diffusivity and flame emissivity, one can imagine situations where hydrogen is either more or less hazardous than gasoline.

The low boiling temperature of hydrogen can cause additional safety hazards either by frostbite or condensation of air if cold surfaces are exposed. A LH_2 transfer system will be a closed system with no vapors escaping during operations such as vehicle Dewar refilling. The gasoline transfer system is an open system with potentially hazardous vapors being vented into the atmosphere and in many instances liquid is spilled on the vehicle, the ground, etc. making smoking more of a hazard with gasoline than with hydrogen. The safety requirements of the 1970 National Fire Code for Gaseous Hydrogen (GH_2) Systems (82) are built around three axioms: adequate ventilation, leak prevention and elimination of ignition sources. According to this code GH_2 systems must be located above ground, but not beneath electrical power lines, nor near piping or storage of flammable liquids, nor within five feet (1.5 m) of adjoining property, nor within 15 feet (4.6 m) of public sidewalks. Safe distances between LH_2 storage and inhabited buildings have been the subject of a number of studies which have resulted in widely varying recommendations for storage of large quantities (> 10 kg).

METAL HYDRIDE STORAGE SYSTEM FOR VEHICLES

Description of System

Considerable research has gone into the metal hydrides, of which the most pertinent work is that done at Brookhaven National Laboratory (see previous chapter). Reilly and Wiswall (83)(84)(85) studied the reactions of hydrogen with vanadium and alloys of magnesium-copper and magnesium-nickel. These alloys, and the other metals listed in Table 5.17 react reversibly with hydrogen to form metal hydrides with low dissociation temperatures.

TABLE 5.17: METAL HYDRIDES WITH RELATIVELY LOW DISSOCIATION TEMPERATURES

Hydride	Dissociation Temperature, °F	Hydrogen, weight percent
LiH	1380	12.5
MgH_2	550	7.6
UH_3	806	1.25
Mg_2Cu-H_2	460	2.7
Mg_2Ni-H_2	480	3.6
$PdH_{0.6}$	302	0.57

Source: UCRL-51228

There are many other hydrides with dissociation temperatures greater than 1000°F; which therefore are of little interest. Of those listed, palladium is too expensive and uranium is too heavy, but magnesium is in plentiful supply, inexpensive, and of relatively low density. Hence, the magnesium hydrides are of the greatest interest here.

The use of magnesium alloys as a practical means of storing hydrogen has already been proposed for vehicular applications (86)(87). The reactions found to occur have been given as:

$$Mg_2Ni + 2H_2 \rightleftharpoons Mg_2NiH_4 + 13{,}700 \text{ Btu/lb } H_2$$

$$2Mg_2Cu + 3H_2 \rightleftharpoons 3MgH_2 + MgCu_2 + 15{,}600 \text{ Btu/lb } H_2$$

$$Mg + H_2 \rightleftharpoons MgH_2 + 16{,}000 \text{ Btu/lb } H_2$$

With pure magnesium, this reaction is slow; but with as little as 5% nickel in a magnesium-nickel alloy, the kinetics are excellent above 400°F. For these hydrides, the temperature dependence of the dissociation pressures is shown in Figure 5.16.

The above reactions indicate that heat is either liberated when the reaction occurs or must be supplied to release the hydrogen. If the hydride were stored in porous or powdered form, the hydrogen could be evolved at the same rate at which heat is supplied. The converse is also true. Hence, the hydride storage bed can be regenerated by supplying hydrogen at conditions above the dissociation pressure and temperature.

FIGURE 5.16: DISSOCIATION PRESSURE vs TEMPERATURE FOR THREE METAL HYDRIDES

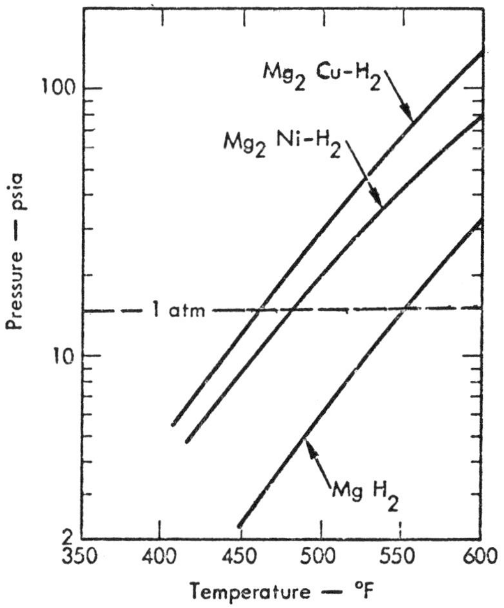

Source: UCRL-51228

Theoretically, this evolution-regeneration cycle could be repeated indefinitely, but accumulations of impurities and caking might raise practical difficulties. To explore this, limited laboratory tests have been made, demonstrating that a Mg_2NiH_4 bed can be depleted and regenerated about five times with no observable change in characteristics.

In order to compare the hydride storage requirements with those of 100 pounds (16.3 gallons) of gasoline, unalloyed MgH_2 will be considered as the base case, neglecting weight and volume changes due to the copper or nickel catalyst. For equivalent energy storage, 37.1 pounds of hydrogen are needed, requiring 488 pounds of MgH_2.

Using data supplied by Winsche et al (86), it is found that this corresponds to a hydride volume of 8.92 cubic feet, based on an assumption that the hydride contains 40% voids to allow hydrogen gas permeation. This is competitive with liquid hydrogen, which requires 8.37 cubic feet. Uncertainties exist, however, since the design of a practical system has yet to be proposed, and there are some serious problems.

To illustrate how a hydride might function as a fuel source, Figure 5.17 is one concept of a complete system. The hydride bed is heated by the hot exhaust gases by a suitable heat exchanger fed directly from the exhaust manifold.

FIGURE 5.17: THEORETICAL CONCEPT FOR A MAGNESIUM HYDRIDE FUEL STORAGE SYSTEM

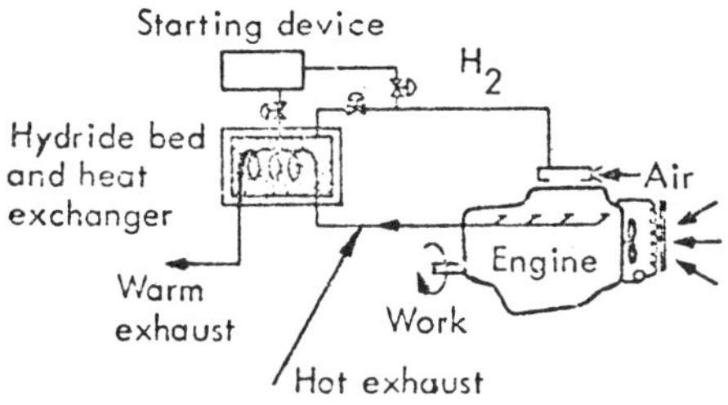

Source: UCRL-51228

Some means must be provided for starting the engine. This could be a small electrical heater, for heating a small portion of the hydride to release enough gas for starting. It might be possible to use the hydrogen evolved from the sensible heat of the hydride bed in cooling from the previous operation. Or there might be a small high-pressure reservoir of hydrogen, or a small, separate bed of low dissociation temperature hydride such as that of $LaNi_5$, which dissociates at about 30°C.

The most important question is whether sufficient energy will be available in the exhaust gases to dissociate the hydride during continuous operation. This question can only be answered concretely by experiment, but a reasonable estimate can be made by carrying out a cycle analysis.

Typical temperatures and pressures, efficiency, and effect of compression ratio can be determined by calculation. Several assumptions are made. Following standard procedures, the ideal Otto cycle is used, which consists of atmospheric pressure intake, isentropic compression, constant volume combustion, and isentropic expansion. Effects of variation of specific heat of both mixture and combustion products are included, but the energy lost due to dissociation during combustion is neglected, along with any effects of the clearance gases.

The cycle analysis is done for only the single case of stoichiometric hydrogen/air mixture. The results are shown in Table 5.18. It can be seen that, depending on the compression ratio, there are only 17,000 to 18,900 Btu/lb of hydrogen exhaust energy at about 3000°F, of which about 90% is available for hydride dissociation at 450°F. Note that considerable energy is lost in the cooling water, some of which might be utilized by a suitable redesign if the engine temperature were kept low enough to avoid preignition.

TABLE 5.18: EFFECT OF COMPRESSION RATIO ON ENERGY DISTRIBUTION IN A HYDROGEN-FUELED INTERNAL COMBUSTION ENGINE

	Otto Cycle State Points*						Energy Balance—Btu/lb H_2				
	Compression		Combustion		Exhaust			Rejected Energy			Brake
CR	P_2	T_2	P_3	T_3	P_4	T_4	Useful Work	Radiation	Cooling	Exhaust	Efficiency %**
6	174	745	867	5,540	92	3,440	11,400	4,000	17,200	18,900	22.2
8	257	870	1,180	5,640	88	3,170	13,120	3,830	16,450	18,100	25.2
10	349	987	1,490	5,730	84	3,030	14,050	3,750	16,050	17,650	27.3
12	446	1,080	1,810	5,780	81	2,890	15,030	3,640	15,630	17,200	29.2
14	549	1,165	2,140	5,870	79	2,810	15,450	3,570	15,480	17,000	30.0

*P_1 = 14.7 psia; T_1 = 150°F. Other values are also in these units.
**Based on lower heating value of 51,500 Btu/lb for H_2.

Source: UCRL-51228

It appears, then, that the concept in Figure 5.17 is feasible in principle, but practical design considerations may be limiting. The two basic problems are to design a highly efficient heat exchanger and minimize the required additional weight and volume so as to meet practical vehicular requirements. The solution of these problems will require additional chemical research to find catalyzed hydrides of light metals (such as lithium) with lower dissociation temperatures, innovative heat exchanger concepts, and overall system engineering to make better use of the cooling system energy and/or use of combinations of hydrides.

There are, of course, other detailed problems such as sealing of the hydride from the exhaust gases, sequential dissociation of the hydride, material compatibility, modular design to permit easy recharging, and a simple method for engine startup. The heat transfer problem is critical, but no data exist for physical and thermal properties of the hydrides of interest. Hence, the required additional volume and weight to accommodate heat transfer cannot be estimated at this time.

In addition to the problems within the vehicle, there always remain the logistics of handling and recharging the hydride. When the magnesium is recharged, 16,000 Btu/lb of hydrogen are released; at 37.1 pounds per vehicle, this amounts to 595,000 Btu. Some means must be devised for dissipating this energy or using it purposefully, and the use certainly depends on the type of commercial system employed. Recharging will be done either at local service stations or at some central facility with the service stations acting only as collection and exchange points (as is done currently with batteries). The latter seems preferable. By concentrating the recharge energy at a central location, the energy from 100 million exhaust pipes is being effectively concentrated, for whatever use can be made of it. In other words, the potential exists for increasing the overall national energy conversion efficiency.

Safety Considerations

Safety considerations could be limiting. Magnesium hydride, like all hydrides, becomes highly pyrophoric when recycled several times. Handling large quantities of granular magnesium in depleted beds is very hazardous but could be done safely under controlled conditions. A more serious situation arises, however, with respect to accidents involving vehicles with partially depleted beds. Magnesium fires burn intensely and are extremely difficult to extinguish. Hence, the use of hydrides may have serious shortcomings for reasons other than technical problems relative only to vehicle performance.

There are the general safety problems to be considered in handling either gaseous or liquid hydrogen. Many people have pointed out, however, that hydrogen is no more hazardous than gasoline. For example, escaping hydrogen gas diffuses rapidly into the atmosphere rather than remaining as a slowly evaporating liquid. It is also worth noting that manufactured natural gas used in some areas is handled casually even though such gas contains as much as 50% hydrogen. Furthermore, hydrogen is used widely for industrial purposes, and large quantities of liquid hydrogen are shipped and stored daily.

It is significant that there have been no catastrophic accidents involving hydrogen since the Hindenberg dirigible fire of 1937. Nevertheless, the fears generated by this accident still persist, accounting for the continuing general attitude that hydrogen is more hazardous than other fuels. Hydrogen is undoubtedly a hazardous material unless properly handled but no more so than other fuels.

The flammability limits and relative densities of the gaseous fuels are listed in Table 5.19. Hydrogen has very wide flammability limits, and ignition can occur with very low energy input.

TABLE 5.19: FLAMMABILITY LIMITS AND RELATIVE DENSITIES OF GASEOUS FUELS

Fuel	Flammability limits in air—% by vol	Density at STP (air=1.0)
Hydrogen	4.0-75.0	0.069
Gasoline vapor	1.5- 7.6	3.90
Butane	1.9- 8.5	2.07
Propane	2.2- 9.5	1.562
Methane	5.3-14.0	0.55

Source: UCRL-51228

Although combustion of a flammable hydrogen-air mixture can occur easily, a very high energy input is required to detonate an unconfined explosive mixture. Note, however, that the lower flammability limit for hydrogen is about twice that of gasoline, butane, and propane. Hence, with respect to leakage, these other fuels seem more hazardous than hydrogen. Furthermore, hydrogen diffuses rapidly in the atmosphere while the other fuels, being heavier than air, settle in low areas, remaining there as an explosive hazard until dissipated by diffusion.

Finally, hydrogen is nontoxic and not considered a health hazard. Thus, despite public resistance, it would seem that proper handling, storage, and use procedures can be developed to accommodate hydrogen as a national fuel without seriously compromising public safety.

Weight, Economy and Range

Since the brake thermal efficiencies for an engine using any of these fuels will not vary widely, any large differences in vehicle performance will result from the weight increases needed for fuel and the required tankage. Hence, it is important to estimate the total weight (and volume) requirements so as to place the hydrogen-fueled system in perspective.

Table 5.20 summarizes the characteristics of the various systems. Calculations of tankage weights and volumes for pressurized gas storage assume the use of a spherical pressure vessel fabricated from high strength steel with a 100,000 psi yield strength and a safety factor of 10.

TABLE 5.20: CHARACTERISTICS OF FUEL SYSTEMS FOR THE INTERNAL COMBUSTION ENGINE

Fuel	Weight (lb)		Vol (ft^3)		Total	Total	
System	Fuel	Tank	Fuel	Tank	Weight (lb)	Volume (ft^3)	Basic Problems
LH_2/air	37.1	37.1	8.4	3.0	74.2	11.4	Boil-off, safety
LH_2/LOX	334.0	107.0	12.6	3.5	441.0	16.1	Volume, safety
MgH_2/air	488.0	122.0	8.9	?	610.0	$>$8.9	Design feasibility, volume, safety
Propane	95.7	68.0	3.5	0.6	163.7	4.1	Availability, pressure, safety
Butane	97.0	23.0	2.8	0.5	120.0	3.3	Availability, pressure, safety
Gasoline	100.0	25.0	2.2	~0.1	125.0	2.3	Availability, pollution

Source: UCRL-51228

These weights, so calculated are probably high, because advanced technologies involving filament winding and lightweight composite materials are now available. Cryogenic storage vessel requirements are computed from data compiled by Russell (88). These data represent modern aerospace technology, and the estimates derived here are for a spherical container with a 1% boil-off loss per day. It should be noted that a spherical container is the most efficient, because heat transfer is minimized due to the low ratio of surface area to volume and because the stress-state is minimized for a given internal pressure. Any other vessel shape will likely require greater weight and volume.

Since the magnesium hydride system requires incorporation of a heat exchanger, a meaningful estimate of the required volume cannot be made at this time. The additional weight, however, is arbitrarily assumed to be 25% of the hydride weight, or 120 pounds. Likewise, a gasoline fuel tank is estimated to weigh 25 pounds.

The effect on vehicle performance can be easily calculated as follows. It can be shown that the power required to propel a vehicle of weight (W) up a grade of angle (ϕ) at a constant velocity (V) is approximately:

$$P(hp) \cong 2.67 \times 10^{-3}(W \sin\phi + 0.02W + 0.03V^2)V$$

W being expressed in pounds and V in miles per hour. The percent increase in power required for an increase in weight is, therefore:

$$\frac{\Delta P}{P} = \left[\frac{(\sin\phi + 0.02)}{\sin\phi + 0.02 + 0.03V^2/W}\right]\frac{\Delta W}{W}$$

This expression is used to evaluate the systems listed in Table 5.20 for various size vehicles, the results being shown in Figure 5.18. Note that the right-hand ordinate in the figure is the percentage increase in fuel consumption. The shaded regions are drawn arbitrarily to indicate uncertainty in the calculations of required weights. It should be noted that, although the weight required for the gasoline system introduces only a 2% loss in range, the introduction of emission control devices is expected to cause a 6 to 12% reduction in fuel economy and, hence, range.

Obviously, liquid hydrogen with air, on a weight basis, is the best choice, while the MgH_2 system appears the least attractive. It is important, however, to consider the volume requirements and basic problems inherent in each system (Table 5.20).

For estimating the incremental fuel consumption, which is a direct measure of the loss in vehicle range for each system, Figure 5.18 is useful. It should be noted also that boil-off losses are not included for the liquid systems. Boil-off would add to the fuel loss, and so these two systems should be charged with this penalty. In any event, considering the importance of finding suitable alternative fuels for transportation needs, a maximum 9% loss in vehicle range does not seem too significant.

HYDRIDE-DEHYDRIDE POWER SYSTEM USEFUL FOR REFRIGERATION

Various types of power cycles using a number of working fluids have been previously proposed. In one long-known and widely used system, heat energy in a working fluid is used in a Rankine cycle to drive a turbine to produce power. The cycle is completed through the sequence of condensing the working fluid,

FIGURE 5.18: PERFORMANCE PENALTY FOR ADDITIONAL FUEL STORAGE WEIGHT

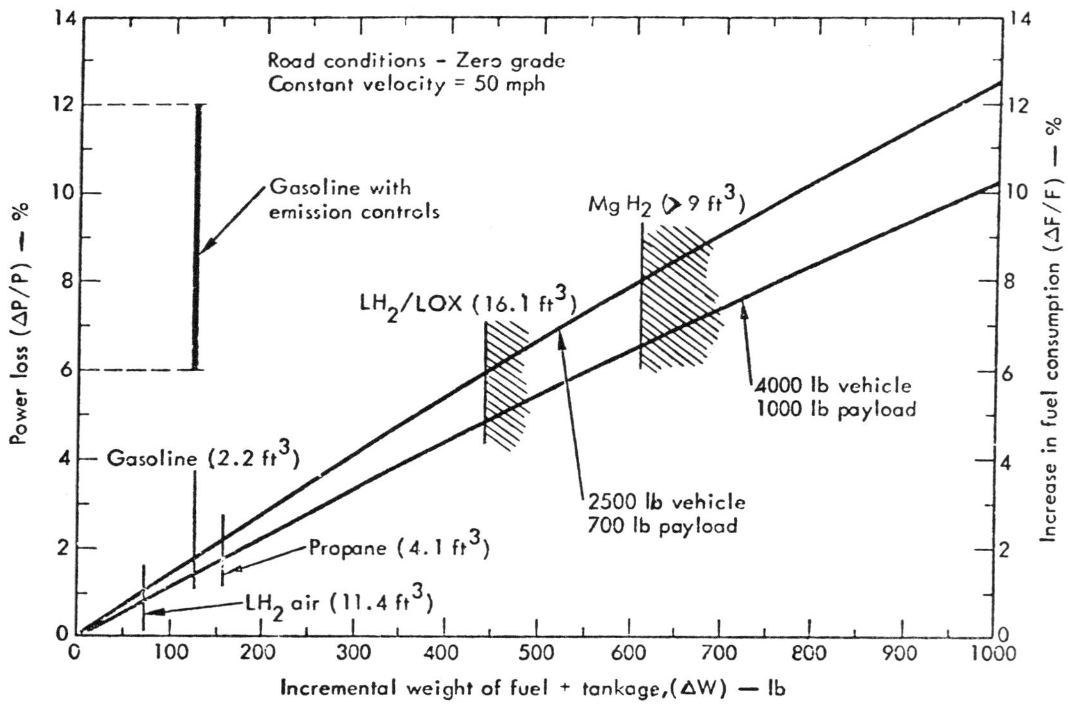

Source: UCRL-51228

compressing the liquid, and evaporating the liquid to a high pressure vapor state. Various fluids which may be easily condensed near ambient temperature, such as water, isobutane and chlorinated hydrocarbons, can be used as the working fluid.

In another long-known and widely used power system, gases are used as the working fluid in a Brayton cycle to drive a turbine. In a closed system, the cycle is completed by effecting recompression of the gas through the use of a compressor and heat exchangers. This type of power system differs basically from the Rankine cycle system by the nature of the working fluid and the manner by which it is recompressed. This type of power system is well known in the field of thermodynamics and is used in many high-temperature energy conversions, such as in jet engines where an open system is used and in high-temperature gas-cooled nuclear power plants.

Various types of refrigeration cycles using a number of working fluids have been heretofore proposed. In one long-known and widely used system, refrigeration is secured by vapor-compression of the working fluid with subsequent cooling to the liquid state, expansion of the liquid through an expansion valve with refrigeration being produced in that portion of the cycle where evaporation occurs. Working fluids commonly used in this vapor-compression cycle are chlorinated hydrocarbons.

In the air-standard refrigeration cycle, the phases of the cycle are essentially the reverse of those encountered in the Brayton cycle. Air is initially compressed and is then cooled before being expanded to further lower its temperature. This type of cycle is used in low-pressure air liquefaction plants, and in other liquefaction devices, such as the Collins helium liquifier.

In the absorption refrigeration cycle, absorption of the working fluid occurs at a low pressure and temperature with release at a relatively higher pressure and temperature. Subsequent cooling to the liquid state and expansion through an expansion valve produce refrigeration. Ammonia-water and lithium bromide-water are common working fluid-absorber combinations.

Various types of working fluids for use in power and refrigeration cycles have been proposed. In the Stirling cycle, hydrogen has been proposed as a working fluid which is directed in sequence to a compressor, heat exchanger and expansion device in a manner analogous to the Brayton cycle.

In U.S. Patent 3,370,420, working fluids which thermally dissociate upon heating, and reversibly recombine in the gas phase upon cooling, are proposed for several closed cycle thermodynamic processes (Brayton, Stirling and Rankine). The patent asserts that any nondissociating gas, when used as a working fluid, severely restricts the efficiency of a power cycle.

A good discussion of the various advantages and disadvantages which have characterized many types of working fluids previously used in power cycles is contained in U.S. Patent 3,152,357. All the described fluids are types which undergo a change of state from liquid to vapor in the course of the power cycle. An ideal working fluid is described as one having good thermal stability, high cycle efficiency, lower corrosiveness, favorable critical properties and low toxicity. A proposal for developing power from a system comprising a multiplicity of power fluids is described in U.S. Patent 3,395,103. Direct use of a relatively low temperature energy source, such as geothermal water, to heat and vaporize a working fluid for expansion through a turbine for the development of power has been hindered by the large cost of the heat exchanger required in order to effect heat transfer. The process proposes to eliminate a portion of this cost by adding a second power plant of the Rankine cycle type and using the residual heat in the hot primary fluid to heat and vaporize a different working fluid having a lower boiling point than that used in the first cycle.

Recently, the properties and useful application of various metal hydrides have been considered in power and refrigeration cycle systems. A considerable amount of work in this respect has been done under the auspices of the U.S. Atomic Energy Commission. The ability of hydrides to chemically store hydrogen at a relatively low temperature and pressure in a concentrated form, and then to release the hydrogen at an elevated temperature and pressure has been recognized, and a number of hydride-forming materials have been identified.

In U.S. Patent 3,508,414, a method of storing hydrogen is described in which gaseous hydrogen is absorbed by titanium-iron alloys. When such a hydride containing 2 weight percent hydrogen is maintained at a temperature of 25°C, hydrogen is released at a constant rate until less than 1.0 weight percent of the hydrogen remains in hydride form.

In U.S. Patent 3,315,479, a method of storing hydrogen by formation of nickel-magnesium hydride is discussed. Similar formation of copper-magnesium hydrides is discussed in U.S. Patent 3,375,676. U.S. Patent 3,516,263 further discusses the formation of titanium-iron hydrides, and points out that a particular type of pressure vessel may be used to contain the hydride, and to heat the hydride to develop hydrogen pressures exceeding 10,000 psi.

By alternating the formation and decomposition of the metal hydride, workers at the Brookhaven National Laboratory have proposed, in Report No. 15844, April 1971, to use the alternate decomposition and regeneration of the hydride as a gas circulation pump. Such systems have also been proposed for refrigeration.

More recently, in U.S. Patent 3,504,494, a closed cycle method for intermittently producing high energy steam has been described. In the system reported, a power cycle, followed by a recharging cycle, is utilized. In the power cycle, a first hydride bed is heated to desorb hydrogen gas therefrom. The gas flows to a second hydride bed where the hydrogen can be absorbed at a lower temperature than the temperature of desorption from the first bed. Absorption of the hydrogen by the second bed releases the heat of absorption which is used to convert water to steam.

The steam is used for power production, and the residual heat remaining in the steam after such power production is used for heating the first hydride bed and enhancing the desorption of hydrogen therefrom. After complete desorption of the hydrogen from the first bed and condensation of the residual steam, the recharging cycle is started.

In the recharging cycle, the second hydride bed is heated by a heat source which can be a low energy isotope source, a chemical heater, an electrical heater or other suitable source of thermal energy. The second bed thus desorbs hydrogen, and the first bed is cooled so that it can absorb the hydrogen desorbed from the second bed preparatory to recommencing the power cycle after recharging. An integral and essential part of this review of prior art is the lack of any finding substantiating the fact that both power and refrigeration can be practically generated from low temperature energy sources except in a conventional manner as referred to above in the examples of the use of isobutane and chlorinated hydrocarbons. Although the prior art recognizes the possible use of hydrides in a gas circulation pump, the art does not contemplate the economic or practical or technical feasibility of large-scale commercial power generation and refrigeration, and the need for energy conservation achieved through extraction of energy from low-temperature energy sources.

A highly useful combination power and refrigeration system and methods for developing shaft power and refrigeration simultaneously are presented by L.E. Terry and R.J. Schoeppel; U.S. Patent 3,943,719; March 16, 1976. Power and refrigeration are continuously and efficiently generated directly from hydrogen in a unique hydride-dehydride-hydrogen (subsequently referred to as HDH) cycle.

Hydrogen is used as the working fluid, and is chemically compressed in the course of the HDH cycle, but is not thermally dissociated. The hydrogen has the advantage, as a working fluid, of having good thermal stability, high cycle efficiency, low corrosiveness, favorable critical properties and low toxicity. Waste or low grade heat can be utilized for freeing chemically bound hydrogen from the hydride during the course of the HDH cycle.

Broadly described, the combination power and refrigeration system of this process comprises a plurality of hydride-dehydride reactors, means for periodically supplying hydrogen gas, and alternately heat, in out-of-phase staggered cycles to each reactor, means for removing heat from the reactors, means for continuously converting the pressure energy of pressurized hydrogen gas developed by the reactors to shaft work or power and refrigeration, and means for recycling relatively low pressure hydrogen to the reactors.

The means for periodically supplying heat to the reactors preferably comprises a low-grade thermal source, such as geothermal water, solar energy or stack gases, and the energy-to-power converting means is preferably a turbine. The energy-to-refrigeration converting means is preferably a heat exchanger.

The process also includes a method for simultaneously developing shaft power and refrigeration from a low thermal energy source, which ideally includes the steps of reversibly combining hydrogen with a hydride-forming material at a relatively low temperature and pressure to chemically store hydrogen gas, heating the hydride at constant volume to effectively chemically compress the hydrogen gas, dehydrogenating the hydride material by transferring heat thereto at an elevated preferably constant temperature to release hydrogen gas under elevated constant pressure, reversibly and adiabatically expanding the pressurized hydrogen gas through a device to develop power and supercool hydrogen gas, cooling the spent hydride under constant volume to effect chemical decompression so as to effectively recondition the hydride-forming material, and then cyclically repeating these steps.

The preferred practice of the method entails carrying out the described steps in staggered or out-of-phase relationship using a plurality of discrete beds of hydride-forming material to develop a substantially continuous source of hydrogen under pressure for supply to the power device. In certain embodiments of the process, the hydrogen gas produced upon dehydrogenating the hydride material is used for heating. It is an important aspect of the process that the heat source used for supplying heat to the hydride reactors can suitably be a low-grade source of thermal energy, and such sources are preferred, although high temperature sources can be used.

In this respect, the hydride-forming material utilized is preferably one which forms the hydride at relatively low temperatures and releases the hydrogen gas at suitably high pressures at a temperature which is not higher than about 450°F.

The process provides a practical means for achieving energy conversions in which a low grade source of thermal energy is utilized in a more efficient and economical manner than can be achieved with any presently existing power cycle involving a working fluid which must be heated to achieve a change of state or upgrade its energy content prior to passage into or through both power and refrigeration producing devices.

The patent provides a description and drawings of the apparatus by which the process is accomplished including the use of one or more fluid phases, one or more fluidized beds, or one or more stationary hydride beds, or gaseous hydrides, means for transferring heat to or from the hydrides, and means for expansion of the contained hydrogen for the generation of power and refrigeration.

REFERENCES

(1) Hausz, W., G. Leeth and C. Meyer, "Eco-Energy," General Electric TEMPO, Report 729206.

(2) "Rocket Power to Light the Cities," *Skyline Magazine,* volume 30, n 4 (1972).

(3) Weil, K.H., "The Hydrogen I.C. Engine—Its Origins and Future in the Emerging Energy-Transportation-Environmental System," 1972 Intersociety Energy Conversion Engineering Conference, Paper No. 729212, San Diego, California.

(4) Gregory, D.P., "A Hydrogen-Energy System," American Gas Association (August 1972).

(5) Scott, R.B., et al, *Technology and Uses of Liquid Hydrogen,* Pergamon Press (1964).

(6) Drell, I.L. and F.E. Belles, "Survey of Hydrogen Combustion Properties," NASA 1383 (1958).

(7) Rosen, B., V.H. Dayan and R.L. Proffit, "Hydrogen Leak and Fire Detection," NASA SP5092 (1970).

(8) Coward, H.F. and G.W. Jenks, "Limits of Flammability of Gases and Vapors," U.S. Bureau of Mines, Bulletin 503 (1952).

(9) Austin, L.G., "Fuel Cells," NASA SP120 (1967).

(10) Martini, W.R., "Developments in Stirling Engines," Paper presented at the ASME Winter Annual Meeting, New York (August 2, 1972).

(11) Meijer, R.J., "Philips Stirling Engine Activities," SAE International Automotive Engineering Congress, Paper No. 650004, Detroit, Michigan (January 11 to 15, 1965).

(12) Gordon, S., and B.J. McBride, "Computer Program for Calculation of Complex Chemical Equilibrium Compositions, Rocket Performance, Incident and Reflected Shocks, and Chapman-Jouguet Detonations," NASA SP-273 (1971).

(13) Bergman, P.D., K.D. Plants, J.J. Demeter and D. Bienstock, "An Economic Evaluation of MHD-Steam Powerplants Employing Coal Gasification," BM-RI-7796, Bureau of Mines, U.S. Department of Interior (1973).

(14) Heywood, J.B. and G.J. Womack, *Open-Cycle MHD Power Generation,* Pergamon Press (1969) pp 212-217, 240-241, 781-784.

(15) Erren, R.A. and W.H. Campbell, "Hydrogen: A Commercial Fuel for Internal Combustion Engines and Other Purposes," *J. Inst. of Fuel (London),* volume 6, No. 29 (June 1933) pp 277-290.

(16) King, R.O. and S.V. Hayes, "The Oxidation, Decomposition, Ignition, and Detonation of Fuel Vapors and Gases; Pt. XXX: The Hydrogen Engine and Detonation of the End Gas by the Igniting Effect of Carbon Nuclei Formed by Pyrolysis of Lubricating Oil Vapor," *Can. J. Techn.,* volume 34 (1957) pp 442-454.

(17) Downs, D., A.D. Walsh and R.W. Wheeler, "A Study of the Reactions that Lead to 'Knock' in the Spark-Ignition Engine," *Trans. Roy. Soc. London,* volume 243 (July 1951) pp 517-520.

(18) Anzilotti, W.F. and V.J. Tomsic, "Combustion of Hydrogen and Carbon Monoxide as Related to Knock," *Fifth (Inter.) Symposium on Combustion,* Reinhold, N.Y. (1955) pp 356-366.

(19) Anzilotti, W.F., J.D. Rogers, G.W. Scott and V.J. Tomsic, "Combustion of Hydrogen as Related to Knock; Parallel Behavior of Hydrogen and Paraffinic Fuels," *Ind. Eng. Chem.* volume 46, No. 6, pp 1314-1318 (June 1954).

(20) Finegold, J.G., F.E. Lynch, N.R. Baker, R. Takahashi and A.F. Bush, "The UCLA Hydrogen Car: Design, Construction, and Performance," SAE Paper No. 73507 (May 1973).

(21) Billings, R.E. and F.E. Lynch, "History of Hydrogen-Fueled Internal Combustion Engines," Publ. No. 73001, Energy Research, Provo, Utah (1973).

(22) Finegold, J.G. and W.D. Van Vorst, "Engine Performance with Gasoline and Hydrogen: A Comparative Study," paper presented at The Hydrogen Economy Miami Energy (THEME) Conference, Miami Beach, Florida (March 18-20, 1974).

(23) Billings, R.E. and F.E. Lynch, "Performance and Nitrogen Oxide Control Parameters of the Hydrogen Engine," Publ. No. 73002, Energy Research, Provo, Utah (April 1973).

(24) Stebar, R.F. and F.B. Parks, "Emission Control with Lean Operation Using Hydrogen-Supplemented Fuel," SAE Paper No. 740187 (March 1974).

(25) Swain, M.R. and R.R. Adt, Jr., "The Hydrogen-Air Fueled Automobile," *Proceedings, 7th Intersociety Energy Conversion Engineering Conference,* San Diego, California pp 1382-1387 (September 1973).

(26) Murray, R.G., R.J. Schoeppel and C.L. Gray, "The Hydrogen Engine in Perspective," *Proceedings, 7th Intersociety Energy Conversion Engineering Conference,* San Diego, California (September 1972).

(27) Holvenstot, C.E., "Performance of Spark Ignition 4-Cycle Engines on Various Fuels," ASME Paper No. 61-OGP-14 (April 1961).

(28) King, R.O., S.V. Hayes, A.B. Allan, R.W.P. Anderson and E.J. Walker, "The Hydrogen Engine: Combustion Knock and the Related Flame Velocity," *Trans., Engineers Inst. of Canada,* volume 2, No. 4, pp 143-148 (December 1958).

(29) Ricardo, H.R., "Empire Motor Fuels Committee Report," *Proceedings, Inst. of Auto. Eng.,* volume 18, 1923, pp. 327-341.

(30) Burstall, A.F., "Experiments on the Behavior of Various Fuels in a High-Speed Internal Combustion Engine," *Proc., Inst. of Auto. Eng.,* volume 22, 1927-28, pp 358-392.

(31) King, R.O., W.A. Wallace and B. Mahapatra, "The Oxidation, Ignition and Detonation of Fuel Vapors and Gases, Pt. V. The Hydrogen Engine and the Nuclear Theory of Ignition," *Can. J. of Research,* volume 26, section F, pp 264-276 (1948).

(32) King, R.O. and M. Rand, "The Oxidation, Decomposition, Ignition, and Detonation of Fuel Vapors and Gases, Pt. XXVII. The Hydrogen Engine," *Can. J. Techn.,* volume 33, pp 445-469 (1955).

(33) Lynch, F.E., "Backfire Control Techniques for Hydrogen-Fueled Internal Combustion Engines," *Proceedings, The Hydrogen Economy Miami Energy (THEME) Conference* (March 1974).

(34) Karim, G.A. and S.R. Klat, "The Knock and Autoignition Characteristics of Some Gaseous Fuels and Their Mixtures," *J. Inst. of Fuels (London),* volume 39, pp 109-119 (March 1966).

(35) Karim, G.A. and H.C. Watson, "Experimental and Computational Considerations of the Compression Ignition of Homogeneous Fuel-Oxidant Mixtures," SAE Trans. No. 710133 (1971).

(36) Oemichen, M., Paper No. 68, *Verein Deutsche Ingenieur* (1942).
(37) Thomas, S.R., Jr., "Experience in Operating a Hydrogen Fueled Internal Combustion Engine," paper presented at Spring Meeting, Central States Section/The Combustion Institute, University of Wisconsin (March 1974).
(38) Ricardo, H.R., *The High-Speed Internal Combustion Engine* (4th edition), Blackie & Son, London, p 35 (1953).
(39) Schoeppel, R.J., "Design Criteria for Hydrogen-Burning Engines," Final Report No. APTD-0901 to EPA (Contract EHS 70-103), School of Mechanical and Aerospace Eng., Oklahoma State University (October 1971), (NTIS No. PB205815).
(40) Reilly, J.J., K.C. Hoffman, G. Strickland and R.H. Wiswall, "Iron Titanium Hydride as a Source of Hydrogen Fuel for Stationary and Automotive Applications," Report BNL-18651, Brookhaven National Lab. (April 1974).
(41) Wiswall, R.H., Jr. and J.J. Reilly, "Metal Hydrides for Energy Storage," Report BNL-16889, Brookhaven National Lab. (September 1972).
(42) Helmore, W. and P.H. Stokes, "Hydrogen-Cum-Oil Gas as an Auxiliary Fuel for Airship Compression Ignition Engines," Report No. E3219 Brit. Roy. Aircraft Establ. (Oct. 1930).
(43) Gerrish, H.C. and H.H. Foster, "Hydrogen as an Auxiliary Fuel in Compression-Ignition Engines," Report No. 535, National Advisory Committee for Aeronautics (NACA) (1935).
(44) Escher, W.J.D., "On the Higher Energy Form of Water (H_2O^*) in Automotive Vehicle Advanced Power Systems," *Proc., 7th Intersociety Energy Conversion Engineering Conference,* San Diego, California (September 1972).
(45) Karim, G.A. and M.E. Taylor, "Hydrogen as a Fuel and the Feasibility of a Hydrogen-Oxygen Engine," SAE Paper No. 730089 (January 1973).
(46) Cameron, H.M. and N.E. Morgan, "Development of Hydrogen-Oxygen Fueled 3-Kilowatt Internal Combustion Engine," AIAA Paper No. A64-756 (September 1964).
(47) Morgan, N.E. and W.D. Morath, "Development of a Hydrogen-Oxygen Internal Combustion Engine Space Power System," NASA CR-255 (July 1965).
(48) Underwood, P. and P. Dieges, "Hydrogen and Oxygen Combustion for Pollution Free Operation of Existing Standard Automotive Engines," *Proceedings, 1971 Intersociety Energy Conversion Engineering Conference,* Boston (August 1971).
(49) Sorensen, H., "The Boston Reformed Fuel Car–A Low Polluting Gasoline Fuel System for Internal Combustion Engines," *Proc., 7th Intersociety Energy Conversion Engineering Conference,* San Diego, California (September 1972).
(50) Breshears, R., H. Cotrill and J. Rupe, "Hydrogen Injection for Internal Combustion Engines," presented at EPA Alternate Automotive Power Systems Coordination Meeting, Ann Arbor, Michigan (May 1974).
(51) Mucklow, G.F., "The Effect of Reduced Intake-Air Pressure and of Hydrogen on the Performance of a Solid Injection Oil Engine," *R.A.S. Journal,* volume 31, No. 193, pp 17-59 (January 1927).
(52) Adt, R.R., D.L. Herschberger, T. Kartage and M.R. Swain, "The Hydrogen-Air Fueled Automobile Engine (Part 1)," *Proc., 8th Intersociety Energy Conversion Engineering Conference,* Univ. of Pennsylvania, Philadelphia, pp 194-197 (August 1973).
(53) Adt, R.R., H. Greenwall and M.R. Swain, "The Hydrogen/Methanol-Air Breathing Automobile Engine," *Proceedings, The Hydrogen Economy Miami Energy (THEME) Conference,* pp S10-37-48, Miami Beach, Florida (March 1974).
(54) Hattingen, U. and W. Jordan, "Wasserstoff als Zusatzkraft," *VDI Nachrichten,* No. 21/24, pp 4-5 (May 1974).
(55) Egerton, A., F.L. Smith and A.R. Libbelohde, "Estimation of the Combustion Products from the Cylinder of the Petrol Engine and Its Relation to 'Knock'," *Phil. Trans., Roy. Soc. of London,* volume 234 A, p 466 (1935).
(56) Eccleston, D.B. and R.D. Fleming, "Clean Automotive Fuel: Engine Emissions Using Natural Gas, Hydrogen-Enriched Natural Gas, and Gas Manufactured from Coal (Synthane)," Tech. Prog. Report-48, U.S. Bureau of Mines, Bartlesville, Oklahoma (February 1972).
(57) Hoffman, K.C., "The U.S. Energy System–A Unified Planning Framework," doctoral dissertation, Polytechnic Institute of Brooklyn (June 1972).

(58) Hoffman, K.C., editor, "Energy Systems Analysis and Technology Assessment Programs," Annual Report FY 1973, BNL-17959 (June 1973).

(59) Associated Universities, Inc., Upton, New York, "Reference Energy Systems and Resource Data for Use in the Assessment of Energy Technologies," AET-8 (April 1972) Available from U.S. Department of Commerce, National Technology Information Service, #PB-221422/9, Springfield, Maryland.

(60) Dupree, W.G., Jr. and J.A. West, "United States Energy Through the Year 2000," a U.S. Department of the Interior Report (December 1972).

(61) Strickland, G., J.J. Reilly and R.H. Wiswall, Jr., "An Engineering-Scale Energy Storage Reservoir of Iron Titanium Hydride," The Hydrogen Economy Miami Energy Conference, Miami, Florida (March 18-20, 1974).

(62) Strickland, G. and J.J. Reilly, "Operating Manual for the PSE&G Hydrogen Reservoir Containing Iron Titanium Hydride," BNL-18725 (February 1974).

(63) Proceedings of the 7th Intersociety Energy Conversion Engineering Conference (IECEC), held at San Diego, California (September 25-29, 1972).

(64) Feirtag, M., "Cars in Search of the Future," *Technology Review* (October/November 1972).

(65) Gregory, D.P. and R.J. Dufour, "Utilization of Synthetic Fuels Other than Hydrogen," Institute of Gas Technology Report on Project 8936 for ORNL (May 1972).

(66) Schuck, O.H., "The Future of Transportation," *AIAA Student Journal* (April 1972).

(67) Reese, R.M. and A.D. Carmichael, "A Proposed Hydrogen-Oxygen Fueled Steam Cycle for the Propulsion of Deep Submersibles," presented at the 6th IECEC, Boston, Mass. (August 3-5, 1971).

(68) Escher, W.J.D., "Helios-Poseidon, A Macro System for the Production of Storable, Transportable Energy from the Sun and the Sea," Escher Technology Associates Report PR-8 (February 1972).

(69) Silverstein, A. and E.W. Hall, "Liquid Hydrogen as a Jet Fuel for High-Altitude Aircraft," NACA Lewis Flight Propulsion Laboratory Report RM E55C 28a (April 15, 1955).

(70) Haldane, J.B.S., "Daedalus, or Science and the Future," a paper read to the Heretics on February 4, 1923, Cambridge, Mass.

(71) Lynch, R.R., "Preliminary Study of the Effects of Advanced Technology Applications on Fuel Conservation," (Proposed NASA Contractor Report by LTV Aerospace Corp.) Unpublished report as of November 1972.

(72) Kirkham, F.S., "Hydrogen Fueled Transport Aircraft–Mach 3 to 8," NASA Langley Research Center, unpublished technical paper (1972).

(73) Kolk, F.W., "Noise–A Triumph of Ignorance," *Astronautics & Aeronautics* (Oct. 1972).

(74) Witcofski, R.D., "Hydrogen Fueled Hypersonic Transports," presented at the American Chemical Society Annual Meeting, Symposium on Non-Fossil Chemical Fuels, Boston, Massachusetts (April 13, 1972).

(75) Rubert, K., "Hypersonic Research Engine," (series of updated project reviews August 1969 through October 1972), initially in Proceedings of the Short Course: Reusable Launch and Reentry Vehicles for Space Flight, University of Tennessee Space Institute, volume 2–Propulsion (August 1969).

(76) "Synerjet: Beyond the Shuttle," *Spaceflight* (British) (November 1972).

(77) Kirkpatrick, E.M., "Production of Liquid Oxygen from Atmospheric Air," U.S. Patent 2,960,834; November 22, 1960.

(78) Yaffee, M.L., "DOD, Airlines Face Energy Crisis," *Aviation Week and Space Technology,* p 54 (November 20, 1972).

(79) Gregory, D.P. and J. Wurm, "Production and Distribution of Hydrogen as a Universal Fuel," Paper No. 729208, *Proceedings of the 7th Intersociety Energy Conversion Engineering Conference,* San Diego, California, p 1329 (1972).

(80) Code of Federal Regulations, Department of Transportation, Hazardous Material Regulations, Title 49, Parts 170 to 190 (Rev. January 1, 1969) U.S. Government Printing Office, Washington, D.C.

(81) Martin, F.A., "The Safe Distribution and Handling of Hydrogen for Commercial Applications," Paper No. 729209, *Proceedings of the 7th Intersociety Energy Conversion Engineering Conference,* San Diego, California, p 1335 (1972).

(82) National Fire Protection Association, Gaseous Hydrogen Systems at Consumer Sites, NFPA 50A, National Fire Codes, volume 2, Boston, Massachusetts (1969-1970).
(83) Reilly, J. and R.H. Wiswall, *Metal Hydrides,* Brookhaven National Laboratory, Upton, New York, Report BNL-50082(S-70), pp 35-38 (1967).
(84) Reilly, J. and R.H. Wiswall, "The Reaction of Hydrogen with Alloys of Magnesium and Copper," *Inorgan. Chem.* 6, 2220 (1967).
(85) Reilly, J., R.H. Wiswall and S. Aronson, *Metal Hydrides,* Brookhaven National Laboratory, Upton, N. Y., Report BNE-50149(S-71), pp 34-39 (1968).
(86) Winsche, W., R.H. Wiswall, J. Reilly, T. Sheehan, C. Waide and K. Hoffman, *Metal Hydrides as a Source of Fuel for Vehicular Propulsion,* Brookhaven National Laboratory, Upton, N.Y., Rept. BNL-11754 (1967).
(87) Hoffman, K., J. Reilly, R. Wiswall, T. Sheehan and W. Winsche, "Metal Hydride Energy Storage Systems," *Intersociety Energy Conversion Conf.,* Boulder, Colorado (1968), IEEE Publ 68C-21-Energy, pp 981-985.
(88) Russell, C.R., *Elements of Energy Conversion,* Pergamon Press, London, England (1967), pp 378-388.

NONTECHNICAL ASPECTS OF A HYDROGEN ECONOMY

The source of material for this chapter was the report, N74-11728-735. A complete bibliography of the report starts on page 283.

SAFETY IMPLICATIONS

Legal Aspects

Concern for safety in the United States energy system is an example of what may be called the principle of federated regulation. Federated in this context means a sharing of power among the Federal, state and local governments and private units. Regulation means a rule or order guiding action and issued by a public (government) or private executive.

Authority for safety is scattered between public and private decision makers and among Federal, state and local responsibility. It is also divided between safety programs designed to prevent accidents, and programs designed to repair losses resulting from accidents. Despite this diversity, two dominant features have stood out. First, because of a strong feeling in policy for the reserved police power of the states over health, safety and general welfare, the states rather than the Federal government have had the primary and overwhelming responsibility; and second, because of equally strong adherence to economic doctrines of laissez faire, many safety decisions are fundamentally private.

An example of the latter is that while a state or locality may require a certain level of safety by an energy company, the basic safety code imposed is no more than an industrial code, self-generated by the industry itself. Two recent Federal acts discussed below may radically change this picture in the very near future.

Federal Responsibility: While Federal responsibility over energy safety has been minimal, there are several areas in which the Federal government plays a dominant safety role. The first area is that of wholly federally owned energy facilities. As of June 1972, the Federal government owned about 42% (24.1 million kilowatts) of all the hydroelectric generating capacity in the contiguous United States.

Since the Tennessee Valley Authority is a federally owned and created energy facility, safety is primarily a Federal responsibility. Where energy takes on an interstate character, the Federal government will assume dominant regulatc·y responsibility. Of all the oil pipelines in the United States, 85.1% are regulated by the Interstate Commerce Commission. The Federal Power Commission has been authorized to regulate the interstate aspects of electrical and natural gas transmission, and thus has primary responsibility. The Atomic Energy Commission since 1946 has held pervasive authority over atomic energy. The enabling legislation is replete with Congressional concern for protection of the public health and safety. While no case has yet held that the Federal act has ousted state authority, there are some indications that such may result.

Two recent Federal Safety Acts, The Federal Coal Mine Health and Safety Act of 1969 and the Occupational Safety and Health Act of 1970 (OSHA), have substantially altered the traditional roles for safety legislation between the Federal and state governments. The 1969 Coal Mine Act introduced two brand new features into the United States energy system: first, it wrote into Federal law a comprehensive safety code for this energy industry; and second, it established the first entry by the Federal government into the repair aspect of safety legislation (thought heretofore to be an exclusive state prerogative)—Workmen's Compensation.

Note that this legislation vitally affects the United States energy system as coal is now the main source of energy for electric generation plants, and is projected to become increasingly important as supplies of other fossil fuels diminish. As revolutionary as the 1969 Coal Mine Act was, the 1970 OSHA is positively radical. The act potentially (with some exceptions for governmental employers otherwise treated) covers all employers whose business affects interstate commerce. (As a footnote, virtually any business can be found to affect interstate commerce.)

The Secretary of Labor is given the power to rule safety by administrative standards. While Congress did not preempt the states from retaining any authority over occupational health and safety, the political dynamics are such that the OSHA eventually will be the primary force in the employee-safety field preempting contrary state action for every industry in America. This is surely a radical first in American safety legislation.

State Responsibility: By a large margin, the primary and overwhelming responsibility for safety legislation in the United States has been in the states. This is due to a very strong belief that, in our Federal system, the states have the traditional reserved power to protect and promote the health, safety and general welfare of its citizenry. Every state has enacted broad-ranging safety codes, but weak programs and insufficient enforcement led to the 1970 OSHA. State energy codes are concerned mostly with job safety, equipment and the construction and use of materials in and around the home.

Housing codes, however, are generally adoptions of privately generated codes. Another area of state energy safety code responsibility is the rather limited professional and trade codes, for example, the licensing of electricians. Trends indicate that there will be increased licensing and professional requirements for trades involved in the United States energy system.

Another area of state responsibility for safety is that of state remedies for accidents. Every state has enacted a Workmen's Compensation system. These laws provide a low-level, no-fault recovery for workers injured as a result of a job connected accident. For persons not workers (i.e., the public) injured by an energy related accident, the states also provide a common law tort remedy. These laws generally require proof of fault such as negligence or recklessness on the part of an agent of an energy supplier. Strict liability or a no-fault liability may be applied to extra-hazardous activities involving energy.

Federal Preemption: A Developing Area — Of particular significance in the area of state safety responsibility is a recent first impression case decided by the United States Court of Appeals for the Eighth Circuit, Northern States Power Company versus Minnesota. There, the central question posed was whether the United States government had the sole authority under the doctrine of preemption to regulate radioactive waste released from nuclear power plants to the exclusion of any state authority.

The Court reached an affirmative result and the case makes several important points. The case was pervaded with controversy because of the ever-increasing demand for power tempered by ecological concerns and the ever-sensitive Pandora's box of problems in the area of Federal-State relations. The state, of course, asserted its traditional power under the Tenth Amendment to protect and promote the health, safety and general welfare of its citizens. Significantly, the Atomic Energy Act, as construed by the Court, did not expressly exclude state authority to control radioactive waste.

The Court concluded that Congress had impliedly preempted the states from any regulation of health and safety in the level of radioactive effluents discharged from nuclear energy plants. This was felt to be so because Congress had vested the AEC with authority to resolve the proper balance between desired industrial progress and adequate health and safety standards, and that only through the application and enforcement of nationally uniform standards, promulgated by a national agency, could the dual objectives of industrial progress and adquate health and safety be met.

The Court noted that should the states be allowed to impose stricter standards on the level of radioactive waste, they might conceivably be so over-protective as to unnecessarily stultify industrial development. This result was reached despite the fact that the conventional approach had been to recognize the legitimate interests of the states to protect the health and safety of its citizens.

Previously, courts had refused to find Federal preemption over state health and safety laws absent of clear and unmistakable expression. Preemption, on the other hand, was much more readily found in cases not involving public health and safety.

Here, for the first time, it is suggested that where first, national interests are involved in a search for energy, and where second, there is thus a collision with the traditional power residing in the states to protect the health and safety of its citizens, and where, third, Congress has not expressly preempted state law, then finally, the Courts may find preemption in general Congressional legislation affecting energy use. This is truly startling.

Thus it may be seen from a comparison of the Federal-state areas of responsibility for safety that a trend is developing as follows: increased Federal responsibility in both safety codes and workmen's compensation and the increased Federal regulation of safety will be to the preemption of state authority to set higher or different standards.

It is significant to note that the Northern States Power Company case mentioned above involved atomic energy, one of the proposed prime energy sources to be used in the production of hydrogen. Should hydrogen enter the energy system, comprehesive Federal regulation of hydrogen and a comprehensive safety code can be expected. Also, should a conflict result between Federal and state authority over safety and health conditions connected with the use of hydrogen, it can be expected that the courts, even in the absence of legislative expression, will find that Congress has preempted state authority.

Political Aspects

Widespread public use of gaseous hydrogen in small units such as kitchen ranges, residential and commercial space heaters or refrigerators would require development of appliance safety standards and of additives for safety purposes. It must also be remembered that safe is a relative, judgmental term. Hydrogen must be evaluated relative to other fuels, not on an absolute basis, since all known fuels can be used only if a degree of risk is accepted.

Additives: Because people cannot ordinarily see or smell hydrogen in its pure gaseous form, whether it is or is not burning, an additive would be required to provide a warning and leak detection method. Present users of hydrogen, such as NASA, typically require very high purity levels and, thus have not made use of odorants or illuminants as safety devices, relying instead on elaborate precautions and leak detection mechanisms. This would not be practicable for household use, but odorants and illuminants would serve well in the public market.

Natural gas is now odorized in over half the states because of similar problems of leak detection, and several codes such as the NFPA or ASA provide standards. Natural gas odorants may be naturally occurring components of the gas (such as hydrogen sulfide or mercaptans) or may be added during processing.

Hydrogen could be used in residential or commercial applications for virtually any role now filled by natural gas: space heaters, water heaters, furnaces, kitchen ranges or refrigerators. For these purposes, odorants and illuminants currently used in natural gas could be used in the hydrogen distribution system.

However, it would be desirable to direct research toward development of a new variety of additives. Two attractive applications of hydrogen, fuel cells and catalytic burning, would be reduced in efficiency or impossible if current additives were used. The fuel cell, which could be used in local electric power substations, operates most efficiently with pure hydrogen and pure oxygen. It is possible that odorizers could be added locally, after first supplying the fuel cell with pure hydrogen, so that both fuel cell efficiency and customer safety could be provided.

Such additions may, however, prevent the use of hydrogen in a catalytic burner. Unfortunately, additives used today would poison the catalyst. Development of different additives or different catalyst materials should be undertaken.

An illuminant would be necessary for open-flame hydrogen burners because hydrogen has a colorless and almost invisible flame in ordinary light. This makes it difficult to know whether a flame is present, thus increasing the danger of personal injury or ignition of materials (paper, dish towels, etc.) which might unknowingly be placed too close to a flame. Use of a suitable illuminant, or a more complicated and probably more expensive flame sensor, could reduce this danger.

Appliance Safety Standards: Safety standards and codes for natural gas appliances are well developed and could be adapted to hydrogen requirements. Some experience with hydrogen safety is in fact already available from the history of "town gas" or "producer gas," once widely used domestically and still common in Europe. Most such gas is a mixture of several elements and is typically approximately 50% hydrogen. Although town gas does not have exactly the same characteristics as would a hydrogen supply of the type contemplated it does provide some illustration of the fact that a properly designed distribution and appliance system can be safely operated with a hydrogen-rich gas.

Adaptation of existing codes to accord with hydrogen's characteristics would probably accompany the development of hydrogen as a major new fuel. It would be necessary to carry out conversions on all burners, and in many cases whole new appliances would be either necessary or highly desirable for reasons of both safety and efficiency. This large-scale conversion coupled with the wide-spread public image of hydrogen as a dangerous substance seems likely to produce pressures for the development and elaboration of a comprehesive safety code on the Federal level.

The potential use of hydrogen-fueled furnaces or decentralized space heaters will require revisions in codes which may become extensive. Since the combustion of hydrogen results in water vapor and excess air, with only traces of nitrogen oxides, it would not be necessary to vent the combustion exhaust to the outside of the building. Current codes require venting of combustion products and various construction and safety requirements are premised on that venting. Changes in building and safety codes can become objects of major political combat in local areas, a fact which may either slow down the introduction of hydrogen as a fuel or provide more impetus to a comprehensive Federal code.

Industrial Uses and Safety: Development of large-scale industrial use and transportation-storage of hydrogen is already present. NASA requirements for liquid hydrogen in the space program have resulted in considerable experience in transporting and storing large quantities. NASA has developed extensive safety procedures and training methods for personnel (1).

Storage tanks, mobile transportation tanks and connector fittings have also been developed. There is no doubt that handling massive quantities of high-purity hydrogen in liquid form is dangerous but manageable, as NASA has not experienced a fatality in connection with liquid hydrogen. Other organizations, such as the Linde Corporation, manufacture and transport or store hydrogen in both liquid and gaseous form and have found the safety problems real, but surmountable.

Large-scale use of hydrogen, particularly in its more difficult liquid form, will require elaboration of these experiences into standardized safety and handling requirements. It is not improbable that Federal or state licensing requirements will be imposed upon personnnel engaged in the transport and storage of liquid hydrogen. Large-scale storage will give rise to requirements for fire protection such as those governing existing oil tank farms. Liquid hydrogen is already transported in over-the-road trailers and by rail tank car, with apparent safety.

Pipeline transmission of gaseous hydrogen has also been a standard practice for several years in the United States. One network of hydrogen pipelines in Germany dates from 1938 (2). Some safety problems with materials, such as the recently recognized hydrogen environment embrittlement, may become more significant with time and large-scale use. Furthermore, any system tends to be more error-prone as it becomes more complex and extensive. Still, there is no apparent reason to believe that safety in hydrogen pipelines poses any insurmountable difficulties.

Assessment of Safety Factors

Technical: There is no point in denying the danger implicit in widespread use of hydrogen as an energy medium. Compared to natural-gas-type mixtures, hydrogen is much more prone to leak (even through air-tight fittings), can be ignited by very low energy (even invisible static sparks), and is flammable over a very wide range of hydrogen-air mixtures. In liquid form, its extremely low temperature (below 20°K) can liquify any air which comes into contact with it, its tank or lines. This produces fire and explosion hazard. Furthermore, a hydrogen-oxygen flame is invisible and a hydrogen-air flame is nearly invisible in daylight.

There are, however, some counter-balances to what appears to be a very damning list. It is true that hydrogen is more prone to leak than any other gas and that its diffusion rate is extremely high. But, since hydrogen will leak (diffuse) so very rapidly it may never reach a concentration sufficient to sustain ignition.

Because it is only one-fourteenth as dense as air, hydrogen rapidly departs the scene of spills or leaks; gasoline, propane or fuel oil vapors, on the other hand, are heavier than air and tend to concentrate around a leak. Even natural gas diffuses only about one-third as rapidly as hydrogen. Very large hydrogen spills or leaks disperse so rapidly that the hazard would be past within minutes. A 500-gallon spill of liquid hydrogen diffuses to below flammability in about one minute.

Ordinary flames such as those from hydrocarbons release a great deal of radiant heat, while a hydrogen flame produces relatively little. This is a hazard in fighting hydrogen fires, because the flame is so nearly invisible and its radiant energy so low that one may actually walk into the flame before realizing it is there. This difficulty may be avoided because a hydrogen fire in all probability may not be a pure hydrogen fire, but may involve foreign luminous substances. The lower radiant energy is potentially an advantage, since fire fighters could work closer to the flame or contain a blaze by removal or cooling of close-by flammable materials.

Hydrogen thus does present some special requirements for flame detection, materials handling, leak detection and safety in general. The fact that hydrogen is in widespread use in industry and the space program, as well as its common use over long time periods in town gas shows that it can be used safely.

Union Carbide Corporation has been reported as believing that hydrogen is safer to handle than is gasoline or propane (3). It may be confidently expected that a decision to develop large-scale hydrogen fuel systems would be accompanied by the development of adequate safety devices and techniques.

Public Image: The literature on hydrogen is replete with references to the 1937 explosion of the German dirigible Hindenburg at Lakehurst, New Jersey. In fact, the term "Hindenburg Syndrome" is widely used as shorthand to indicate public identification of hydrogen as an exceptionally dangerous substance. There is an additional segment of the population who identify hydrogen with "that stuff bombs are made of" and are likewise concerned about the safety of hydrogen.

Derek Gregory (2) and many others have pointed out that the Hindenburg incident is greatly over-dramatized. In fact only 36 people were killed and 65 survived, a ratio and an absolute number which compares quite favorably with the results of a great many airplane crashes since 1937. The dirigible had made 18 round trips between its German home port and either the United States or Brazil without accident. While sabotage has been alleged, several safety provisions were violated at the time of the explosion.

It may even be argued that the rapid burning characteristics of hydrogen and its low radiant energy led to the survival of more passengers than would otherwise have been expected. But, of course, the important point is that a public image was established and still, at least in part, persists. It is also true that there is no relationship whatever between a hydrogen-fueled kitchen range and a thermonuclear bomb; but some people may think that there is. These facts are important to remember, for opponents of hydrogen as an energy carrier may present an arsenal of objections based upon the images of the Hindenburg and the H-Bomb.

There is a conventional public education and public relations method which should be able to counter unfounded public fears about hydrogen. It will be necessary to show that hydrogen dangers can be met and that safety standards can obtain acceptable risk levels.

Given the general concensus that petroleum fuels are becoming rapidly depleted, and that their prices will rise, and given the powerful and still growing concern for environmental protection, American society may be faced with a tradeoff in which a real risk of some increased danger will be exchanged for having energy in the quantities desired. This is not to say that hydrogen safety should not and cannot be developed until the incremental risk is rather low; it is, however, to argue that hydrogen simply cannot be dismissed because it is marginally less safe now than are other fuels presently in use.

LEGAL IMPLICATIONS

Energy Law

This section investigates the law of the United States energy system. The survey will cover the past and present in hopes of detecting trends so that some forecasts can be made about law's impact should hydrogen be introduced as an energy carrier. A constant in the system has been a demand for abundant, ever increasing quantities of low-cost energy. Abundance was assumed as a fact and a value, and law reflected this. The law utilized to establish the rules of the road for the energy system was basically the English common law, but along the way some new and distinct developments were made. In short, law was called upon to assist in building the system, but was also affected by the system it created.

Property Concepts and Ownership: English common law concepts of property were borrowed. The concept of private property is prominent in the system, which is largely privately owned. There is, however, some public ownership. The Federal government, of course, owns a part of the energy system, for example the Tennessee Valley Authority and the numerous hydroelectric generation plants. Also, local governments, cities, counties and districts do have a substantial ownership role, primarily in the ownership of local electrical utilities. Singularly absent from any ownership role are the states.

The predominance of private ownership results in three features as follows: ownership is becoming highly concentrated, raising antitrust problems; ownership in private hands is assisted by the law of condemnation; and the public interest demanded comprehensive regulatory schemes over private property.

Antitrust and Energy Monopoly: During the 1960's a growing list of oil companies started acquiring interests in competing fuels. The result was intensive horizontal integration in the fossil fuel industry. For example, by 1970, of the 25 largest oil companies, 18 had involvements in oil shale, 11 in coal, 18 in uranium, and 7 in tar sands. This trend created what is now commonly called the "Energy Company." The Energy Company phenomenon has led to a renewed application of the antitrust laws to control the concentration of energy resources within a company.

It should be noted, however, that it was long ago recognized that there must be many instances for natural monopolies, to wit the public utility. While this, of course, is quite contrary to basic notions of free enterprise competition, natural monopolies are very much a legal part of the American scene. Even so, a natural monopoly may be subject to antitrust action. In the recent case of Gulf States Utilities Company versus FPC, the Supreme Court held that the FPC's broad regulatory authority over electric utility companies included the responsibility of considering possible anticompetitive aspects in authorizing a utility security issue.

Responsibility was necessary, the Court felt, because it provided the first line of defense against anticompetitive practices which might later become subject to antitrust proceedings. Also, in another 1973 Supreme Court case, Otter Tail Power Company versus United States, the Court held that even though a utility may be quite highly controlled by a regulatory agency, it is nevertheless subject to independent scrutiny under the antitrust laws. This result was reached because repeals of the antitrust laws, by implication from a regulatory scheme, are to be strongly disfavored and have only been found, said the Supreme Court, in cases of plain repugnancy between the antitrust laws and the regulatory provisions.

On another antitrust front, the recent suit filed by the Florida Attorney General charged 15 major United States oil companies with conspiring to violate the antitrust laws by creating a nationwide fuel shortage. The suit alleges that the oil companies have engaged in an illegal monopoly and in unreasonable restraints of interstate commerce and trade.

It also alleges that the oil companies conspired to reduce competition; raise, fix and stabilize prices and exclude nonmajor oil companies from significant aspects (e.g., refining and exploration) of the industry. The relief sought is that the companies be forced out of crude oil exploration and production. Furthermore, in July 1973 the Federal Trade Commission accused the eight largest United States oil companies of monopolistic refining and marketing practices which have increased their profits, thus forcing American motorists to pay inflated prices.

Paralleling the Florida suit, the FTC charged that the big oil firms have controlled the market from oil well to gas pump for the last 25 years, making it almost impossible for new companies to enter the refining business. The FTC action, in issuing an antitrust complaint, was termed unusual since it usually issues a proposed complaint, giving the parties involved a chance to settle the matter out of court.

All this renewed interest in application of the antitrust laws is sure to have a profound effect. The horizontally and/or vertically integrated energy company may be broken up into a severely atomized system. With capital investment costs approaching the limits of all but the largest company, it is anticipated that a vigorous application of the antitrust law to reduce company size may significantly hinder private entry into the hydrogen economy.

Eminent Domain: Of major importance is the ability of a company or government to condemn land necessary to carry on an energy activity. The common law of property has lent the venerable concept of eminent domain. Its application in the United States, controlled by the Fifth Amendment, provides that private property shall not be taken for public use without just compensation. Just compensation is defined as the full monetary equivalent of the property taken.

The owner should be placed in the same position, monetarily, as he would have been had his property not been taken. The monetary equivalent is determined by the concept of market value: the owner is entitled to the fair market value of his property at the time of the taking, and that the value is normally to be ascertained by what a willing buyer would pay in cash to a willing seller.

Future energy needs in the nation indicate that substantial amounts of private property may have to be condemned, for which just compensation must be paid. Proposals that hydrogen production plants be sited on land or that major new underground pipeline distribution systems be constructed will require that lands so taken be condemned, requiring substantial amounts of compensation. Offshore plant sites, thus derive an economic advantage, a factor which may join with environmental pressures and cooling water needs to create a rush to the sea.

Regulatory Law

Federal Regulation: Federal regulatory power is based upon the power conferred on Congress to regulate interstate commerce. The public interest in energy is generally defined as two-fold: First, to insure an efficient and dependable energy system to meet the needs of a growing national economy; and second, to serve the needs of consumers by insuring abundant, low-cost, safe energy. These purposes may, however, conflict.

Regulation of a company or industry to insure its viability and growth in an expanding economy requires promotion of the specific company or industry. The extent to which the company or industry is promoted may result in diminished concern for the needs of the consumers. Similarly, the reverse is true; the extent to which the regulatory agency sees its mandate as the protection of the consumer may result in harm to the company or industry.

Federal regulation is not as comprehensive as one would think. In surveying the total energy resources in the United States, coal, petroleum, natural gas, nuclear power, and hydroelectric power can be listed. The Federal regulatory agency involved in coal resources is the United States Bureau of Mines. Its authority is generally limited to the safety aspects of coal mining.

The Federal government has made no attempt to regulate the price of coal, nor the siting or licensing of mines. The Federal agency involved in regulating petroleum resources is the Interstate Commerce Commission. The ICC regulates approximately 85% of all the intercity ton miles of pipelined oil. Its main concern is with the orderly movement and safety of the transportation process. The Federal agency involved in regulating natural gas is the Federal Power Commission. The Natural Gas Act of 1938 gave to the FPC broad powers to protect consumers against exploitation at the hands of natural gas companies through a comprehensive and effective regulatory scheme of dual (state and Federal) authority. The Federal Power Commission's authority is limited to:

- transportation of natural gas in interstate commerce,
- its sale in interstate commerce for resale, and
- natural gas companies engaged in such transportation or sale.

The Federal agency involved in the regulation of nuclear energy is the Atomic Energy Commission. The Atomic Energy Act of 1954, amending prior acts, granted to the AEC broad powers over atomic energy to insure its proper development consistent with a concern for adequate health and safety standards.

Hydroelectric power resources are regulated by the Federal Power Commission. Its authority over nonfederal hydroelectric projects dates back to the agency's beginning in 1920. Of a total generating capacity of 57.4 million kilowatts, 24.1 million kilowatts is federally owned, 19.8 million privately owned, and 13.5 million nonfederal, publicly owned. Of the 33.3 million kilowatts of nonfederal capacity 29.4 million kilowatts (88%) is operating under FPC major licenses.

State Regulation: As noted above, Congress has relied upon the states in many instances to supplement the Federal government's regulatory power. Even when Federal authority has been exercised, the usual pattern is a dual system of Federal and state authority. State authority is perhaps most pervasive in the coal mining industry where comprehensive safety codes, siting requirements and pollution standards abound. State residual authority in the petroleum industry is likewise quite comprehensive.

A well-known illustration is the powerful Texas Railroad Commission. In the areas of natural gas and electricity the states retain the residual power of economic regulation, mostly by fixing consumer prices. Perhaps the most highly state-regulated aspect of the energy system is of electric power generation, distribution and sale.

Competition for energy among the states gives rise to a diversity of state resolutions, thus impelling the Supreme Court to find in many cases that the area is federally regulated. This insures regulation in the national interest but diminishes the regulatory power of the states.

For a number of years, spokesmen for the energy industry have noted the need for a national energy policy and coordination among the various Federal and state regulatory agencies. Prompt action seems certain.

As recently as April, 1973 the establishment of a cabinet-level agency entitled the Department of Energy and Natural Resources, which would be responsible for the balanced utilization and conservation of our nation's energy and natural resources, was proposed. The agency would consolidate many of the regulatory functions scattered throughout the Federal structure. Thus, it may be that the Federal and state regulatory structures will be, by and large, consolidated into a single Federal cabinet-level department.

Surely, as the demand for energy increases and the supplies decrease, this agency will be given more and more functions of research, development, funding and regulation of the entire energy picture. In just two years, the name of the proposed agency had been changed from a Department of Natural Resources to a Department of Energy and Natural Resources thus indicating an expansion of conceived powers. It seems rather clear that decisions in a hydrogen economy will be almost entirely federalized.

Environmental Law

One of the most dramatic and recent developments on the American scene has been the growth of environmental concern. It has been suggested that environmental activism began in the 1950s, when the focus was on atmospheric nuclear testing, moved to pesticides in the sixties, and finally in the seventies to national policy.

On January 1, 1970 the National Environmental Policy Act was signed into law. NEPA has been called the most important piece of environmental legislation ever written and will surely have a major impact on energy decisions in the future.

Federal Law: The most important Federal environmental law, the National Environment Policy Act, declares that it is the policy of the Federal government...

> to use all practicable means and measures, including financial and technical assistance, in a manner calculated to foster and promote the general welfare, to create and maintain conditions under which man and nature can exist in productive harmony, and fulfill the social, economic, and other requirements of present and future generations of Americans.

To implement that policy, the act directs that, to the fullest extent possible, all policies, regulations and laws shall be interpreted and administered in accord with the purposes of the act. Also required is the famous NEPA environmental impact statement. The statement must survey the following five items:

- the environmental impact of the proposed action
- any adverse environmental effects which cannot be avoided should the proposal be implemented
- alternatives to the proposed action
- the relationship between local short-term uses of man's environment, and the maintenance and enhancement of long-term productivity
- any irreversible and irretrievable commitments of resources which would be involved in the proposed action should it be implemented.

The significance of NEPA can be seen by a comparison of two cases, one pre-NEPA and the other post-NEPA, both involving the Atomic Energy Commission. In 1969 (prior to NEPA) a United States Appeals Court found that the AEC need not consider the thermal pollution impact of a proposed nuclear power plant. Only two years later, after the enactment of NEPA, another court found that the AEC's regulations were inadequate because they failed to fully consider environmental factors, including thermal pollution.

It must be emphasized again that NEPA requires all agencies of the Federal government (including courts and agencies involved in the energy system) to interpret and administer all policies, regulations and law in accord with the environmental purposes of the act. The environmental attractiveness of hydrogen as a fuel and/or energy carrier would fit in well with the NEPA requirement. Production methods, however, may present significant problems.

Supplementing NEPA are various Federal acts designed to protect specific areas of the environment. The Clear Air Act of 1970 has had recent emphasis. In Sierra Club vs. Ruckelshaus it was held that the phrase "to protect and enhance the quality of the nation's air resources" means that there can be no (not even the slightest down turn to standardized clean air) degradation of preexisting clean air. The Federal Water Pollution Control Act Amendments of 1972 establish a complicated and comprehensive Federal permit system under Federal standards to insure clean water.

For example, the Act requires that by no later than July 1, 1977, permit holders must use the best practicable (waste) control technology currently available. Its breadth of coverage ranges from domestic water sources to the high seas. The Rivers and Harbors Act of 1899 is the most famous of the Federal criminal environmental protection acts.

State and Local: Every state and locality has power to protect the environment. Because of the great diversity, no attempt will be made to catalog or organize the laws; rather, some highlights will be mentioned. Perhaps the most significant recent development has been the inclusion of environmental protection in state constitutions. Florida, Illinois, Michigan, New York, Pennsylvania, Rhode Island and Virginia all have recently adopted constitutional provisions concerning the environment. Some of these enact constitutionally based provisions giving the citizen a right to a clean environment.

The basis for state and local regulation of the environment is the traditional idea of police power possessed by state and local governing bodies. The conventional rule has been that these powers are not to be superseded by a Federal act unless there was a clear Congressional intent. However, the trend seems to be that Congress (with the assistance of the courts) will find, in ever increasing areas of environmental control, a need for a federalized environmental law.

Private: The private environmental law remedies that can be asserted by groups and individuals include:

- constitutional theories
- public trust doctrine
- statutory remedies
- common law tort remedies

The most frequently asserted constitutional theory is the Ninth Amendment idea that there is a zone of reserved rights which the government must guard and protect from intrusion. Development of the constitutional right to a clean environment at the Federal level seems to have little potential, although, as noted above, a state right may exist.

An offshoot of the constitutional right argument is the Public Trust Doctrine. The doctrine specifies that there are certain common properties (river beds, seashore, lands) held in the public trust by the government for the use of the general public. When such common properties are reallocated to private uses and where there is not substantial evidence that the private use has a considerable degree of public benefit, the government has defaulted on its trust responsibility.

While the Trust Doctrine has been applied in a rather limited number of areas, it may play an increasing role in the private protection of the environment as increased energy needs threaten the naturalness of the public domain.

The third area of private environmental law is statutory remedies. The trend seems to be for the states to provide an increasing amount of private statutory remedies while the Federal government is limiting the availability of private remedies.

The fourth area of private remedies for environmental deprivation lies in the venerable theories of the common law. A part of the common law is tort law or 'he hit me' law. Trespass is an ancient action for damages or an injunction where there is a deliberate intrusion on land. Another tort theory is nuisance. Nuisance is perhaps the oldest and most used of the common law remedies to control pollution.

It generally requires an unreasonable interference with the use and enjoyment of the land. What is unreasonable is determined by a balancing of the benefits to the community at large and the defendent in particular against the inconvenience costs to the plaintiff. Usually the nuisance theory has been used to adjust the intangible invasions of land such as noise, dust, odor and the like.

The next theory is negligence. In negligence the landowner must show that the polluter has failed to exercise reasonable care in preventing pollution of plaintiff's property. The last theory is that of strict liability for ultrahazardous activities, defined as those activities which can be characterized as abnormally dangerous such as blasting.

International Law

Since several of the suggested methods for the production of hydrogen would use large quantities of water and would produce substantial quantities of reject heat, it has been suggested that the plants be located offshore. Proposed offshore locations vary from the shoreline, to the territorial sea, to the contiguous zone, to the continental shelf sea area, to the high seas. Each one of these siting locations itself, along with the production process by-products (primarily salts and reject heat) and the use of large quantities of water raise serious international law questions.

The threshold question is, who owns the seas? The traditional answer to this question has been, no one. Under the basic principle of the freedom of the seas, the seas, the water itself, marine life, and the minerals (in and under) belong to no one; or, in other words, they were treated as res nullius. Since the seas and its resources belong to no one, it follows that any one could use them in any manner, for any purpose.

There are, however, growing signs of an inclination to treat the oceans and other resources shared in common as res communis, belonging to the community rather than res nullius (4).

This shift in international law was created by a two-fold desire: to exploit the resources of the seas, and to control the pollution of the seas and its resources so that they may continue to be available for exploitation.

The old pristine idea of res nullius no longer exists. It is settled that the seas (and its resources) are divided up into several bands. Working outward from the shoreline, first there is a band known as the territorial sea, then a band known as the contiguous zone, then a band known as the continental shelf, and finally, the high seas.

It is clear that the territorial sea is part of the territory of the coastal state, that is, the coastal state owns it. Since ownership is a valuable right (the coastal state owns the natural resources and can keep others out), coastal states want as much territorial sea as possible. There is no general agreement in international law as to the width of the territorial sea.

While the United States claims that its territorial sea is an internationally acceptable three miles, other countries make claims of a width of a territorial sea anywhere from 6 to 12 to 25 and indeed up to 200 miles. Nevertheless, should the hydrogen generation facility be located in the United States' territorial sea, it would be subject to all the domestic laws of the United States as well as some minimal international law requirements such as nonobstruction of navigation as well as developing international standards on pollution.

The next band of the seas is that area known as the contiguous zone. It is 12 miles in width and includes within it the territorial sea, if that is less than 12 miles. The contiguous zone is not owned by a coastal state but it can exercise considerable authority in it as needed to prevent the infringement of its customs, fiscal, immigration, or sanitary regulations within its territory or territorial sea. Thus, should the hydrogen generation plant be located in the contiguous zone, it would be subject to all domestic American laws concerning pollution, as well as international rules against interference with navigation and developing international pollution controls (5).

The next band is known as the continental shelf (6). It is defined generally as the seabed extending from the shoreline to a depth of 100 fathoms. The Continental Shelf Convention reserves to the coastal state the exclusive right to explore and exploit the natural resources on the seabed and in the subsoil of the continental shelf. In so doing, however, the coastal state must not unjustifiably interfere with navigation, fishing or the conservation of living resources of the sea.

All appropriate measures must be taken to protect the living resources of the sea in areas (known as safety zones) surrounding installations or devices on the continental shelf. It seems fairly clear, then, that a hydrogen production facility could, consistent with international law, be sited on the American continental shelf (assuming its configuration to be analogous to an offshore oil drilling rig). International law constraints would severely restrict the size and the plant's ability to cause overboard pollution.

Finally comes the last band of the seas, the high seas. The international law of the high seas is codified in the convention on the High Seas (7). Its basic principle is the freedom of the high seas; the high seas are open to all nations, and no one may validly subject any part of it to ownership. While the Convention makes no attempt to provide an exhaustive list of the various permutations of the basic principle, it, most significantly, cautions in Article 2 that such freedoms are to be exercised by all states with reasonable regard to the interests of the other states in the exercise of the freedom of the high seas. Article 25, in part, of the convention also specifies that

> All states shall cooperate with the competent intertional organizations in taking measures for the prevention of pollution of the seas or air space above, resulting from any activities with radioactive materials or other harmful agents (7).

The question then becomes: (assuming that a hydrogen production facility were to be sited on the high seas) could it be argued, under Articles 2 and 25, that the quantities of reject heat and salts resulting from the production process would be an unreasonable use of the high seas and/or a pollution of the seas by other harmful agents? Such may well be the case (8). This interpretation is in accord with the United Nation's 1970 Declaration of Principles Governing the Seabed and Ocean Floor:

> States shall take appropriate measures for and shall cooperate in the adoption and implementation of international rules, standards and procedures for, inter alia: (a) The prevention of pollution and contamination, and other hazards to the marine environment, including the coastline, and of interference with the ecological balance of the marine environment;
> (b) The protection and conservation of natural resources of the area and prevention of damage to the flora and fauna of the marine environment (9).

Also supportive of that interpretation is Article 5 of the Convention on the Continental Shelf which, while applicable only to the continental shelf, provides in part:

> The coastal state is obligated to undertake, in the safety zones (around installations or devices on the continental shelf), all appropriate measures for the protection of the living resources of the sea from harmful agents (6).

The basic premise of the report to the United Nation's 1972 Conference on the Human Environment was that a certain level of development not yet reached in the developing countries is a prerequisite for a decent environment and all developed nations must help the have-nots (10). Thus, if a hydrogen production facility were to be sited on the high seas, it may well run into severe objection on environmental grounds and/or strong demands from the developing countries of the world for a share of the energy produced (11)(12).

ENVIRONMENTAL IMPLICATIONS

Hydrogen is produced primarily from an abundant and noncontaminating raw material, water. When combusted, it again becomes water and rejoins the water cycle. However, as a synthetic fuel, it must be considered whether hydrogen can be produced by a base power source which is environmentally sound. Further, any fuel must be transported and stored, and possible adverse effects of this sector must be examined.

Simply put, there are two possible environmental contaminants which result from hydrogen combustion, NO_x and water. Water is not normally considered to be a contaminant, but there are some conditions under which hydrogen combustion might produce sufficient water to be a problem.

For example, tightly insulated homes using ventless hydrogen space heaters might require dehumidification. It is possible that areas of high use, such as a hydrogen-fueled city, might develop a water island effect similar to the heat island which develops around cities due to reject heat. Although this problem needs study, it is expected that a very minor impact will be found. Table 6.1 shows the relative amounts of water produced per Btu by three energy sources.

TABLE 6.1: WATER PRODUCTION FOR VARIOUS ENERGY SOURCES

Fuel	Water, lb/Btu
Gasoline (C_8H_{18})	74×10^{-6}
Methane (CH_4)	105×10^{-6}
Hydrogen	174×10^{-6}

Source: N74-11728-735

Although some added water product will be found, the difference is certainly not great. When hydrogen is burned with pure oxygen, water would be the only combustion product. If burned with air at high temperatures (above approximately

1750°K) atmospheric nitrogen would become involved and some amount of NO_x would be produced. Such high temperatures would be rare in residential use and found only in specialized commercial uses, where controls should be possible. These hydrogen burn characteristics compare favorably with the usual hydrocarbon combustion which produces not only water, but a witch's brew of unburned hydrocarbons, NO_x, CO, CO_2, SO_2 and much more.

One fact is inescapable: Americans cannot have energy and zero environmental impact. Heat engines must necessarily produce reject heat; transportation of energy requires vehicles and highways, pipelines, transmission lines or other structures which may produce visual pollution if nothing else. The production of base power in large quantities involves, at the very least, land use problems. Hydrogen and electricity may be compared in that both may be produced from a variety of base power sources.

A system may be designed which minimizes environmental damage at the source by making a flexible choice of the environmentally optimal generation method. Hydrogen has an advantage over electricity in the areas of transportation and storage. Overhead electrical transmission lines range from unsightly to ugly, a fact which is rapidly driving electrical transmission underground. But underground electrical transmission is currently an order of magnitude more expensive than pipeline hydrogen transmission of equivalent energy.

Storage of electricity in large amounts is not yet possible, except through the use of indirect methods such as the expensive and environmentally undesirable technique of pumped storage. Storage of hydrogen in large quantities is possible, either underground with very minimal environmental impact or above ground, but with much less land use impact.

Electricity and hydrogen are both clean-burning in end use. It is primarily between the generating plant and the end use that hydrogen has environmental and cost advantages. Electric power transmission is conventionally done by means of overhead lines, but environmental objections are driving the cost of overhead transmission rapidly upwards and may soon make it virtually prohibitive.

However, long-distance underground electrical transmission is more than ten times as expensive as overhead transmission. Hydrogen, by contrast, can be transmitted any necessary distance without significant power loss by means of underground pipeline, at a cost which would probably be about 10% of the cost for underground electrical transmission. Since hydrogen could be used as a natural gas substitute in most applications, and could be used in local fuel cell or turbine substations to regenerate electricity, hydrogen pipelines may replace electrical transmission lines as the nation's long-distance energy carrier.

As noted above, both electricity and hydrogen are limited for mobile power use, as in automobiles, primarily because of storage problems. The weight, bulk, or expense of storage batteries for electricity plus the limited range attainable has placed definite practical limitations on the much-discussed electric automobile, despite its otherwise favorable environmental character. Hydrogen suffers from similar storage limitations; the low density of hydrogen dictates that an automobile with any range would need to carry a very bulky tank which may be difficult to handle safely. A cylindrical tank 20 by 60 inches would hold about 35 gallons of liquid hydrogen, equivalent to approximately 12 gallons of gasoline in energy content, and would be much more awkward to handle than today's conventional "fill 'er up" at the corner station. Primarily for this reason, the use of hydrogen in automobiles is not considered a likely alternative in the near future.

However, the concomitant effects of environmental controls and gasoline high prices or scarcity may override these constraints. If the storage limitation can be resolved by a modified automobile design or by acceptance of limited ranges (as in the urban car intended for short-range travel only), and if the physical and safety limitations of handling liquid hydrogen for this purpose can be circumvented environmentalists may rejoice in almost ideal auto exhaust content.

At the same time, the auto-driving public could still use the private, flexible transport to which they have become so practically and emotionally attached. Politically and socially, this is a solution devoutly to be desired.

In short then, hydrogen per se is environmentally an almost ideal fuel. It is significantly cleaner than hydrocarbon fuels, and its only combustion by-products should be easily controlled. Hydrogen transmission by underground pipeline offers aesthetic advantages over conventional electrical transmission. Hydrogen's environmental impact is likely to be highly positive, allowing cleaner cities and factories. The adverse impact, if any, will be connected chiefly with the method by which hydrogen is originally produced.

SOCIOECONOMIC IMPLICATIONS

United States material wealth and national power have rested upon a relative abundance of natural resources, an industrious and capable population, an abundance of energy and no doubt an occasional bit of luck. Darmstadter (13) has presented data establishing a close relationship (correlation coefficient of +0.87) between high energy consumption rates and high gross national product per capita for a large sample of nations.

Cook (14) and others have noted the same basic relationship. American society is an affluent (even effluent) society in which the availability of massive quantities of energy at low prices has been a major supporting element.

Social Costs and Values

Social factors are in part responsible in creating the current energy shortage. Legislative and other restrictions on strip mining, offshore oil exploration, nuclear plant siting and safety, automobile emissions standards and many others have all played a part. These actions have helped to control a growing pollution problem and preserve environmental purity levels, but is it worthwhile to do these things, in view of the other costs?

Caldwell (15) concludes than an ecologically right position may be politically wrong; but conversely, a political right may be an ecological wrong. Lueck (16) has pointed out that the problem has not been clearly defined. Is social cost the difference between Plant A with 100% effective pollution controls and Plant B with zero pollution controls? Or is social cost the difference between the cost of pollution controls in Plant A and social benefits obtained by those controls?

If the money cost is higher than the social benefit, should pollution be permitted? It is concluded that "there is no way in the market on purely economic grounds to compare the cost of pollution control measures with the social costs which are incurred if the measures are not taken."

Any economic activity may be analyzed from a number of points of view. One viewpoint may be labeled environmental economics. Perhaps the most familiar environmental economics approach is the cost method. It postulates that economic activity involves costs. Costs are then divided into two types: internal and external. Internalized costs include such familiar items as capital, labor, materials, overhead, depreciation, etc., in short, those things that are real costs in an economic activity.

On the other hand, external costs (also known as social costs) include such things as pollution, unsightliness, noise, odors and the like. These costs are not born by the economic activity but are the price that society in general must pay. While this may seem unfair, it can readily be seen that it would be most difficult and indeed almost impossible to come up with an index of such costs. Economists, until most recently, have, therefore, tended to ignore external costs.

Another environmental economics analysis is the "goods" method. It states that any economic activity involves the use of goods. Goods are of three types, economic goods, public goods and free goods. Economic goods include such familiar items as raw materials, equipment and the like. Air and water are the prototypical free goods because they exist in such superabundancy that they are free to everyone.

Thus, no one pays for the air that he breathes or that his car uses or the cooling water that a factory uses. A public good may be described as a formerly free good but which has become subject to competition for its use beyond its free availability and thus comes under public control usually under a licensing system. A good example is the wave bands on the radio spectrum (which were formerly a free good) regulated and allocated by a comprehensive Federal government licensing scheme.

Economic goods have an economic value and are hence controlled by the market mechanism; they can be bought and sold for a price. Public goods and free goods, on the other hand, are outside the market place and do not have an economic value; they do not have a price and cannot be bought or sold. While it is highly unlikely that public or free goods will ever become economic goods (that is to say they will never be bought and sold in the market place) there is a changeover point at which a free good may become a public good. That point is reached when competition for its use sets in, because it is so heavily used/or polluted, that it loses its ability of self renewal.

The changeover point is reached when the free good becomes scarce. It should be noted, however, that the change is not a total break of free to the public but is gradual, resulting in less free and more public.

Antipollution laws, from an environmental economics point of view, are in reality, attempts to internalize pollution costs or to change free goods into public goods. While it may never be possible to place an economic value on social costs or to charge for air or water, the concern for the environment will not diminish.

The push for environmental quality will result in increased costs as more and more external (social) costs are forced to be internalized; and stricter standards for air and water purity decrease their free nature and increase their public character. The impact of this on a hydrogen economy is obvious: hydrogen energy will simply be more costly. An appreciation of environmental economics was ever present throughout the study and is evident in this report.

Energy and the Social Future

A different energy system, particularly if it involves restricting energy, will have a massive impact upon society. As one example, if energy use is restricted but antipoverty programs are continued, there must be a redistribution of wealth by taking from the rich to give to the poor a greater share of a relatively static supply. The steady-state economy may obtain an equilibrium no-growth point, but social dynamics continue to require readjustments within the overall equilibrium. This would present American social and political institutions with unprecedented value and practicality questions.

Even a shift from one form of energy to another (as from coal-electric to nuclear-hydrogen) will affect the relative positions of social vested interests, the economic viability of whole cities, states and regions and certainly occupations and dwelling patterns. Forecasting those changes is difficult, and with present knowledge probably impossible if a high degree of accuracy is required. Battelle Memorial Institute has found that energy forecasts made during the 1960s all lacked adequate consideration of environmental factors and that few dealt with possible constraints on access to foreign supplies of energy (17).

The impact on both economics and petroleum requirements by the NEPA automobile emissions controls has been quite large and was unforeseen. Safety and environmental requirements have combined to constrain development of nuclear power plants to a pace much slower than expected.

Economic Costs

Reference has been made to the threat of shortage in natural resources. Petroleum and many minerals are becoming scarce, thus forcing us to turn to lower quality or higher cost; or to accept higher social costs in greater environmental damage or increased reliance on external supplies. But there is another shortage, very real and as yet largely unappreciated by the public, of economic resources.

In order to provide the energy requirements so freely forecast for the next fifty years, society must generate enormous amounts of capital investment. It will be necessary to develop new supplies through improved techniques, research and development. It will be necessary to construct refineries, tankers, pipelines, railroads, superports, nuclear fuel enrichment and fabrication facilities, etc.

It will also be necessary to pay interest on borrowed capital, to provide environmental protection facilities, and to develop marketing/distribution systems. The price tag for these far-flung investments will be enormous. Chase Manhattan Bank estimates that the noncommunist world will need to find one trillion dollars in capital for the petroleum industry alone in 1970-1985. The National Petroleum Council has estimated that the 1971-1985 capital cost of the United States' energy industry will reach $450 to $550 billion; Chase Manhattan estimates approximately $400 billion, a figure some eight times the capital investment made in the 1955-1970 period. Hottel (18) and others have cited a need for oil and LNG tankers which varies from some 350 tankers of 250,000 tons each up to as many as 1,000 tankers, all very large, by 1985. Capital cost estimates for these tankers and their associated superports vary widely, but $40 billion should be a conservative estimate.

The scale of nuclear energy capital needs is likewise large, and has been getting larger rather than smaller. The fast breeder reactor demonstration plant at Oak Ridge which the AEC, the TVA and private industry have agreed to build by 1980 is estimated to cost $640 million (19). This unit cost may be reduced for later installations, but the capital needs for widespread nuclear development will climb high into the billions.

If the National Petroleum Council's estimate of approximately $500 billion capital investment required by 1985 for development of the United States' energy system is accepted, it is clear that enormous amounts of social resources will be absorbed in this one area. It is not at all certain that the private market has this much capital available.

It is very probable that the nation's energy industry will turn to the federal and perhaps the state governments for assistance, perhaps even as partners as in the case of the fast breeder reactor demonstration plant noted above. It is also within the realm of probability that government interests and private capital shortages may produce further development of government energy organizations on the TVA model.

It is not seen how the hydrogen economy can be more than a marginal aid in this situation; it may perhaps relieve some foreign exchange burdens. Since some conversion costs would be incurred in transition to hydrogen, there may be no net economic gain.

The probability of very large-scale energy development under control of the Federal and/or state governments suggests that there is an opportunity for the introduction of large-scale, integrated hydrogen systems. Predictions of growth in all areas of American life suggest that, since a great deal of added construction, even a number of new cities, will be needed, it will be possible to construct a planned, integrated model city based on hydrogen energy. This option should be considered by Federal energy policy groups. Future growth in Hawaii, for example, might combine a solar, wind, or geothermal power source with hydrogen production to reduce Hawaii's costly dependence on imported hydrocarbons.

Hydrogen and Society

Hydrogen and electricity both require some base power source for their production. Each is flexible, and they are nicely complementary for some purposes (e.g., hydrogen as a storage form for electricity). A major social advantage of hydrogen over hydrocarbons is environmental in that it could be used in any quantity without creating waste management or pollution burdens.

Esthetically, pipeline hydrogen avoids the scenic clutter of overhead electric lines. Perhaps most importantly, hydrogen fuel can be substituted for many petroleum fuels. This relieves the social tension inherent in managing a resource which is growing scarcer, releases remaining petroleum for those chemical uses in which it is vital, and allows greater national self-sufficiency in energy supplies. Rapidly declining natural gas reserves will mean that soon natural gas users must either:

- stop their activity,
- pay much higher prices and accept risks of supply interruption due to dependence on foreign sources, or
- switch to synthetic gas

In some cases, use of electricity under restraints of technical and environmental nature may be possible. Hydrogen, a synthetic fuel, could be produced and transmitted by pipelines already in existence (with slight modifications or additions) which might otherwise lose their function. Saving this enormous capital and energy investment for continued use is a social benefit of great value.

Emissions controls and petroleum scarcity/cost are rapidly developing towards a point where the private, personal automobile may become a luxury for the few. The enormous social overhead invested in the automobile and its associated appurtenances (highways, repair facilities, etc.) should be conserved if possible.

Further, since so many social, life style, and economic patterns are keyed to the private automobile, (over 80% of the United States work force commutes to work by automobile) there would be major social dislocations inherent in significant restrictions on automobile use.

A wholesale switch to a radically different energy form would require a revolution in the auto industry; this alone suggests that such variations as the electric automobile are socially expensive. Electric cars have energy storage restrictions, and are energy-expensive. A synthetic fuel similar to present hydrocarbon fuel, such as hydrogen-produced ammonia, could be more readily assimilated and at lower social costs.

It is not clear that such a minor transition, from gasoline to the functionally similar hydrogen, would be overly expensive when compared to the costs of continued hydrocarbon use. It would be necessary to make major modifications in storage and handling but perhaps minimal changes elsewhere.

Fortune (20) estimates that the difference in on-the-car emission controls between 1975 and 1976 alone will amount to $3 billion. David (21) says that between 1975 and 1985 the environmental-protection costs for automobiles will approximate $100 billion. Such costs do not include the social costs of using irreplaceable petroleum fuels, balance of payments deficits, and United States energy dependence on external supplies. It may be that hydrogen is thus worthwhile as a replacement fuel.

POLITICAL IMPLICATIONS

There are essentially two ways to analyze the political effects of hydrogen as an energy medium: that of considering the political role of energy in general, or that of attempting to isolate particular effects of hydrogen as a unique form of fuel. It seems appropriate to consider, first, certain aspects of the politics of energy and then to ascertain any particular impact of hydrogen as fuel.

Politics and Energy

American use of energy has meshed with several strong values in the American social, political, and economic system to produce a peculiar set of inputs and outputs. The society has made several demands related to energy provision and utilization: that progress must exist in an observable fashion, generally in the form of an ever-rising standard of living measured chiefly in material goods; that the premise of a consumer-controlled, market economy based upon a system of private property and private competitive enterprise must be maintained as a value and as a symbol; and that the public image of the United States as a great nation be preserved.

All of these requirements produce an implicit energy policy: the supply of energy must be adequate at all times, which means that it must constantly grow larger; the price of energy must be kept at a level which does not impede its use to support progress; and energy production should be carried out primarily by private enterprise, although government regulation is allowed as a (regrettable) necessity due to the interstate character of energy supply, natural monopolies, and other factors.

In support of these demands, the United States historically has possessed an abundance of readily available raw materials, a high-grade technology capable of performing the necessary engineering, and an abundance of empty space in which things could be thrown away after use. The economic system has generally performed quite effectively in providing the rising living standards, the growth capital, and the cheap energy required.

The image of the United States as the number one nation has been buttressed by the above factors plus the enormous military power and political-economic influence which an energy-rich society made possible.

However, service to the energy demands and values associated with them has brought a confrontation with issues not previously raised in this country, or, in some senses, in any other country. Use of an energy-rich economic mix and virtually unquestioning identification of growth with progress has resulted in an enormous economic expansion, an energy shortage and the emergence of environmental protection concerns.

The United States adopted a set of petroleum policies which were nominally intended to encourage exploration for and development of domestic petroleum resources. This was intended to create sufficient domestic petroleum production so that, in the event of war, the United States would not be dependent upon external sources for vital fuel supplies.

Extensive development of domestic petroleum fields plus import quotas on foreign oil, however, resulted in what someone has called a "drain America first" policy. But this also resulted in a growing dependence upon Middle East oil. It is one of the costs society bears in return for what was considered a benefit, the operation of an energy-intensive system largely dependent upon the activities of private, profit-oriented businesses to whom low costs and high profits are not only desirable but necessary. A governmental energy supply agency might have followed a policy of drain the Middle East first and restrict domestic supplies. They are not presented here as good versus bad choices, but simply as political choices of the type any society makes.

In similar fashion, America has followed policies which subsidized energy costs in order to encourage economic expansion. One form of subsidy was failure to count environmental costs as part of the cost of production and consumption. Air, water, and other environmental factors were viewed as free goods—an infinite stock. There is no doubt but that these policies have been effective in attaining the overall goal of progress but also helped produce the current energy and environmental problems.

The United States has a rapid and increasing rate of consumption of nonrenewable domestic petroleum and natural gas. It is common, but extremely misleading, to evaluate resource reserves in terms of how many years' supply is represented at current rates of consumption because those rates will increase drastically unless action is taken to stop the growth of consumption. If environmental disruption has created such concern at current rates of consumption, surely that concern will grow as consumption grows.

International Politics

It is probable that the single greatest political impact of a well-developed hydrogen energy system would be international rather than national. Its impact would be more in terms of what it might prevent than what it would cause. Every study surveyed has been in agreement that, for at least the next several decades, the oil resources of the Middle East will play a crucial role in the world's political and economic life.

The Middle East is the only place which can possibly supply the vast oil demand of the great industrial countries in Europe, in Japan and in the United States. Even if hypothetical and optimistic domestic oil discoveries are made in the next decade, there is no short-term way to avoid this dependence on Middle Eastern oil.

The dimensions of that dependence, in national security terms, are staggering. The National Petroleum Council's 1972 projections indicate that by 1985, the United States may import as much as 50% of its total petroleum consumption or, in other words, may import as much in 1985, as was consumed in 1970.

Hottel (18) considers an optimistic case in which the 1985 projected oil deficiency of 16 million barrels per day would require approximately 350 oil tankers of 250,000 tons each (classed as supertankers but smaller than ships being built) to supply that need. This requires handling an average of one giant tanker every two hours, and means an enormous capital investment in both tankers and superports.

It also poses probably the insoluble problem of air and naval defense of many enormous ships (high-value targets) which must follow restricted routes and schedules.

Another international crisis which is growing quite rapidly is the imbalance of international payments represented by oil imports. By 1985, America's balance of payments deficit per year for petroleum alone may range from \$30 billion to as high as \$70 billion. Hottel (18) has cited a study by the Continental Oil Company in which the potential total flow of funds from the United States, Europe and Japan to the eleven members of the Organization of Petroleum Exporting Countries (OPEC) may amount by 1985, to as much as \$500 billion per year. But a balance of payments deficit is more than simply a problem of finding the money with which to pay.

It raises such questions as whether the United States can or should allow Saudi Arabia to purchase, as an example, an American oil company. European and Japanese fears of American purchase of their domestic industries may be replaced by American fears of foreign ownership of essential domestic industries.

Indeed, since all the industrial countries are more or less dependent on the Middle East for oil (Japan perhaps most remarkably so, being without domestic supplies) worldwide balance of payments problems may be forced upon all countries as the Middle-Eastern nations seek foreign investment opportunities. On the other hand, if supplies become restricted, the United States may find itself in uncomfortable competition for energy with its current allies.

In this context, hydrogen presents some extremely attractive possibilities for both international economic balance and national security. It is worth noting that since Russia and China will probably continue to be much more self-sufficient in petroleum than the United States, they will gain relatively as American self-sufficiency declines.

The United States found in Viet Nam that its military operations are extremely fuel-dependent, particularly as distance from this country increases logistic requirements. In fact, defense is already over 50% dependent on foreign oil sources. Vansant (21) has shown that in 1968, the United States military procured approximately 58.5% of its fuel from domestic sources and the remainder directly from foreign sources.

However, since the domestic supply already included approximately 20% foreign supplies, the military's 58.5% included nearly 12% which originated overseas. This means that in 1968, the United States' military was already some 53% dependent on foreign oil. As the fuel import ratio increases in the future so will this depenency ratio.

Civilian economic reasons also exist for minimizing dependence on petroleum imports since a blackmail cut-off of supplies by the OPEC countries, or a wartime interruption, or large-scale sabotage would create severe shortages in the civilian economy.

The National Petroleum Council suggests that by 1978, a maximum of two million barrels per day of oil could be saved, under emergency conditions, if previously planned voluntary and mandatory restrictions could be invoked. Saving a maximum of two million barrels per day is not very comforting when compared to their demand estimates that approximately six million barrels per day would be the minimum imported quantity of oil in 1980. This would leave an absolute deficit of at least four million barrels per day or some 20% of the total oil consumption estimated for that date, even with the recommended controls in effect.

Developing integrated hydrogen energy systems for military use would both reduce United States' import requirements and perhaps provide increased military capabilities. Vansant (22) and Hallet (24) have both studied the potential use of synthetic fuels, such as hydrogen, for military purposes. Since hydrogen-fueled aircraft have already flown several test missions and there seem to be few significant technical barriers to overcome, this avenue should be explored.

As an example of a potential weapons system, consider a nuclear powered aircraft carrier equipped with high-performance hydrogen-powered aircraft. Since the carrier power plant seldom operates at peak power, it could routinely be used to generate hydrogen through electrolysis for on-board storage and use by its complement of aircraft.

The range and time-at-sea capability for the system would be greatly extended since the carrier could constantly replenish the aircraft fuel stores. This should also result in cost savings in terms of fuel depots and perhaps even in the very expensive necessity of maintaining a large fleet of carriers on station and in port.

Additionally, it would result in a balance of payments savings and render the military less dependent on foreign fuels. Similar possibilities for improved capability and economic savings exist in use of hydrogen as a fuel for civil or military aircraft.

Domestic Politics

Domestically, hydrogen offers several advantages. Since a new fuel is essential, political costs are reduced to the degree that the system is technically, economically, and politically compatible with the previous system. Hydrogen can fill nearly every role played by natural gas, and can do so in much the same way.

In other words pipeline companies should welcome rather than resist hydrogen as a fuel; utilities can use hydrogen as a peak-shaving auxiliary, as a boiler fuel substitute, or as a natural gas substitute for eventual residential distribution; and even the energy companies may fit hydrogen into their nuclear and chemical divisions. Workers and unions in energy-related areas should face minimal employment and retraining adjustments.

There is the further political boon that hydrogen, widely cited as an eco-energy, genuinely is an environmentally desirable fuel. Its combustion product is water, which makes it eminently recyclable since its source is also water. The environmental problems associated with hydrogen are nearly all related to the primary source of power to produce the hydrogen, not to hydrogen itself. Since the source is likely to be coal or nuclear power, there is no totally new political or economic factor to be adjusted by the political system.

Politically and economically, hydrogen and electricity are both natural competitors and naturally complementary. Either is convertible into the other; either is environmentally clean (absent consideration of the original power source); and either is quite flexible in application. If nuclear power plants must be sited remote from load centers, and particularly if environmental pressures force electric transmission underground, hydrogen fuel should be a clear favorite over electricity. Since it is technically and economically improbable that hydrogen/electricity is an either/or situation, a satisfactory mix of the two can develop over time with minimal disruption by utilities and energy companies.

REFERENCES

(1) Rosen, B., Dayan, V.H., and Proffit, R.L., "Hydrogen Leak and Fire Detection," NASA SP 5092, 1970.

(2) Gregory, D.P., "A Hydrogen Energy System," American Gas Association and Institute of Gas Technology, August, 1972.

(3) *Business Week,* September 23, 1972, quoting Union Carbide's Product Manager for Hydrogen, Stanford H. Henry.

(4) Utton, A.E., "International Environmental Law and Consultation Mechanisms," 12 Colum. *J. Transnat'l L.* 56, 60 (1973).

(5) Convention on the Territorial Sea and the Contiguous Zone (done April 29, 1958), 15 UST 1606, 516 UNTS 205, in force September 10, 1964.

(6) Convention on the Continental Shelf (done April 29, 1958), 15 UST 471, 499 UNTS 311, in force June 10, 1964.

(7) Convention on the High Seas (done April 29, 1958), 13 UST 2312, 450 UNTS 82, in force September 30, 1962.

(8) Brown, E.G., "International Law and Marine Pollution: Radioactive Waste and 'Other Hazardous Substances'," 11 Nat. Res. J. 221 (1971).

(9) G.A. Res. 2749, par. 11, 25 U.N. GAOR Supp. 28, at 24, U.N. Doc. A/8028 (1970), 64 Dept. State Bull. 155 (1971).

(10) 3 Environment L. Rev. 695 (1972).

(11) Edwards, D.J., "A Proposal for Participating in Natural Resource Development Starting with the High Seas," 11 Nat. Res. J. 636 (1971).

(12) Haight, G.W., "The Seabed and the Ocean Floor," 3 Intl. L. 642 (9169).

(13) Darmstadter, J., *Energy in the World Economy,* Johns Hopkins Press for Resources for the Future, 1971.

(14) Cook, "The Flow of Energy in an Industrial Society," reprint from *Scientific American,* September, 1971.

(15) Caldwell, K., *Environment: A Challenge for Modern Society,* Natural History Press, 1970.

(16) Lueck, D.W., "Energy and the Environment," General Electric Co. TEMPO GE 72 TMP-57, 1972, p. 10.

(17) Pacific Northwest Labs, Battelle Memorial Institute, "A Review and Comparison of Selected U.S. Energy Forecasts," December, 1969 (Prepared for Office of Science and Technology in the Executive Office of the President).

(18) Hottel, H.C., "Challenges in Production of Fossil Fuels," *Chemical Engineering Progress,* June, 1973.

(19) *Professional Engineer,* March, 1973, pp. 22-23.

(20) *Fortune,* June, 1973, p. 121.

(21) David, E.E., "An Energy Policy from the Federal Standpoint," *Chemical Engineering Progress,* June, 1973.

(22) Vansant, C., *Strategic Energy Supply and National Security,* Praeger, New York, 1971, p. 51.

(23) *Houston Post,* July 25, 1973.

(24) Hallet, N.C. et al, *Study, Cost and System Analysis of Liquid Hydrogen,* NASA, Ames Research Center, NASA Report n. 73-226 (1968).

HYDROGEN TECHNOLOGY EXPERTS

A NASA contractor report, N76-10323, was the source of the information in this chapter. For a complete bibliography, see page 283.

The purpose of this chapter is to present, in easy reference form, sources of reliable information relating to the various areas of hydrogen technology. The experts, who are points of contact, could be queried by researchers in similar fields of research and development work for unpublished information and more of the details which have not appeared in published reports, journal articles, or society presentations. In addition, further inquiries could be made regarding active current programs where the published results may not appear for several months. No attempt was made to ascertain that persons listed are actually available for extensive consultation.

It should be understood that a list of this type is dated; people retire or change organizational affiliation, fields of interest change, and programs change emphasis and terminate with time. Thus, the list is subject to updating and revision. Retirees who are still active professionally are listed at their last affiliation unless they indicated a preferred alternate address.

Selection for inclusion was based upon the independent judgments of the expert's peers and his appropriate publications in the literature. This is recognized as being an incomplete listing. Nevertheless, it should provide sufficient contacts to those who seek related information.

HYDROGEN PROPERTIES

Experts on thermodynamic, transport, physical, and phase equilibria properties are listed in this section. The experts selected for this section were limited to the United States. However, it should be pointed out that there are many recognized experts on hydrogen properties in other countries, especially the USSR. Extensive coverage of hydrogen thermophysical properties is given in an ASRDI companion document: Hydrogen Technological Survey—Thermophysical Properties, edited by R.D. McCarty (NASA SP-3089).

Expert Organization	Expertise
AHLERS, Guenter Bell Laboratories Murray Hill, New Jersey	Specific heat measurements on solid hydrogen and ortho-para rate of conversion measurements on solid hydrogen.
DILLER, D.E. Cryogenics Division National Bureau of Standards Boulder, Colorado	Viscosity, index of refraction, specific heat, PVT, and thermal conductivity measurements on hydrogen.
GOODWIN, R.D. Cryogenics Division National Bureau of Standards Boulder, Colorado	PVT properties, equation of state and thermodynamic functions for hydrogen.
GRILLY, E.R. Los Alamos Scientific Laboratory Los Alamos, New Mexico	PVT properties of liquid and solid hydrogen at pressures to 3,500 atm (current); hydrogen, deuterium, and tritium properties, ortho-para conversion of tritium.
HIZA, M.J. Cryogenics Division National Bureau of Standards Boulder, Colorado	Solid-vapor and liquid-vapor phase equilibria of hydrogen mixtures (H_2—CH_4, H_2—C_2H_6, H_2—C_2H_4).
McCARTY, R.D. Cryogenics Division National Bureau of Standards Boulder, Colorado	Thermodynamic transport properties of hydrogen, oxygen, and other cryogens. Editor of ASRDI Hydrogen Technology Survey. Computer programs for thermophysical properties.
MEYER, Horst Department of Physics Duke University Durham, North Carolina	Equation of state, compressibility, thermal expansion, dielectric constant, NMR, and acoustical measurements on solid hydrogen and deuterium. NMR measurements on liquid hydrogen and deuterium.
MILLS, R.L. Los Alamos Scientific Laboratory Los Alamos, New Mexico	High pressure hydrogen PVT measurements to 20 kilobars. Melting curve and volume change properties of hydrogen. Separation of ortho-para states of hydrogen and deuterium.
MULLINS, J.C. Department of Chemical Engineering Clemson University Clemson, North Carolina	Thermodynamic properties of para-hydrogen from 1-22 K.
RODER, H.M. Cryogenics Division National Bureau of Standards Boulder, Colorado	Thermal conductivity measurements, PVT measurements, and calculation of thermodynamic properties for hydrogen. Calculation of thermodynamic properties for oxygen. Edited ASRDI Oxygen Technology Survey—Thermophysical Properties, Vol 1, NASA SP-3071. Edited ASRDI Oxygen Technology Survey—Density and Liquid Level Measurement, Vol 5, NASA SP-3083.

Expert / Organization	Expertise
STEWART, J.W. Physics Department University of Virginia Charlottesville, Virginia	Conducted early experimental measurements on the compressibility of solid hydrogen (measured ΔV vs pressure at 4 K). Also made early measurements of the dielectric constant of hydrogen gas in the range of 20 to 300 K.
STREET, W.B. U.S. Military Academy West Point, New York	Phase equilibria of hydrogen mixtures (He-H_2, D_2-Ne, and H_2-Ne).
WEBER, L.A. Cryogenics Division National Bureau of Standards Boulder, Colorado	PVT properties and equilibrium thermodynamic properties for hydrogen and oxygen.
WOOLLEY, H.W. National Bureau of Standards Guest Worker, Heat Division Equation of State Section Gaithersburg, Maryland	Thermodynamic properties, coexistence curves-application of scaling theory to the critical region, extrapolation of properties to high temperature and high density regions, calculation of ideal gas properties, ortho-para conversion.
ZIEGLER, W.T. School of Chemical Engineering Georgia Institute of Technology Atlanta, Georgia	Thermodynamic properties of para-hydrogen from 2 to 22 K. Phase equilibria studies for binary systems H_2-CF_4 and H_2-$CClF_3$.

HEAT TRANSFER

The experts listed in this section have performed experimental work to determine the heat transfer characteristics of hydrogen in various boiling modes, states, and environments. Some of the individuals also have expertise in the design of hydrogen or helium heat exchangers.

Expert / Organization	Expertise
CLARK, J.A. Department of Mechanical Engineering University of Michigan Ann Arbor, Michigan	Heat transfer to liquid hydrogen at low and high gravity conditions, both single and multiphase fluids. Temperature measurement of liquid and multiphase hydrogen at low and high gravity conditions. Drop tower and centrifuge experiments ($a/g = 10^{-3}$ to 10^5).
DANEY, D.E. Cryogenics Division National Bureau of Standards Boulder, Colorado	Experimental investigation of turbulent natural convection of liquid hydrogen and liquid deuterium within enclosed vessels, with particular application to moderators and targets.
FLEMING, R.B. General Electric Company Research and Development Center Schenectady, New York	Flow distribution and ultrahigh effectiveness of cryogenic heat exchangers. Participated in the development of the perforated plate heat exchanger. Considerable experience with liquid helium heat exchangers.

Expert **Organization**	**Expertise**
HENDRICKS, R.C. NASA—Lewis Research Center Cleveland, Ohio	Experimental and analytical convective heat transfer—two phase, near critical, super critical—and pool boiling (1 to 10 g's). Choke flow, transfer, and applied thermophysical properties programs.
HSU, Y.Y. Reactor Safety Research Division Energy Research and Development Administration Washington, D.C.	Performed experiments to determine the heat transfer characteristics of liquid hydrogen in the sub and super critical regions; all types of boiling were investigated. Also studied critical discharge parameters.
SIMONEAU, R.J. NASA—Lewis Research Center Cleveland, Ohio	Heat transfer work with hydrogen near the thermodynamic critical region (film boiling). Also experimental work with two-phase choked flow of hydrogen.
SINDT, C.F. Cryogenics Division National Bureau of Standards Boulder, Colorado	Experimental work to determine the convective and nucleate-boiling heat transfer coefficients for normal-boiling liquid, triple-point liquid, and slush hydrogen. Comparison of experimental data to classical heat transfer correlations.

HYDROGEN PRODUCTION

There are five large hydrogen liquefaction plants currently operating in the United States. The plants are listed in this section, along with the following information: the company operating the plant, the approximate capacity of the plant, the corporate safety director for hydrogen liquefaction, the plant operations manager and superintendent for each facility.

Also listed are experts on hydrogen liquefaction and purification, the ortho-para conversion of hydrogen and its isotopes, and individuals who have performed various types of experimental work on slush and gel hydrogen. Also listed are individuals who are currently trying to produce metallic hydrogen, people who are studying electrolysis systems for producing hydrogen, and people who are studying various thermochemical processes and cycles to produce hydrogen.

Hydrogen Liquefaction Plants in the United States

AIR PRODUCTS AND CHEMICALS, INCORPORATED
Allentown, Pennsylvania

Leonard Ball
Corporate Safety Director over all operations (air separation and H_2 liquefaction plants).

Hugh Wynn
Manager of all liquid hydrogen plant operations.

Long Beach, California LH_2 Plant
Capacity: ~30 tons/day
Plant Manager: N.S. Schoenherr

New Orleans, Louisiana LH_2 Plant
Capacity: ~30 tons/day
Plant Manager: C.A. Loyd
Cryogenics Superintendent: J.L. Hubbard

UNION CARBIDE CORPORATION, LINDE DIVISION
New York, New York
Hydrogen safety at the two Union Carbide liquefaction plants is a prime responsibility of the respective plant managers.

Ontario, California LH_2 Plant
Capacity: ~30 tons/day
Plant Operations Manager: I.E. Barber
Superintendent of Hydrogen Plant: K.S. Wehr

Ashtabula, Ohio LH_2 Plant
Capacity: currently 7½ tons/day
Plant Manager: V.E. Kraszeski
Superintendent of Production: D.R. Davis

AIRCO INDUSTRIAL GASES
Division of Air Reduction Company, Inc.
Murray Hill, New Jersey

E.F. Szymanski
Division Safety Director

Pedricktown, New Jersey LH_2 Plant
Capacity: ~6 tons/day
Plant Superintendent: W.H. Whitener
Assistant Plant Superintendent: McHough Griffin

Hydrogen Liquefaction and Purification

Expert Organization	Expertise
JOHNSON, J.E. Product Manager of Hydrogen Linde Division of Union Carbide New York, New York	Process design, energy optimization, and operations analysis of hydrogen liquefaction plants.
JOHNSON, V.J., Guest Worker Cryogenics Division National Bureau of Standards Boulder, Colorado	Process design, gas liquefaction, and operation of liquid hydrogen plants. Chief of the gas liquefaction at the National Bureau of Standards. Former director of the NBS Cryogenic Data Center.
STROBRIDGE, T.R. Cryogenics Division National Bureau of Standards Boulder, Colorado	Thermodynamic analysis of liquefaction cycles—design. Plant operating experience.
VANDER AREND, P.C. Cryogenic Consultants, Inc. Allentown, Pennsylvania	Extensive experience in hydrogen liquefaction, process design, and plant operations. Supervised the initial process design and start-up of several hydrogen liquefaction plants.
VOTH, R.O. Cryogenics Division National Bureau of Standards Boulder, Colorado	Thermodynamic analysis of hydrogen liquefaction cycles and design.

Ortho-Para Conversion

Expert / Organization	Expertise
BROWN, L.F. Department of Chemical Engineering University of Colorado Boulder, Colorado	Ortho-para conversion and catalysts to accelerate conversion of hydrogen and its isotopes.
KEELER, R.N. Head, Physics Laboratory Lawrence Livermore Laboratories Livermore, California	Ortho-para conversion, catalysts to accelerate conversion, and isotope separation.
LAPIN, Abraham Air Products and Chemicals, Inc. Allentown, Pennsylvania	Extensive experience in the area of ortho-para conversion of hydrogen; both rate of conversion and catalysts for conversion. Supervised the experimental work on the super catalysts developed at Air Products Company.
MILLS, R.L. Los Alamos Scientific Laboratory Los Alamos, New Mexico	Ortho-para conversion of hydrogen and deuterium. Separation of ortho-para states of hydrogen and deuterium.

Slush Hydrogen

Expert / Organization	Expertise
DANEY, D.E. Cryogenics Division National Bureau of Standards Boulder, Colorado	Experimental work pumping slush hydrogen with an S-IVB liquid hydrogen centrifugal pump; pump performance and cavitation characteristics. Quality determination of slush using four different methods.
HYDE, E.H. NASA—Marshall Space Flight Center Marshall Space Flight Center, Alabama	Fabrication and operation of a 1,500-gallon slush generator with weigh capability. Facility includes a 4-inch vacuum-jacketed transfer line, and a 23,000-gallon storage vessel. Slush production and flow instrumentation. Production of slush hydrogen using the freeze-thaw method in a 1,500-gallon generator.
KELLER, C.W. Staff Engineer Lockheed Missiles and Space Company Sunnyvale, California	Conducted a feasibility study on the application and impact of using slush hydrogen in a liquid hydrogen fueled vehicle. Conducted an experimental program to simulate and verify a vehicle loading process using slush hydrogen.
MANN, D.B. Cryogenics Division National Bureau of Standards Boulder, Colorado	Supervised the experimental work at the National Bureau of Standards to determine production techniques and parameters, storage characteristics, pumping characteristics, and flow characteristics of slush hydrogen. Also supervised the instrumentation program for slush hydrogen.

Expert Organization	Expertise
SINDT, C.F. Cryogenics Division National Bureau of Standards Boulder, Colorado	Experimental work to determine production techniques, flow characteristics, storage and ageing characteristics, and mixing parameters of slush hydrogen. Gelling of slush hydrogen using light hydrocarbons.

Gel Hydrogen

Expert Organization	Expertise
TARPLEY, W.B. Organic Recycling, Incorporated West Chester, Pennsylvania	Gelling of LH_2 using carbon black, cryogenic silica, lithium borohydride, and lithium aluminum hydride.
VANDER WALL, E.M. Chemical Processing Research Laboratory Aerojet Liquid Rocket Company Sacramento, California	Gelling of liquid and slush hydrogen using light hydrocarbons, carbon particles, and silicates.

Metallic Hydrogen

Expert Organization	Expertise
BROWN, G.V. Chief Magnetics and Cryophysics Branch NASA—Lewis Research Center Cleveland, Ohio	Directing in-house and contract work toward fabricating presses to achieve the 3 megabar range in order to produce the transition from molecular to metallic phase.
HAWKE, R.S. Lawrence Livermore Laboratory Livermore, California	Magnetic compression experiments using liquid and solid hydrogen to produce metallic hydrogen.
HOMAN, C.G. Watervliet Arsenal Watervliet, New York	Metallic hydrogen experiments; 500 kilobars to 1 megabar.
ROGERS, F.J. Lawrence Livermore Laboratory Livermore, California	Theoretical studies on metallic hydrogen.
RUOFF, A.L. Cornell University Ithaca, New York	Experimentalist, producing metallic hydrogen by two methods; compressing solid hydrogen to 3 megabars and the electrolysis of lithium hydride.
SPAIN, I.L. Chemical Engineering Department University of Maryland College Park, Maryland	Experimental program to develop ultra-high pressure techniques for compressing hydrogen to pressures in the range of 1 megabar.

Expert **Organization**	**Expertise**
VAN THIEL, M. Lawrence Livermore Laboratory Livermore, California	Shock wave experiments on solid hydrogen up to 1,000 kilobars.

Electrolysis Systems

Expert **Organization**	**Expertise**
KINCAIDE, W.C. Product Manager of Electro-Chemical Systems Teledyne Isotopes, Incorporated Timonium, Maryland	Research and development work on small and large electrolysis systems; experimental work with hydrogen generators with capacities from 1 to 200 std. liters/minute, and large custom plants with capacities up to tons/day.
SRINIVASAH, S. Brookhaven National Laboratory Electrochemistry Research Division Upton, New York	Systems studies of electrolysis plants. The program plans for a small prototype plant of ~200 kilowatts, and finally construction of a demonstration electrolysis plant of up to 26 megawatts capacity at a utility site.

Thermochemical Production of Hydrogen

Expert **Organization**	**Expertise**
FLOWERS, A. Manager of Synthetic Fuels Headquarters, American Gas Association Arlington, Virginia	Manager of a program being conducted by the Institute of Gas Technology to screen thermochemical cycles to determine which are most feasible for large-scale production of hydrogen.
FUNK, J.E. Dean, College of Engineering University of Kentucky Lexington, Kentucky	Theoretical studies and experimental programs investigating the thermo-chemical processes for producing hydrogen. Investigating techniques to evaluate thermo-chemical cycles and determine thermal efficiency. The work also focuses on the process irreversibilities and the need for thermal regeneration.
KRIKORIAN, O.H. Lawrence Livermore Laboratory Livermore, California	Presently conducting experimental work on the feasibility of producing large quantities of hydrogen gas using a thermo-chemical process; selenium is combined with oxides to form selenides, which are then hydrolyzed with acid to produce H_2Se gas, which is then decomposed to form free H_2 gas and selenium.
NORAD, D.L. Chief, Laser and Energy Systems Branch NASA—Lewis Research Center Cleveland, Ohio	Involved in a joint AEC-NASA effort to study production of hydrogen from nuclear heat using both coal conversion and thermo-chemical processes. Also investigating complete hydrogen energy systems including production, transmission, storage, distribution, end use, and scenarios for implementation.

Expert Organization	Expertise
PANGBORN, J.B. Institute of Gas Technology Chicago, Illinois	Presently investigating the theoretical and experimental aspects of thermo-chemical hydrogen production. Parameters include reaction steps, operating conditions, large-scale hydrogen production facility. Parameters include reaction steps, operating conditions, and economy of operation, with a long-term goal of a large-scale hydrogen production facility.
RUSSELL, J.L., Jr. Manager of Special Products General Atomics, Incorporated San Diego, California	Computer search, experimental, and engineering work on thermo-chemical water-splitting cycles.
WENTORF, R.H., Jr. General Electric Company Research and Development Center Schenectady, New York	Exploration of potential processes for economical production of hydrogen from water, using fossil fuel, nuclear, or solar energy, and of the use of hydrogen as an ingredient in closed-loop energy storage or delivery methods.

MATERIALS OF CONSTRUCTION

This section contains a listing of individuals having expertise about the various types of materials used in the construction of hydrogen systems. There has been a vast amount of work on hydrogen embrittlement, fracture mechanics, and insulation; however only a small fraction of the experts representing key organizations are listed in this document. A listing of experts on polymeric composites was also included because of the increased use of these composites in superconducting systems. Since multilayer and powder insulations are used almost exclusively for liquid hydrogen systems, only experts for these types of insulations are listed. Experts on other types of low performance insulation systems were not considered.

Metals

Hydrogen Embrittlement:

Expert Organization	Expertise
BEACHEM, C.D. Naval Research Laboratory Washington, D.C.	Test methods. Microscopic mechanisms, stress corrosion cracking mechanisms, and fracture mechanics analysis.
CHANDLER, W.T. Rocketdyne Division of Rockwell International Canoga Park, California	Hydrogen environment embrittlement. Effect of high-pressure hydrogen on metals and alloys.
GRAY, H.R. Research Metallurgist NASA—Lewis Research Center Cleveland, Ohio	Hydrogen embrittlement of structural materials; iron base alloys, nickel-cobalt base alloys, and titanium base alloys. Hot salt stress corrosion. Responsible for the materials aspect of the NASA—interagency hydrogen energy systems technology study (HEST).

Expert **Organization**	**Expertise**
GROENVELD, T.P. Battelle Memorial Institute Columbus, Ohio	Hydrogen embrittlement of iron and steels, also a few superalloys.
HARRIS, J.A. Pratt and Whitney Aircraft Florida Research and Development Center West Palm Beach, Florida	Investigating the effect of external hydrogen environment embrittlement on the mechanical properties of high nickel alloys and superalloys.
LOGINOW, A.W. U.S. Steel Company Research Laboratories Monroeville, Pennsylvania	Environmental cracking of steels: stress corrosion cracking, cathodic crack propagation, and gaseous hydrogen cracking at pressures to 14,000 psi.
NELSON, H.G. Materials Science Branch Ames Research Center Moffet Field, California	Presently studying the influence of environmental factors on slow crack growth and crack initiation on iron, nickel, titanium, and aluminum base alloys. Attempting to correlate transport kinetics to the environmental influences.
TROIANO, A.R. Case Western Reserve Cleveland, Ohio	Stress corrosion cracking in high strength and austenitic stainless steels, particularly oil and pipeline tubing.

Fracture Mechanics:

Expert **Organization**	**Expertise**
BROWN, W.F., Jr. Chief of Fractures Branch of Materials Processing Laboratory NASA—Lewis Research Center Cleveland, Ohio	Mechanical properties testing of aluminum, annealed titanium, nickel steels, copper alloys, and superalloys.
LANDON, P.R. Lawrence Livermore Laboratories Livermore, California	Specifications for, and welding of, stainless steels. Stability of stainless steels and beryllium. Failure analysis.
REED, R.P. Cryogenics Division National Bureau of Standards Boulder, Colorado	Elastic and elastic-plastic fracture tests and fatigue crack growth rate from 4 to 300 K on structural alloys and welds of cryogenic materials.
WESSEL, E.T. Westinghouse Electric Corporation Research and Development Center Pittsburgh, Pennsylvania	Development and application of advanced fracture mechanics concepts. Currently testing superalloys, superconducting materials, copper, and brazed, soldered, and welded joints at cryogenic temperatures.

Polymeric Composites

Expert Organization	Expertise
HERTZ, Julius Convair Division of General Dynamics Corporation San Diego, California	Fabrication, development, and testing of high modulus advanced polymeric composites; graphite-reinforced epoxies and polyimides.
KASEN, M.B. Cryogenics Division National Bureau of Standards Boulder, Colorado	Mechanical properties testing of high modulus, advanced composites in the temperature regions 4 to 20 K. Current work is with composites useful in superconducting systems.
MORRIS, E.E. Structural Composites Industries Azuza, California	Filamentary reinforced plastic composites, conventional and advanced. Structural fibers are glass, Kevlar 49, graphite and boron with epoxy and polyimide resins. Current interest is high-strength and low-thermal conductivity composites for cryogenic and superconducting systems.

Insulations

Expert Organization	Expertise
BARBER, J.R. Head of Fluid Systems NASA—Lewis Research Center Cleveland, Ohio	Contract manager dealing with the advancement of multilayer insulation (MLI) technology. Research and evaluation of various types of multilayer spacer materials. Measurements of MLI performance with boundary temperatures of 40° to 700°R.
CLAPP, M.B. Chicago Bridge and Iron Company Oak Brook, Illinois	Knowledgeable about insulation requirements and criteria of large cryogenic storage vessels. Responsible for the preliminary design of large cryogenic vessels and liquefaction systems.
CONTE, R.R. Minnesota Valley Engineering Company New Prague, Minnesota	Multilayer insulation of cryogenic vessels to 9,000-gallon capacity. Helium boiloff gas radiation shield cooling techniques. High vacuum technology and fabrication techniques.
DIEKMAN, Don Silbrico Corporation Hodgkins, Illinois	Manufacture and installation of perlite in large field-erected cryogenic storage vessels. Job site installation to specifications on moisture, particle size, and density.
HYDE, E.H. NASA—Marshall Space Flight Center Marshall Space Flight Center, Alabama	Superfloc development and testing. Project monitor for NASA—Marshall multilayer and polyphenyl oxide (PPO) foam programs.
KROPSCHOT, R.H. Chief, Cryogenics Division National Bureau of Standards Boulder, Colorado	Research and development work on all types of cryogenic insulations; powder, opacified powders, foams, microspheres, and multilayer insulations. Review articles on cryogenic insulation systems.

Expert Organization	Expertise
MATSCH, L.C. Linde Division of Union Carbide Corporation Tonowanda, New York	Research, development, and testing of all types of multilayer insulations. Development and testing of opacified powders. Ortho-para conversion and catalysts for conversion. Hydrogen getters and sorbents.
PARMLEY, R.T. Lockheed Missiles and Space Company Palo Alto, California	Design, analysis and testing of multilayer insulations and systems. Research and development of microsphere insulation. Design and environmental testing of 46-inch diameter flight weight spherical LH_2 dewar for space tug using microsphere insulation.
RUCCIA, F.E. Arthur D. Little Company Cambridge, Massachusetts	Development and testing of opacified powders. Basic studies of high performance insulation systems. Feasibility studies of insulation systems for long-term storage of LH_2 in space. Design, development, and fabrication of a double-guarded, flat-plate calorimeter for study of multilayer insulation parameters. Design of high performance insulation systems for the Apollo program instrumentation and sensor systems.
TATRO, R.E. Convair Division of General Dynamics Corporation San Diego, California	Research, design, development, and production of superfloc multilayer insulation. Analysis, design, predicted performance, and environmental testing of 87-inch diameter LH_2 flight weight vessel with superfloc insulation.

ENGINEERING SYSTEMS AND SUBSYSTEMS

Individuals who have expertise in some of the more important subsystems of a hydrogen facility are listed in this section. Experts in the areas of large storage vessels, vacuum-jacketed transfer lines, heat exchangers, liquid hydrogen pumps, refrigerators, nonlubricated compressors, and hydrogen disposal systems are listed.

A subsection on nonlubricated compressors was included because these types of compressors are often used in hydrogen systems, and a vast amount of operating experience has been accumulated in using these types of compressors at the helium plants in the U.S. Experts for other subsystems such as vaporizers, purifiers, small storage vessels, portable Dewars, high-pressure gas banks, cryogenic bearings, hydrogen turbines, and the other types of compressors will possibly be added to a future, more comprehensive register.

Large Storage Tanks and Vessels (Field-Erected)

Expert Organization	Expertise
BODLEY, R.W. Graver Tank and Manufacturing Company East Chicago, Indiana	Technical information on the design, fabrication, and erection of large cryogenic vessels.

Expert Organization	Expertise
CLAPP, M.B. Chicago Bridge and Iron Oak Brook, Illinois	Leader of an engineering group responsible for estimating costs for inquiries on cryogenic vessels and liquefaction systems, and the preliminary design of the above. Knowledgeable about all technical aspects of cryogenic systems and structures.
MURPHY, J.C. Manager of Engineering Systems Group Pittsburgh—Des Moines Steel Company Neville Island Pittsburgh, Pennsylvania	Design, fabrication, and erection of large cryogenic vessels (LH_2, LO_2, LN_2, and LNG).

Vacuum-Jacketed Transfer Lines

Expert Organization	Expertise
BOWERS, W.M. Liquid Metal Engineering Center Atomics International Division of Rockwell International Canoga Park, California	Experience with liquid hydrogen fueled rocket engine test facilities at Rocketdyne. Responsible for maintenance and vacuum integrity of the transfer and run lines. A current member of the ANSI-B31.10 Code for Cryogenic Piping Systems Committee.
EDESKUTY, F.J. Los Alamos Scientific Laboratory Los Alamos, New Mexico	Considerable experience with vacuum-jacketed transfer lines ranging in size from 2 to 16 inches in diameter. Experience with Invar welding techniques, annular end seals, expansion joints, multilayer insulation, and sorbents.
HOWARD, F.S. NASA—Kennedy Space Center Kennedy Space Center, Florida	Responsible for maintenance and integrity of the 10-inch vacuum-jacketed LH_2 transfer lines at launch complex 39A and 39B. Initiated design of an annular bellows seal modification at spool piece joints. Supervised rebuilding of the V.J. transfer lines for the Skylab mobile launch platform.
MARTINDALE, D.L. General Manager, Cryolab Division of CTI, Incorporated Los Osos, California	Extensive experience in cryogenic systems. As manager of cryogenic systems for Ametek/Straza, was responsible for supervision of design, development and manufacture of cryogenic systems, components and related products. General manager and project engineer for the cryogenic ducting system with accessories for the Saturn/Apollo project.
MOORE, A.L. Director of Engineering Ametek/Straza Corporation El Cajon, California	Experience in the design, manufacture, and test of vacuum-jacketed transfer lines. Extensive experience with transfer line components: ducting, spool joints, annular end seals, expansion joints, sorbents, and insulation.

Expert Organization	Expertise
MORDHORST, G.D. Cryogenic Engineering Company Division of CTI, Incorporated Denver, Colorado	Active in research, design, and fabrication of vacuum-insulated transfer lines and vacuum-insulated vessels.
SPENCE, R.E. Manager of Standard Products CVI, Incorporated Columbus, Ohio	Experience in the fabrication of vacuum-insulated transfer lines for cryogenic systems. Also valves, field joints and couplings. A member of the ANSI-B31.10 Code for Cryogenic Piping Systems Committee.

Heat Exchangers

Expert Organization	Expertise
COWANS, K.W. Kinergetics, Incorporated Tarzana, California	Super-high-efficiency, countercurrent heat exchangers for all cryogenics applications.
FLEMING, R.B. General Electric Company Schenectady, New York	Flow distribution and ultra-high effectiveness of heat exchangers. Helped develop the perforated-plate heat exchanger. Liquid helium heat exchangers.
JACOBS, R.B. R.B. Jacobs Associates, Inc. Boulder, Colorado	Active in heat transfer and heat exchanger design for helium and hydrogen systems; liquefiers, refrigerators, and related equipment.

Liquid Hydrogen Pumps

Expert Organization	Expertise
BROOKS, W.S. Manager SSME Turbomachinery Engineering Rocketdyne Division of Rockwell International Canoga Park, California	Design and development of turbomachinery for pumping LH_2 and LO_2 in rocket engines.
CAINE, G.H. Sunstrand Corporation Rockford, Illinois	Research and development on induction motors (1 watt to 50 HP) for cryogenic pump applications. Cryogenic pump design and performance studies for LO_2, LH_2, and LHe. Low NPSH LH_2 boost pump development.
CONNELLY, R.E. NASA—Lewis Research Center Cleveland, Ohio	LH_2 and LO_2 turbopumps. Project manager of Mark-9 LH_2 turbopump for Rocketdyne. Inducer development work on the F-1 LO_2 pump. Basic research on inducer cavitation of LH_2 turbopumps. Project manager of the Nuclear Engine for Rocket Vehicle Application (NERVA) LH_2 turbopump.

Expert Organization	Expertise
FURST, R.B. Rocketdyne Division of Rockwell International Canoga Park, California	Fluid mechanics design of pumps. Hydrodynamic development of F-1 and J-2 engines for the Saturn program. Authored NASA Monograph on centrifugal pumps.
GINSBURG, Ambrose Chief, Fluid Systems Components Division NASA—Lewis Research Center Cleveland, Ohio	Responsible for the management and supervision of research and development work on liquid hydrogen turbopumps and components.
GROSS, L.A. NASA—Marshall Space Flight Center Marshall Space Flight Center, Alabama	Development, performance and testing of liquid hydrogen pumps.
HAGER, J.A. Manager SSME Fuel Turbomachinery Rocketdyne Division of Rockwell International Canoga Park, California	Design and development of liquid hydrogen pumping for turbomachinery.
HARTMAN, M.J. Chief, Fan and Compressor Branch NASA—Lewis Research Center Cleveland, Ohio	Conducted design and performance studies on axial and centrifugal flow liquid hydrogen pumps. Developed a rationale design approach to predict cavitation performance of hydrogen pump inducers.
HUPPERT, Murl Aerojet Liquid Rocket Company Sacramento, California	Authored NASA Axial Flow Pump Monograph. Research and development of cryogenic pumps. Pump inducer design, liquid hydrogen pump design for J-2 engine and the NERVA pump.
MOORE, R.D. Engine Research NASA—Lewis Research Center Cleveland, Ohio	Performed experimental work on cavitation parameters of pumps using liquid hydrogen.
ROTHE, Kurt Manager of Rotating Machinery Rocketdyne Division of Rockwell International Canoga Park, California	Extensive experience in turbomachinery field. Manager of turbomachinery for Rocketdyne work on the Apollo and Space Shuttle programs. Supervised selection of the turbomachinery for the space shuttle main engine (SSME). Cavitation studies with LH_2. Pump work with LH_2, LO_2, RP-1, and F_2.
SPRATLEY, M.L. SSME Development Engineer Rocketdyne Division of Rockwell International Canoga Park, California	Design and development of liquid hydrogen pumping turbomachinery.
STINSON, H.P. NASA—Marshall Space Flight Center Marshall Space Flight Center, Alabama	Performance, cavitation, testing and development of liquid hydrogen pumps.

Expert Organization	Expertise
WISLICENUS, G.F. 4641 E. Coronado Drive Tucson, Arizona	Extensive experience with turbomachinery of all types; cavitation and performance characteristics of liquid hydrogen pumps. Former consultant to Rocketdyne.

Refrigerators

Expert Organization	Expertise
COLLINS, S.C. Naval Research Laboratory Washington, D.C.	Research and development of small helium refrigerators. Development of small compressors for refrigerators. Research and development work on LO_2 generators. Small hydrogen liquefiers and expansion engines.
STROBRIDGE, T.R. Cryogenics Division National Bureau of Standards Boulder, Colorado	Cryogenic refrigerators of all sizes (1 watt to 10 kilowatts). Studies of efficiency and performance parameters. Field experience.
STUART, B.W. Cryogenic Technology, Incorporated Waltham, Massachusetts	Research and development of small and large cryogenic refrigeration systems (2 to 200 watts). Hydrogen recondensing and liquefaction refrigerators. Closed cycle refrigeration systems for bubble chambers and accelerators.
VANDER AREND, P.C. Cryogenics Consultants, Incorporated Allentown, Pennsylvania	Extensive experience in the field of cryogenic refrigerator design. Designed the hydrogen refrigerators for several bubble chambers.

Compressors (Nonlubricated)

Expert Organization	Expertise
DAVIS, Dick, Plant Supt. CHERNOHAN, Glen, Operations Supt. Cities Service Jayhawk Helium Plant Satanta, Kansas	Both diaphragm and dry piston type compressors in use at this facility. Information on maintenance schedules, field experience, diaphragm life, and PTFE piston ring replacement schedule available.
FARRELL, Paul, Plant Supt. JONES, L.B., Maintenance Supvr. Phillips-Greenwood Helium Plant Elkhart, Kansas	Both diaphragm and dry piston type compressors in use at this facility. Operating experience, maintenance schedules, dry piston ring replacement schedules.
PESISCA, Al, Plant Supt. LYNCH, Tom, Maintenance Mgr. Kansas Refined Helium Company Otis, Kansas	Both diaphragm type and dry piston compressors in use at this facility. Information on dry piston ring life, maintenance schedules, typical piston ring life, and valve and diaphragm life is available.
SCHEGEL, Carl, Plant Supt. GAMBLIN, Gurnell, Ass't. Plant Supt. U.S. Government Helium Plant Keyes, Oklahoma	Diaphragm type compressors in use at this facility, vast field experience, maintenance schedules, diaphragm change intervals.

Hydrogen Disposal Systems

Expert Organization	Expertise
GRUMER, Joseph Bureau of Mines Pittsburgh, Pennsylvania	Performed early evaluation work and experimental tests on hydrogen flare stacks and burn ponds. The main concern was the safety aspects of each disposal system. Parameters investigated were completeness of burning, stability of the flames, and flame-back characteristics of the flare stacks.
HOWARD, F.S. NASA—Kennedy Space Center Kennedy Space Center, Florida	LH_2 burn pond operations at Kennedy Space Center. Siphoning problems, safety, and optimum design criteria.
LAPIN, Abraham Air Products and Chemicals, Inc. Allentown, Pennsylvania	Supervised programs to evaluate and conduct experimental testing with hydrogen flare stacks.

TRANSPORTATION SYSTEMS FOR HYDROGEN

Individuals listed in this section are deeply involved in the safe transportation and distribution of liquid hydrogen over highways, railways, waterways, and within NASA installations. These experts are knowledgeable about transport trailer design, department of transportation (DoT) requirements, industrial safety regulations, emergency procedures, accident/incident investigation and analysis, and NASA safety regulations and recommended procedures. Experts from industry, NASA, the Federal regulatory agencies, and several advisory agencies are included.

Commercial Operations

Expert Organization	Expertise
BURKE, Stephen Manager of Chemicals Distribution Air Products and Chemicals, Inc. Allentown, Pennsylvania	In charge of truck-trailer vessel design and any design modifications to the trailers. Knowledgeable of all Department of Transportation (DoT) requirements and special permits to transport LH_2 over the highways.
CAVANNA, C.E. Manager of Engineering LOX Equipment Company Livermore, California	Manufacturers of truck-trailers and rail cars for cryogenic fluids. Extensive design and fabrication experience. Technical information on design, performance, and fabrication of mobile vessels.
HEIBERGER, F.M. Safety Affairs Department Linde Division of Union Carbide Corp. Tarrytown, New York	Corporate safety regulations, emergency procedures, and DoT requirements for truck-trailer and rail transport of liquid hydrogen.
MATTHEWS, R.L. General Manager of Distribution Airco Industrial Gases Murray Hill, New Jersey	Distribution of liquid hydrogen by truck transport, driver training, and DoT requirements and regulations.

Expert Organization	Expertise
TRAMMELL, Byron Manager, Truck Fleet Operations Air Products and Chemicals, Inc. Allentown, Pennsylvania	Supervises truck transport of LH_2 all over the U.S. In charge of purchasing and maintaining all truck-transport equipment, driver qualifications and training, special permits for LH_2 transport, Federal regulations on hours of service limitations, and emergency procedures.

NASA Operations

Expert Organization	Expertise
BAIN, A.L. Systems Engineer NASA—Kennedy Space Center Kennedy Space Center, Florida	Responsible for the design, operation, and maintenance of transportation equipment and hardware at KSC. Responsible for the safe distribution and storage of all propellants and for compliance with DoT regulations, state requirements, and motor carrier safety regulations.
BALL, J.G. Chief, Safety Office NASA—Pasadena Pasadena, California	Responsible for the safe distribution of NASA liquid hydrogen west of the Mississippi. Actively working on hazards analysis and systems analysis and systems analysis of liquid hydrogen; transfer, transport, and storage systems.
HARWOOD, W.R. Manager of Propellants and Pressurants NASA—Headquarters Washington, D.C.	Responsibility for insuring that all NASA facilities have the propellants required; establishes need for and initiates the procurement process. The availability and safe transportation of liquid hydrogen is a major responsibility.
ORDIN, P.M. Program Manager, NASA Aerospace Safety Research and Data Institute NASA—Lewis Research Center Cleveland, Ohio	Knowledgeable in the procedures, guidelines and regulations for liquid hydrogen transportation, distribution and transfer at NASA and NASA contractor facilities. Prepared hydrogen transportation summary including safety program guidelines to ensure proper driver and handling training and safety procedures.
SHAW, R.C. Chief, Cryogenic and Pressure Section NASA—Marshall Space Flight Center Marshall Space Flight Center, Alabama	Responsible for the transport, distribution and storage of cryogens and high-pressure gases at MSFC. Develops safety criteria and specifications for all the propellant transport trailers used at MSFC.

Regulatory Agencies

Department of Transportation

Expert Organization	Expertise
BLACK, W.F. Chief, Hazardous Materials Branch Office of Safety Federal Railroad Administration Washington, D.C.	Knowledgeable about DoT requirements and regulations concerning the transport of liquid hydrogen by rail. Has primary responsibility for the evaluation of hazardous material container design for rail transport.

Expert Organization	Expertise
BOHLMAN, Michael, Lt. Chief of Packaged Cargo Branch of the Cargo and Hazardous Material Division of the U.S. Coast Guard Washington, D.C.	Knowledgeable about shipping LH_2 by waterway transport or barge. No existing regulations on transport of LH_2. Special permits for transport of LH_2 are issued on an individual basis. Technical review of application for LH_2 transport is performed in this office.
MALLEN, Arthur Chief of the Mechanical Engineering Branch of the Office of Hazardous Materials Department of Transportation Washington, D.C.	Knowledgeable about DoT requirements and regulations concerning the transportation of liquid hydrogen by highway and air. Has primary responsibility for evaluation of the hazardous material container design for highway and air transport. Has secondary responsibility for evaluation of rail containers (tank cars for LH_2) for hazardous materials.
RUSSEAU, Gordon Chief of the Office of Hazardous Materials Department of Transportation Washington, D.C.	Knowledgeable about DoT regulation for the transport of LH_2.

Advisory Agencies

Expert Organization	Expertise
GRAZIANO, R.M. Director, Bureau of Explosives Association of American Railroads Washington, D.C.	Twelve-member committee (composed of six railroad representatives and six industry representatives); approves the design of rail tank cars for LH_2. The committee has no regulatory power but does exercise veto power. Primary function of the Bureau is emergency responsibility when hazardous materials are derailed or present a dangerous situation. Influential in exercising judgment on what may be transported by rail.
OLSEN, E.A. Compressed Gas Association (CGA) New York, New York	Provides staff assistance to the CGA Cryogenic and Low Temperature Committee. The duty of the committee is to develop appropriate recommendations upon matters concerning safety in the transportation, storage, handling, and use of gases or fluids at temperatures below -20°F.

INSTRUMENTATION FOR HYDROGEN SYSTEMS

Experts listed in this section have knowledge and experience in measuring temperature, pressure, flow, mass, and liquid level in liquid hydrogen systems. A few of these experts are employed in calibration laboratories for the above instrumentation. A list of experts on combustible gas detectors is also included.

Temperature Measuring Instruments

Expert Organization	Expertise
COLLIER, R.S. Cryogenics Division National Bureau of Standards Boulder, Colorado	Carbon thin-film state and temperature sensors in the cryogenic temperature range.
FURUKAWA, G.T. Manager of the Platinum Resistance Thermometry Laboratory National Bureau of Standards Gaithersburg, Maryland	Provides platinum resistance thermometer calibration service to industry. Sixteen calibration points available between 12 and 90 K. Coauthor of National Bureau of Standards Monograph 126, Platinum Resistance Thermometry, April, 1973.
POWELL, R.L. Cryogenics Division National Bureau of Standards Boulder, Colorado	Thermocouples, resistance thermometers, and vapor pressure bulbs in the cryogenic temperature range. Reviewed most types of temperature sensors and methodology.
RUBIN, L.G. Francis Bitter National Magnet Lab. Massachusetts Institute of Technology Cambridge, Massachusetts	All types of cryogenic thermometry in low and high magnetic fields (to 200 kilo Gauss). Thermocouples, platinum, germanium, carbon, and thermistor resistance thermometers. Hydrogen vapor pressure bulbs and silicon diode thermometers. Review articles on low-temperature thermometry and methodology.
SCHOOLEY, J.F. Chief, Temperature Section National Bureau of Standards Gaithersburg, Maryland	Cryogenic thermometry of all types. Calibration of thermocouples to 1500°C. Development of superconducting thermometric fixed-point devices for the temperature range from 0.5 to 7 K. Calibration of thermocouples, resistance thermometers, and mercury in glass thermometers for national standards laboratories, industry, and hospitals.
SINCLAIR, D.H. NASA—Lewis Research Center Cleveland, Ohio	Aerospace type of temperature instrumentation. Research, testing, and evaluation of carbon, germanium, and mainly platinum-resistance thermometers. Calibration techniques and stability characteristics of carbon thermometers from 4 to 76 K. Extensive summary and review of platinum-resistance thermometers. Development of resistance thermometer methodology.
SPARKS, L.L. Cryogenics Division National Bureau of Standards Boulder, Colorado	Experimental work to establish national standards for thermocouples in the temperature range 4 to 280 K. Types E, K, T, and KP vs Au Fe thermocouples.

Pressure Measuring Instruments

Expert / Organization	Expertise
HARMON, H.S. Lead Engineer of Pressure and Thrust NASA—Marshall Space Flight Center Marshall Space Flight Center, Alabama	Aerospace pressure instrumentation. Pressure instrumentation of the RL-10, J-2, and the SSME. Hydrogen embrittlement of transducer diaphragms.
HAYAKAWA, K.K. Space Division of Rockwell International Downey, California	Extensive experience with pressure transducers, mainly the strain gauge types. Currently working on pressure instrumentation associated with POGO phenomenon (oscillatory behavior of a rocket vehicle resulting from the coupling between the propellant system and the structural vibrational modes of the vehicle). Author of cryogenic pressure measurement section of NASA-CR 120226, Measurement Component Technology.
HOWARD, J.L. Boeing Aerospace Company Seattle, Washington	Director of the Instrument Society of America, Cryogenics Instrumentation Division. Knowledgeable of work being conducted in most areas of cryogenic instrumentation.
INSKEEP, J.Z. California Institute of Technology Jet Propulsion Laboratory Pasadena, California	Extensive experience with all types of pressure transducers in both aerospace and ground installations. Dynamic response of transducer systems and support electrical equipment.
LEDERER, P.S. Assistant Chief for Transducers National Bureau of Standards Gaithersburg, Maryland	Development of reliable and durable calibration and evaluation techniques for pressure transducers. Emphasis is on dynamic calibration. Vibration temperature cycling, and radiation effects on the repeatability, reliability, and durability of transducer calibrations. POGO transducer evaluation.
STOKES, R.W. Los Alamos Scientific Laboratory Los Alamos, New Mexico	Experience with using quartz piezo-electric type transducers in liquid nitrogen. Dynamic response only. POGO studies.
VOTH, R.O. Cryogenics Division National Bureau of Standards Boulder, Colorado	Overpressure and rupture characteristics of dial gauge bourdon tubes using hydrogen gas, and subsequent case pressure relief characteristics; confinement of fragments and loose parts. (High-pressure dial gauges).

Flow Measuring Instruments

Expert / Organization	Expertise
BRENNAN, J.A. Cryogenics Division National Bureau of Standards Boulder, Colorado	Flow calibration facilities for cryogenic flowmeters. Calibration of flowmeters using LN_2 and LNG.

Expert Organization	Expertise
GOLDSTEIN, J.H. NASA—Marshall Space Flight Center Marshall Space Flight Center, Alabama	Research and development with liquid hydrogen flowmeters for aerospace applications; turbine, orifice, head, and venturi types.
HOBART, H.F. Instrument Development and Applications Office NASA—Lewis Research Center Cleveland, Ohio	Research and development work on turbine-type flowmeters and flowmeter bearings. Liquid hydrogen flow calibration facility with accuracy within 1%. Correlation studies of water and LH_2 flow calibrations. Calibration of LH_2 flowmeters up to 2 pounds/second.
JACOBS, R.B. R.B. Jacobs Associates, Inc. Boulder, Colorado	Cryogenic Flowmeters: Designed and developed a vibrating transducer mass flowmeter for liquid hydrogen.
KLEA, J.A. Rocketdyne Division of Rockwell International Canoga Park, California	Research, development, and manufacture of the LH_2 and LO_2 flowmeters for the J-2 and F-1 engines, and the space shuttle main engine (SSME). Turbine-type flowmeters.
UDELL, D.R. Space Division of Rockwell International Downey, California	Experience with cryogenic flow measurement. Author of the Cryogenic Flow Measurement section of NASA-CR 120 227, Measurement Component Technology.
WHEELOCK, H.R. Wyle Laboratories Norco, California	Calibration and testing of flowmeters with LO_2 and LH_2 for the X-15, Titan, Centaur, and Saturn vehicles. Calibration and testing of a dynamic response, ultrasonic-type flowmeter for the shuttle program.

Liquid Level and Quantity Measuring Instruments

Expert Organization	Expertise
COLLIER. R.S. Cryogenics Division National Bureau of Standards Boulder, Colorado	Carbon thin-film state sensors for point-type liquid-level sensing. Radio Frequency (RF) resonance mass tank gauging.
CRUZ, J.E. Cryogenics Division National Bureau of Standards Boulder, Colorado	Time Domain Reflectometer (TDR) continuous liquid-level sensors.
ELLERBRUCH, D.A. Electromagnetics Division National Bureau of Standards Boulder, Colorado	RF frequency technique of tank mass gauging using LH_2, LO_2, LN_2, and LNG. Subcritical and supercritical states, and also zero gravity conditions. Spherical vessels up to 50 feet diameter.
HAMLET, J.F. NASA—Marshall Space Flight Center Marshall Space Flight Center, Alabama	Liquid level sensors of all types, especially capacitance-type bullseye point sensors and capacitance-type continuous sensors.

Combustible Gas Detectors

Expert Organization	Expertise
DAS, K.B. Senior Specialist Engineer Materials and Processes Group Boeing Aerospace Company Seattle, Washington	Responsible for the development of an ultrasensitive hydrogen detection system and the associated methodology. System may be used for leak detection, combustible gas detection, and detection of hydrogen gas within metals. System described in ASTM-STP 543.
DELAUNE, S.D. Supervisor, Marketing Group Mine Safety Appliances Company Pittsburgh, Pennsylvania	Technical information on combustible gas detectors and alarms.
MACINTYRE, J.R. 231 Queensbury Drive Southwest Huntsville, Alabama	Retired systems engineer for General Electric's Space Division. Developed a selective hydrogen detection system using an activated thin-film sensor. The sensor has very large resistance changes for small amounts of hydrogen (0.1 to 4%) in air. The detection system was developed for continuous monitoring applications but could also be used as a hydrogen leak detection instrument.
PEARSE, J.N. Vice President of Product Design and Research Appleton Electric Company Chicago, Illinois	Technical information on combustible gas detectors and alarms.
PHETTEPLACE, C.J. Bacharach Instrument Company Mountain View, California	Technical information on combustible gas detectors and alarms.

HYDROGEN APPLICATIONS

This section lists experts who are working in various areas where hydrogen is used as a fuel or an energy carrier. Listed are experts on rocket propulsion who have done research and development work on various rocket engines and systems during the space program, from its inception up to and including work on the advanced space shuttle. Included are experts who were involved in the early experimental tests using hydrogen for fuel in modified conventional aircraft as well as the individuals pursuing the research and development work on the present hydrogen-fueled supersonic and hypersonic research aircraft.

Individuals who are doing basic research work with metal hydrides and the use of metal hydrides in hydrogen energy storage systems are listed. Also listed are experts on hydrogen-air and hydrogen-oxygen fuel cells and their application in future energy systems. Another category lists experts who are investigating the performance and emission parameters in using hydrogen for fuel in internal combustion engines. The last category is a listing of experts who are currently doing economic and technical feasibility studies on total hydrogen energy systems.

Rocket Propulsion (LH_2-LO_2)

Expert Organization	Expertise
DOMOKUS, S.J. Vice President of Advanced Programs Rocketdyne Division of Rockwell International Canoga Park, California	Involved in the development of rocket systems for many years. Deeply involved in the reaction control systems for the Gemini, Apollo, Navajo, and Saturn 1-B vehicles.
DOUGLASS, H.W. Chief Chemical Energy Division NASA—Lewis Research Center Cleveland, Ohio	Participated in the development of the LH_2-LO_2 technology pertinent to the Space Tug and the Advanced Space Shuttle propulsion systems. Basic research in combustion technology in low-gravity environment; flame behavior, propagation, extinguishment, and fire safety in space.
EIDSON, B.L. General Dynamics Corporation San Diego, California	Worked on the Centaur program since its inception. Feed system design, tankage and outlet design, boost pump design, and integration tests between the fuel system and the RL-10 engine. Main propulsion attitude control systems for the shuttle interim upper stage and the Space Tug.
EK, M.C. Vice President of Engineering Rocketdyne Division of Rockwell International Canoga Park, California	Vast experience in the design, development, and testing of all types of liquid rocket engines.
GOETZ, O.K. NASA—Marshall Space Flight Center Marshall Space Flight Center, Alabama	Systems level work and in-house testing of individual engine components. Presently associated with the space shuttle main engine.
GREGORY, J.W. Chief, Rocket Systems Branch NASA—Lewis Research Center Cleveland, Ohio	Developing a high performance hydrogen-oxygen engine (20,000 pounds thrust) for use on the Space Tug. R&D work being conducted on various new components; LH_2 pump, LO_2 pump, thrust chamber, injector and preburner.
HENNINGS, Glen Chief Power Experiments Branch NASA—Lewis Research Center Cleveland, Ohio	Extensive experience in propulsion systems as chief of the Rocket Systems Division at the Plumbrook Facility of the Lewis Research Center; components work, data acquisition, systems operations, testing of complete engine systems. Fuel systems testing on rockets and turbojets.
LOMBARDO, J.A. Chief, Propulsion Division NASA—Marshall Space Flight Center Marshall Space Flight Center, Alabama	Early jet engine work with Pratt and Whitney Company. Problems, investigations, and analysis work with the RL-10, J-2, F-1, H-1, and the space shuttle main engine (SSME). Heavily involved with H_2-O_2 engine technology.

Expert Organization	Expertise
MULREADY, R.C. Pratt & Whitney Division of United Aircraft Corporation East Hartford, Connecticut	Research and development work on an LH_2 fueled J-57 aircraft engine, a 304 demonstration engine, and the RL-10 rocket engine.
REUEL, N.C. Vice President and Program Manager of SSME Rocketdyne Division of Rockwell International Canoga Park, California	Extensive experience in rocket engine development. Program manager for development of the J-2 engine, and has been associated with every other rocket engine development program at Rocketdyne.
RODGERS, R.N. NASA—Marshall Space Flight Center Marshall Space Flight Center, Alabama	Worked with various aspects of the J-2 engine development program, and currently the SSME development program system tests. Responsibility for integration of engine into vehicle launch procedures.
STOFAN, A.J. Director of Launch Vehicles NASA—Lewis Research Center Cleveland, Ohio	Responsible for the management of the Atlas-Centaur and the Titan-Centaur launch vehicle programs.
STREETMAN, J.W. General Dynamics Corporation San Diego, California	LH_2 and LO_2 feed system design and engine inlet requirements for the Centaur and Atlas vehicles. Integration of the feed system to the engine requirements.

Hydrogen Fueled Aircraft

Expert Organization	Expertise
BREWER, G.D. Advanced Development Projects Lockheed California Company Burbank, California	Recently completed a study for NASA—Ames Research Center on the feasibility of using liquid hydrogen as a fuel for a supersonic transport, and a similar study for NASA—Langley Research Center for subsonic passenger and cargo type of aircraft.
HENRY, J.R. NASA—Langley Research Center Hampton, Virginia	Presently developing design technology to integrate a hydrogen-fueled scram jet engine with the air frame of the vehicle. Mach range of 3 to 11. Extensive experience with jet engine research.
MACKLEY, E.A. Program Manager, Hypersonic Research Engine (HRE) NASA—Langley Research Center Hampton, Virginia	Research and development work on hypersonic propulsion systems and engines using hydrogen for fuel. Simulated structural and engine testing in the NASA—LeRC Plumbrook Hypersonic Tunnel Facility at Mach 5, 6 and 7. Dual mode ram jet engines.

Expert / Organization	Expertise
ORDIN, P.M. Program Manager, NASA Aerospace Safety Research and Data Institute NASA—Lewis Research Center Cleveland, Ohio	Research and development work resulting in the successful flight testing of a workable liquid hydrogen fuel system on a B-57 airplane.
RICH, B.R. Head, Advanced Development Projects Lockheed California Company Burbank, California	Early R&D work on the CL-400 supersonic plane which used hydrogen fuel in a Pratt & Whitney gas turbine engine. Paper studies for hydrogen-fueled supersonic aircraft for NASA—Ames Research Center and subsonic aircraft for NASA—Langley Research Center.
RUBERT, K.F. Department of Mechanical and Aerospace Engineering University of Tennessee Space Institute Tullahoma, Tennessee	Formerly, project manager at NASA—Langley for the research and development work on the hypersonic research engine (HRE). The HRE is a Mach 4 to Mach 8 hydrogen cooled and fueled flight weight dual-mode combustion ram jet.

Hydrogen Superconducting Systems

Expert / Organization	Expertise
ARP, V.D. Cryogenics Division National Bureau of Standards Boulder, Colorado	Feasibility study of refrigerating superconducting transmission lines with subcooled liquid or slush hydrogen.

Metal Hydrides

Expert / Organization	Expertise
BILLINGS, R.E. Billings Energy Research Corp. Provo, Utah	Research and development work on most types of metal hydrides. Production of commercial metal hydride tanks. Use of metal hydrides to store hydrogen fuel for autos. Research and design work on inexpensive electrolysis units for charging metal hydride tanks with hydrogen.
DOUGLASS, D.L. Materials Department University of California Los Angeles, California	Physical chemistry and physical metallurgy of hydride formation and dissociation. Main interest is in the higher capacity magnesium alloy hydrides (Mg-10Al and Mg-25Ni).
LUNDIN, C.E. Denver Research Institute University of Denver Denver, Colorado	Basic research on the stable and newer unstable class of metallic hydrides; kinetics of sorption and desorption, thermodynamic properties, enhancement of properties for an ideal storage material, contamination and poisoning characteristics. Safety aspects and characteristics of iron-titanium hydride; possibilities of a pyroforic dust explosion, effect of shock, impact, static electricity, and flammability of the solid material.

Expert Organization	Expertise
POWERS, G.J. Department of Chemical Engineering Carnegie-Mellon Institute Pittsburgh, Pennsylvania	Conducting a chemical engineering analysis of the heat and mass transfer with reaction in metal hydrides. Applications are automobile hydride storage of hydrogen and solar energy hydrogen hydride storage schemes.
REILLY, J.J. Brookhaven National Laboratory Upton, New York	Physical research on metal hydrides. Studies on new hydride materials and development work on existing materials.
SCHIRBER, J.E. Sandia Laboratories Albuquerque, New Mexico	Theory and experimentation on superconducting palladium and palladium alloy hydrides and deuterides. Other work within the group includes NMR studies, diffusion and permeation studies on rare earth and transition metal hydrides, also theoretical work on the electronic properties of metal hydrides.
VOOK, F.L. Department Head of Radiation and Device Physics Research Sandia Laboratories Albuquerque, New Mexico	Surface theory and experimentation on ion implantation and modification of metal hydride surfaces.

Hydrogen Fuel Cells

Expert Organization	Expertise
ADLHART, O.J. Manager of Energy Systems Englehard Industries Newark, New Jersey	Research and development work on hydrogen-air fuel cells. Author of book entitled, *Handbook of Energy Technology.*
AUSTIN, L.G. Department of Material Sciences Pennsylvania State University University Park, Pennsylvania	Research and consulting work on fuel cells for Army Material Command, NASA, and the National Science Foundation. Author of NASA SP-120, Fuel Cells. Co-author of *Handbook of Fuel Cell Technology.* Areas of expertise: electrocatalysis, theories of electrode behavior, power losses of electrodes, and cell optimization.
BAKER, B.S. Energy Research Corporation Bethel, Connecticut	Research and development work on fuel cells and metal hydrides. Investigation of thermal and chemical compatibility between nickel and silver batteries and metal hydride hydrogen storage systems. H_2-O_2 and H_2-Air Fuel Cell research.
CHAPMAN, L.E. Manager of Engineering General Electric Co. Lynn, Massachusetts	Research and development work on H_2-O_2 type fuel cells for space applications. Fuel cells have a solid polymer type of electrolyte. Life support systems.

Expert Organization	Expertise
DUSENBURY, Bill Branch Chief, Power Generation Branch NASA—Johnson Space Center Houston, Texas	Monitor of NASA research and development contracts for H_2-O_2 fuel cells for spacecraft applications. Subsystem responsibility for flight-type fuel cell systems.
KING, J.M., Jr. Pratt & Whitney Division of United Aircraft Corporation South Windsor Engineering Facility South Windsor, Connecticut	Project engineer responsible for evaluation of advanced applications of fuel cell power plants. Experience in directing the analysis of technical and economic requirements for fuel cells; analysis of alternate fuel cell concepts and in establishing their technological goals.
THALLER, L.H. Fuel Cells Systems Section NASA—Lewis Research Center Cleveland, Ohio	Development work on advanced type of light weight H_2-O_2 fuel cells. Potassium hydroxide and water electrolyte.

Internal Combustion Engines

Expert Organization	Expertise
ADT, R.R., Jr. Mechanical Engineering Department University of Miami Miami, Florida	Performance and emissions studies on two automobile engines that have been modified to run on hydrogen-air mixtures. Engine power output is controlled by varying the fuel-air ratio. Achieving high part load efficiencies and low nitrous oxide levels.
BILLINGS, R.E. Billings Energy Research Corp. Provo, Utah	Investigation of techniques and parameters for using hydrogen as fuel in standard spark plug engines, Wankel and Sachs rotary engines, and also diesel engines. Use of metal hydrides to store hydrogen fuel.
de BOER, P.C.T. Sibley School of Mechanical and Aerospace Engineering Cornell University Ithaca, New York	Experimental program to determine efficiency, performance, and NO_x emissions of internal combustion engines fueled with hydrogen. Direct cylinder injection of hydrogen, compression ignition, rate of combustion, and preignition characteristics. Use of exhaust heat to convert LH_2 to gaseous hydrogen at high pressure.
MCLEAN, W.J. Sibley School of Mechanical and Aerospace Engineering Cornell University Ithaca, New York	Works with analytical models of the hydrogen fueled reciprocating engine. The models are useful for predicting performance and emissions for engines fueled with hydrogen (or alternative fuels) operating over wide ranges of engine design parameters, load and fuel-air mixtures.
SCHOEPPEL, R.J. School of Mechanical Engineering Oklahoma State Engineering School Stillwater, Oklahoma	R&D work using a dual fuel (hydrogen and hydrocarbons) concept for internal combustion engines. Fuels are not premixed and the H_2-HC ratio can be varied full range. Hydrocarbons are carbureted into the engine and the hydrogen is injected into the combustion chamber.

Expert Organization	Expertise
VAN VORST, W.D. Engineering Systems Department University of California Los Angeles, California	Comparison studies between hydrogen and gasoline fueled engines. Using water injection and exhaust gas recycling to control combustion rate of hydrogen fuel. Monitoring exhaust emissions. Two types of speed control: quality governing and ordinary throttling.
WEIL, K.H. Former Head, Department of Mechanical Engineering Stevens Institute of Technology Hoboken, New Jersey	Worked on conversion of existing internal combustion engines to hydrogen operation. Present research and development comprises hydrogen generation, storage, transportation, distribution and energy conversion. Special Consultant for hydrogen systems to National Academy of Sciences and National Research Council, Washington, D.C.

Hydrogen Fuel, General

Expert Organization	Expertise
ESCHER, W.J.D. Escher Technology Associates St. Johns, Michigan	Technological consulting in the area of hydrogen energy-environment. Special interest in the application of hydrogen to solar energy systems and to advanced transportation vehicles.
GREGORY, D.P. Institute of Gas Technology Chicago, Illinois	Studies of total hydrogen energy systems in which hydrogen replaces natural gas systems. Studies of hydrogen as a means of using nuclear energy in chemical, steel production, and the transportation industries. Pipeline transmission of hydrogen gas.
HOFFMAN, K.C. Head of Engineering Systems Div. SALZANO, F.J. Mgr. of Hydrogen Storage and Production Project Brookhaven National Laboratory Upton, New York	Systems analysis of methods for storing energy. Comparison of various methods of storing hydrogen with the main emphasis on using metal hydrides. Physical research of metal hydride materials. System studies for using stored hydrogen gas for peak loading by public utilities. Systems studies of electrolysis plants.
HORD, Jesse Cryogenics Division National Bureau of Standards Boulder, Colorado	Economic and technical feasibility studies of hydrogen-energy systems.
JOHNSON, J.E. Project Manager of Hydrogen Linde Division of Union Carbide Corp. New York, New York	Hydrogen fuel logistics. Methods of manufacture, distribution, and economics of supply.
MICHEL, J.W. Acting Technical Assistant of Advanced Energy Systems Linde Division of Union Carbide Corp. Oak Ridge National Laboratory Oak Ridge, Tennessee	Interest in all aspects of the hydrogen economy. Two experimental programs on fuel production; one on thermochemical methods of producing hydrogen and another program studying methods of producing fuel from garbage and farm waste.

Expert Organization	Expertise
NORAD, D.L. Chief, Laser and Energy Systems Branch NASA—Lewis Research Center Cleveland, Ohio	Studying techniques for using hydrogen for energy storage. Investigating complete hydrogen energy systems including production, transmission, storage, distribution, end use, and scenarios for implementation. Also involved in a joint AEC-NASA effort to study production of hydrogen from nuclear heat (using both coal conversion and thermochemical processes).

SAFETY AND HAZARDS

This section contains a listing of individuals who are very knowledgeable about the hazards and safety requirements associated with hydrogen systems. There is a listing of experts who have had extensive experience in the storage and handling of liquid and gaseous hydrogen. Also listed is a group of people who are knowledgeable about the electrical codes and the electrical equipment to be used in Class 1, Group B hazardous environments. There is a list of experts who have managed, supervised or conducted experimental work on hydrogen fires, explosions, and spills, and studied the parameters involved. Also included is a listing of experts who have been deeply involved in hydrogen safety since the early 1950s; a good share of these individuals have participated in the investigation of hydrogen incidents and accidents in recent years.

Storage and Handling of Hydrogen

Expert Organization	Expertise
BAIN, A.L. Systems Engineer NASA—Kennedy Space Center Kennedy Space Center, Florida	Systems engineer in the support operations at KSC. Extensive experience with the storage, trailer transport, and distribution of LH_2 and LO_2 and the liquid to gas conversion systems at KSC.
BOWERS, W.M. Liquid Metal Engineering Center Atomics International Division of Rockwell International Canoga Park, California	Formerly, a facility engineer for an LH_2 fueled rocket engine test complex at Rocketdyne. Considerable experience with storage, handling and testing with liquid hydrogen.
HENNINGS, Glen Chief Power Experiments Branch NASA—Lewis Research Center Cleveland, Ohio	Extensive experience with the storage, handling and testing with hydrogen, oxygen, and fluorine at the Plumbrook and Lewis Research Center. Supervised liquid hydrogen transfer from 34,000-gallon rail tank cars; and use of the rail tank cars for run tanks during testing of liquid hydrogen pumps. Participated in the design and construction of large hydrogen storage and transport vessels for the Plumbrook facility.

Expert Organization	Expertise
SHAW, R.C. Chief, Cryogenic and Pressurant Section NASA—Marshall Space Flight Center Marshall Space Flight Center, Alabama	Responsible for the transport, distribution, and storage of cryogens and high-pressure gases at MSFC. Responsibility for developing safety criteria and specifications for all the propellant transport trailers used at MSFC.
SIEWERT, R.D. ASRDI Program Manager NASA—Lewis Research Center Cleveland, Ohio	Helped supervise the storage and handling of liquid hydrogen at the Plumbrook facility for 12 years. Participated in the testing of rocket engine pumps at Plumbrook. Responsibility for the safety aspects of storage, transfer, and experimental operations with liquid hydrogen at the Plumbrook Test Facility.
WARD, R.C. Chief, Launch Operations Fuel Systems NASA—Kennedy Space Center Kennedy Space Center, Florida	Early experience with the Atlas program at General Dynamics. Extensive experience in handling RP-1 and liquid hydrogen for the Saturn and Atlas vehicles at Kennedy Space Center.

Electrical Codes and Equipment

Expert Organization	Expertise
HALL, L.J. Engineering Mgr., Engineering Div. The Association of Mill and Elevator Insurance Companies (Mill Mutual) Chicago, Illinois	Chairman of the National Electric Code Panel 14. Extensive experience in the fire prevention field. Evaluation and survey of firms for fire hazards, and for meeting fire prevention standards for hazardous locations.
PEARSE, J.N. Appleton Electric Company Chicago, Illinois	Technical service and advise for Class 1, Group B hazardous location electrical equipment.
SCHRAM, P.J. Underwriters Laboratories, Inc. Northbrook, Illinois	Member of Panel 14, National Electric Code Committee. Manager of the section testing all electrical equipment for hazardous locations at Underwriters Laboratories. Type of equipment includes switching, lighting, and process and control equipment.
SHORT, W.A. Crouse-Hinds Company Syracuse, New York	Technical service and advise for Class 1, Group B hazardous location electrical equipment.

Hydrogen Fires, Explosions and Spills

Expert Organization	Expertise
ALLAN, D.S. Arthur D. Little Company Cambridge, Massachusetts	Responsible for the early work by A.D. Little Company to determine the hazards associated with storing and handling large quantities of

liquid hydrogen. Conducted unconfined LH_2 spill tests to analyze and determine the detonability of hydrogen vapor clouds. Spills of LH_2 with LO_2 were also conducted. Parameters examined were ignition sources, detonability range, electrostatic charge, explosive yield, and thermal radiation.

ATKINS, J.R.
Director of Safety Office
NASA—Kennedy Space Center
Kennedy Space Center, Florida

A study manager for project PYRO; Investigation of LO_2-LH_2, TNT equivalencies, investigation of triggering methods for large spills of LO_2-LH_2, and other studies associated with triggering mechanisms of LO_2-LH_2 mixtures.

FLETCHER, R.F.
NASA—Johnson Space Center
Houston, Texas

Project manager for the Gemini Fireball Program; a series of tests to determine the explosion parameters and TNT equivalencies of LO_2 mixed with RP-1 and LH_2. Investigated the explosion characteristics of LO_2 mixed with RP-1 and LH_2 in rarified environments. Participated as a member of the Panel 7 investigation team for the Apollo 13 Incident. Participated in a NASA shuttle program study to determine if the initiation of the liquid propellants would tend to detonate the solid propellants on the shuttle vehicle.

HORD, Jesse
Cryogenics Division
National Bureau of Standards
Boulder, Colorado

Analysis and synthesis of explosion criteria for liquid hydrogen test facilities.

PROFFIT, R.L.
Program Manager of Environmental Analysis Laboratory
Lockheed Electronics Co.
Las Vegas, Nevada

Developed five different methods for the visualization of hydrogen fires; ultraviolet television, infrared television, infrared photography, ultraviolet photography, and an infrared telescope employing an image converter tube. Participated in the development of the AC/DC hydrogen fire detection system; NASA SP-5092.

REIDER, Roy
Los Alamos Scientific Laboratory
Los Alamos, New Mexico

Performed unconfined experiments wherein hydrogen gas was released in air at a high flow rate (120 pounds/second). Auto-ignition occurred and a deflagration of explosive rapidity occurred. Acoustical, ignition, deflagration and explosive damage data were obtained.

RIEHL, W.A.
Deputy Chief of Nonmetallic Materials
NASA—Marshall Space Flight Center
Marshall Space Flight Center, Alabama

Extensive research and development on the blast hazards of hydrogen in space vehicle liquid propellant systems within NASA-MSFC. Also assisted in directing project PYRO (a joint NASA/USAF liquid propellant explosive hazards program to determine the blast and thermal characteristics of the three liquid propellant combinations in most common use in military missiles and space vehicles). PYRO is the current basis for prediction of liquid propellant blast hazards.

Expert Organization	Expertise
SIEWERT, R.D. ASRDI Program Manager NASA—Lewis Research Center Cleveland, Ohio	Working in the area of fragments and explosions for all types of explosive materials. Program manager for a study grant entitled, The Analysis of the Structure of Blastwaves from Accident Explosions. A member of the technical steering committee for the safety and environmental protection group in JANNAF.
ULLIAN, L.J. Senior Ordnance Engineer Missiles Systems Safety Division Headquarters Air Force Eastern Test Range Patrick Air Force Base, Florida	Participated in range safety work at Patrick Air Force Base. Participated as a Study Manager for project PYRO. Conducted and monitored an experimental program and study of the relationship of the interaction of liquid propellant initiation on the detonation of solid propellants.
VAN DOLAH, R.W. Research Director of Pittsburgh Mining and Safety Research Center Bureau of Mines Pittsburgh, Pennsylvania	Involved in active research, consultation, and accident investigation over a wide range of fires and explosions. Served as a member of the Apollo 204 Incident review board. Consultant to the AEC on hydrogen safety and nuclear reactors. Directed some early small-scale experimental work on hydrogen spills, and subsequent work on LNG spills. Served as Chairman of the Advisory Committee of the Armed Services Safety Explosive Board.
ZABETAKIS, M.G. Mining Enforcement and Safety Administration Arlington, Virginia	Considerable experience in the field of hydrogen fires and explosions. Conducted the research necessary to prepare the Air Force quantity-distance tables for siting hydrogen storage vessels and experiments. Has particpated as an expert in over 100 investigations of fires and explosions (all types) throughout the country.

Hydrogen Safety and Accident Investigations

Expert Organization	Expertise
ATKINS, J.R. Director of Safety Office Kennedy Space Center Kennedy Space Center, Florida	Director of all safety operations (launch and ground support) at Kennedy Space Center. Extensive operational experience with LH_2 and LO_2 safety.
BALL, L.W. Director of Safety and Awareness NASA—Marshall Space Flight Center Marshall Space Flight Center, Alabama	Exercises management controls over the design, manufacturing, and use of hydrogen equipment. Responsible for the operational readiness inspections of hydrogen facilities.
BELLES, F.E. Manager of Research and Development American Gas Association Research Laboratories Cleveland, Ohio	Recently retired from NASA—Lewis Research Center as Director of the NASA Aerospace Safety Research and Data Institute (ASRDI). Presently doing research and development work on natural gas utilization, particularly product

	safety with respect to gas appliances. Area of expertise developed at NASA—Lewis includes combustion, fire hazards, detonation, cryogenic systems, and hydrogen safety. Participated in several NASA investigations involving hydrogen fires.
CHELTON, D.B. Cryogenics Division National Bureau of Standards Boulder, Colorado	Very active in the field of hydrogen safety. Participated as a member of the investigation team for the Cambridge Electron Accelerator explosion and fire. Participated as a member of the Manned Space Center Apollo 13 Incident investigation team. Participated in the investigation of the LNG storage tank explosion on Staten Island. Participated in hydrogen safety reviews of liquid hydrogen bubble chambers and rocket launch facilities. Author of several publications on hydrogen safety and technology.
CONLON, J.W. Chief Operations Safety Office NASA—Johnson Space Center Houston, Texas	Responsible for the overall operational safety of the Space Shuttle operations and hazardous manned testing. A member of the investigation team for the Apollo 204 accident, and also participated in the investigation of the Apollo 13 incident.
EDESKUTY, F.J. Los Alamos Scientific Lab. (LASL) Los Alamos, New Mexico	Member of the Apollo 13 Incident investigation team. Member of the LASL LH_2 safety committee and a member of the LASL meson physics safety committee.
HERNANDEZ, H.P. Lawrence Berkeley Laboratory Berkeley, California	Extensive safety experience in the area of liquid hydrogen bubble chambers. Conducted an external review of the safety aspects of the National Accelerator Laboratory (NAL) 15-foot LH_2 bubble chamber. Participated as a member of the safety investigation team for the Cambridge Electron Accelerator explosion and fire.
MCKINLEY, Clyde Director of Research and Development Cryogenic Systems Division Air Products and Chemicals, Inc. Allentown, Pennsylvania	Extensive experience in hydrogen safety. Participated in the investigations of several incidents at Air Products hydrogen liquefaction facilities at West Palm Beach, Florida.
ORDIN, P.M. Program Manager, NASA Aerospace Safety Research and Data Institute NASA—Lewis Research Center Cleveland, Ohio	Extensive experience in the field of safety with both hydrogen and oxygen. Fluid flow and flight safety studies for NERVA program and for ground and flight operations of early hydrogen aircraft.
PINKEL, Irving—Consultant 4671 West 210th Street Fairview Park, Ohio	Extensive experience in the deisgn and development of LH_2 pumps and fuel systems for hydrogen-fueled aircraft. Principal investigator for the Apollo 204 fire and participated as a member of Apollo 13 accident investigating team. Worked as Chief of the Fluid Systems

Components Division at NASA—Lewis, and retired from NASA—Lewis as Director of the NASA Aerospace Safety Research and Data Institute (ASRDI). Previous experience at Bureau of Mines was involved with hydrogenation experiments and the design of a plant for the hydrogenation of coal.

REIDER, Roy
Los Alamos Scientific Laboratory
Los Alamos, New Mexico

Safety Director for the Los Alamos Scientific Laboratory. Active in LH_2 safety work for many years. Consultant to the advisory committee on reactor safeguards.

SCHLAFKE, A.P., Jr.
Division Chief of Mechanical Engineering
Brookhaven National Laboratory
Upton, New York

Chairman of the Brookhaven Laboratory Cryogenic Safety Committee. Presently Manager of the Cryogenic Support and Design Group for the alternating gradient synchrotron. Participated as a member of the safety investigation team for the Cambridge Electron Accelerator explosion and fire.

SCHMIDT, A.F.
Cryogenics Division
National Bureau of Standards
Boulder, Colorado

Extensive liquid hydrogen safety experience. Consultant to NASA, AEC, and USAF on hydrogen safety. Member of National Bureau of Standards Cryogenics Division Hydrogen Safety Committee.

ULLIAN, L.J.
Senior Ordinance Engineer
Missiles Systems Safety Div. (SEN)
Hdqrs., Air Force Eastern Test Range (ETR)
Patrick Air Force Base, Florida

Participated in range safety work at Patrick Air Force Base. Performed an investigation and study of all full-scale Launch Vehicle Aborts at the Eastern Test Range and the Western Test Range. Involved in the siting and safety planning for firing tests of the first vehicle to use hydrogen. Participated as part of the investigation team for the Apollo 204 fire, the Sycamore Canyon Saturn S-IV fire and explosion, and the Atlas-Centaur launch pad liftoff failure which resulted in a fire and explosion.

ZABETAKIS, M.G.
Mining Enforcement and Safety Administration
Arlington, Virginia

Extensive experience in the field of hydrogen safety, fires and explosions; has participated in the investigation of over 100 fires and explosions (all types) throughout the United States. Authored the book, *Safety with Cryogenic Fluids,* and has participated in many investigations on hazards involving hydrogen safety. Was in charge of the explosion portion of the investigation of the LNG storage tank disaster on Staten Island, and later participated in a grand jury hearing and subsequent House of Representatives hearings. Participated in the investigation of the Minuteman silo fire and explosion. Participated in the investigation of the fire and explosion at the Sacramento test firing stand involving an S-IV test bird.

BIBLIOGRAPHY

The following reports used in the preparation of this book are available from:

National Technical Information Service
U.S. Department of Commerce
5285 Port Royal Road
Springfield, Virginia 22151

AD-787 484 — McAlevy, R.F., III, Cole, R.B., Hollenberg, J.W., Kurylko, L., Magee, R.S., and Weil, K.H., *Hydrogen as a Fuel,* August 1974.

AD-A016 256 — Cole, R.B., Magee, R.S., and Hollenberg, J.W., *Hydrogen Storage and Transfer,* August 1975.

BNL-18224 — Sailor, V.L., *The Hydrogen Economy.*

BNL-18719 — Waide, C.H., Reilly, J.J., and Wiswall, R.H., *The Application of Metal Hydrides to Ground Transport.*

BNL-18720 — Yu, W.S., Suuberg, E., and Waide, C., *Modelling Studies of Fixed-Bed Metal Hydride Storage Systems.*

BNL-18721 — Salzano, F.J., Cherniavsky, E.A., Isler, R.J., and Hoffman, K.C., *On the Role of Hydrogen in Electric Energy Storage.*

BNL-18887 — Reilly, J.J., Wiswall, R.H., Hoffman, K.C., and Waide, C.H., *Metal Hydrides as Hydrogen Storage Media,* May 1974.

BNL-19249 — Salzano, F.J., Editor, *Hydrogen Storage and Production in Utility Systems,* July 1974. Contributors are:

Cherniavsky, E.A. | Isler, R.J.
Dean, J.W. | Kissel, G.
Hoffman, K.C. | Majeski, S.J.

(continued)

Milau, J.S.	Strickland, G.
Miles, M.H.	Suuberg, E.M.
Reilly, J.J.	Waide, C.H.
Sevian, W.A.	Wiswall, R.H., Jr.
Srinivasan, S.	Yu, W.S.

BNL-19266 Burger, J.M., Lewis, P.A., Isler, R.J., Salzano, F.J., and King, J.M., Jr., *Energy Storage for Utilities via Hydrogen Systems,* August 1974.

BNL-20162 Powell, J.R., Salzano, F.J., Yu, W.S., and Milau, J.S., *High Efficiency Power Conversion Cycles Using Hydrogen Compressed by Absorption on Metal Hydrides.*

BNL-50447 Powell, J.R., Salzano, F.J., Yu, W.S., and Milau, J.S., *High Efficiency Power Cycles Using Metal Hydride Compressors,* January 31, 1975.

CONF-730807-3 Michel, J.W., *Hydrogen and Synthetic Fuels for the Future,* August 30, 1973.

LAUR-73715 Stewart, W.F., and Edeskuty, F.J., *Logistics, Economics, and Safety of a Liquid Hydrogen System for Automotive Transportation.*

N74-11728-735 Savage, R.L., Editor, *A Hydrogen Energy Carrier. Volume II: Systems Analysis,* September 1973. Other editors and authors include:

Blank, L.T.	Paleocrassas, S.N.
Cady, T.C.	Powers, S.O.
Chao, R.E.	Riley, R.K.
Cox, K.E.	Salmon, J.D.
Eisenstadt, M.M.	Traina, L.A.
Johnson, R.T.	Williams, R.D.
Koelling, H.A.	Wilson, A.E.
Lee, S.S.	Witwer, J.G.
Murray, R.G.	

N74-16800 Seikel, G.R., Smith, J.M., and Nichols, L.D., *H_2-O_2 Combustion Powered Steam-MHD Central Power Systems,* March 1974.

N76-10323 Ludtke, P.R., *Register of Hydrogen Technology Experts,* October 1975.

ORNL-TM-4305 Escher, W.J.D., *Prospects for Hydrogen as a Fuel for Transportation Systems and for Electrical Power Generation,* September 1972.

ORNL-TM-4307 Johnson, J.E., *The Storage and Transportation of Synthetic Fuels—A Report to the Synthetic Fuels Panel,* September 1972.

SAND-74-8219 Swisher, J.H., Keeton, S.C., West, A.J., and Jones, A.T., *Survey of Hydrogen Compatibility Problems in Energy Storage and Energy Transmission Applications,* September 1974.

UCRL-51228 Austin, A.L., *A Survey of Hydrogen's Potential as a Vehicular Fuel,* June 19, 1972.

The following U.S. patents used in the preparation of this book are available from:

Commissioner of Patents and Trademarks
Washington, D.C. 20231

U.S. Patent 3,922,872 Reilly, J.J., and Wiswall, R.H., Jr.; December 2, 1975; assigned to the United States Energy Research and Development Administration.

U.S. Patent 3,943,719 Terry, L.E., and Schoeppel, R.J.; March 16, 1976.

WIND POWER 1975

by Daniel M. Simmons

Energy Technology Review No. 6

Wind is a free, clean, inexhaustible source of energy. In previous centuries, when dependence on time and deadlines for production and shipments were not as important as they are now, wind served mankind well, despite becalmed and stormy intervals. Wind moved sailing ships of all sizes and drove windmills to grind seeds and other materials, saw wood and to pump water. In later years it was used to produce some electrical current of the DC type to charge storage batteries.

There are three phases to harnessing the power of the wind. The first is the selection of a suitably windy site. This necessitates studies of wind behavior in possible locales, including ocean sites. Secondly a suitably designed windmill and generator to convert the mechanical to electrical energy or to convert seawater to hydrogen and oxygen. The third and rather fundamental difficulty is the available choice of storage batteries or other storage means, so that energy produced at peak times can be used when the wind is not blowing.

The various phases for developing wind power are considered in the first three chapters of this book. The next three describe wind power developments in the U.S., Canada, U.S.S.R., Germany, Denmark, France, U.K., Sweden and other countries in Asia and Africa. The last chapter describes commercially available apparatus.

This Energy Technology Review is based primarily on international studies conducted by industrial and engineering firms or university research teams under the auspices of various governments and governmental agencies.

A partial and condensed table of contents follows here.

1. **WIND BEHAVIOR AND SITE SELECTION**
   Influence of Height Above Ground
   Wind Surveys and Site Selection
   Anemometers and Wind Speed Recorders
   Isovent Cards
   Use of Metereological Data
   Wind Studies in Different Countries

2. **WIND MACHINE DESIGN**
   Major Problems of Wind Power
   Tower Height & Design
   Complete Systems
   Wind Turbines
   Small Wind Electric Plant with Permanent Magnet Generator
   Sailwing Windmills
   Vertical Axis Wind Rotors
   Diffuser-Augmented Wind Turbine
   The J. D. Madaras Project

3. **WIND CONVERSION AND STORAGE SYSTEMS**
   Variable Speed Input vs. Constant Current Output
   Asynchronous AC/DC/AC Converter
   Seawater Electroylsis to H_2 and O_2 and Their Storage
   Compressed Air Manufacture
   Batteries
   Superflywheels
   Alternate Systems
   Flow Charts for Alternate Systems

4. **WIND POWER DEVELOPMENT IN THE U.S. AND CANADA**
   Small Wind Power Systems
   An Economic Study
   Large Scale Applications
   Ocean-Based Wind Machines
   Preliminary Design of a 100 KW Wind Turbine Generator
   Proposed Regions

5. **U.S.S.R. DEVELOPMENTS**
   Large Scale Wind Generator
   Agricultural Usage
   Wind Machine Layouts
   Water Lifting Machines
   Other Wind Powered Machines
   Wind Machines for Making Electricity
   Typically Russian Designs

6. **WIND POWER DEVELOPMENTS IN OTHER COUNTRIES**
   Netherlands and adjacent Low Countries
   West Germany & Denmark
   France & Great Britain
   Sweden
   India
   Israel
   Algeria
   Curaçao

7. **COMMERCIALLY AVAILABLE WIND MACHINES**
   Aermotor
   Automatic Power, Inc.
   Dempster Industries, Inc.
   Edmund Scientific Co.
   Environmental Energies, Inc.
   Heller-Aller Co., Inc.
   Solar Wind Co.
   Winco
   Windworks

ISBN 0-8155-0575-2

300 pages

SOLAR ENERGY FOR HEATING AND COOLING OF BUILDINGS 1975

by Arthur R. Patton

Energy Technology Review No. 7

Solar energy can be used for indirect heating purposes in many ways. The information in this book has been limited to so-called low temperature solar thermal processes. Designs requiring photocells or other thermoelectric generators and lenses or reflecting mirrors plus tracking equipment have been excluded.

Low temperatures are the easiest to obtain, and the necessary collectors are fairly simple in construction. A black surface is used to absorb the sun's rays, this surface is usually covered with glass and the collector is insulated on the back and sides against heat loss. Water or some other heat transfer fluid is passed through the collector and can reach temperatures from 60°C (140°F) to about 95°C (203°F). The thermal energy is then stored in a heat storage system (perhaps based on the latent heat of fusion of selected salts). Coupled to the heat storage system are heating loops to furnish heat by convection and to operate an air conditioning system. In most temperate zones an auxiliary heater, operated with conventional fuels, must also be connected.

Large scale applications designed for schools and similar building are beginning to appear or are in the planning stage. This book describes in detail several large scale feasibility studies with designs suitable for institutions and industrial plants.

Descriptions are based on studies conducted by industrial or engineering firms or university research teams under the auspices of various government agencies. A partial and condensed table of contents follows here.

1. **SYSTEM COMPONENTS**
   Collectors—Flat Plate Construction
   Loss Control Concepts
   Internal Heat Transfer
   Heat Storage Designs
   Solar Powered Air Conditioning

2. **HISTORY OF EXPERIMENTAL SYSTEMS**
   System Configurations
   Air as the Working Fluid
   Water as the Working Fluid
   Integral Collector-Storage Configurations
   Other Configurations

3. **DETAILED DESCRIPTIONS AND LOCATIONS OF EXPERIMENTAL SYSTEMS**
   Heating Systems
   Cooling and Refrigeration Systems

4. **SIMULATED SYSTEMS**
   University of Wisconsin Study
   Albuquerque Model
   System Models
   Systems Evaluation and Cost Analyses
   Marshall Space Flight Center
   Background
   Description of Test System
   Model Techniques
   Simulation Results
   University of Pennsylvania Study

5. **FEASIBILITY STUDIES FOR LARGE SCALE INSTALLATIONS**
   National Science Foundation Study
   Overview
   Evaluation of Present Status of Technology
   Implementation Studies
   MITRE Study
   General Assessments
   Problem Areas
   Improvement Studies
   Economic Viability Calculations

6. **GENERAL ELECTRIC STUDY**
   Scenario Projections
   Systems Descriptions
   Technical Evaluations
   Capture Potential and Projections of Market Penetration
   POCE—Proof of Concept Experiments

7. **WESTINGHOUSE STUDY**
   Energy Consumption
   Operational Requirements
   Capture Potential and Master Plan

8. **TRW STUDY**
   Climatic Region Classifications
   Reference System Designs

9. **COLORADO STATE UNIVERSITY INTEGRATED SYSTEM DESIGN**
   Project Objectives and Data Handling
   Desirable Building Designs
   Lithium Bromide Absorption Cooling Unit

10. **OTHER RESEARCH**

11. **AVAILABLE HARDWARE**

12. **NAMES & ADDRESSES FOR FURTHER INFORMATION**

13. **REFERENCES**

ISBN 0-8155-0579-5 **$24** **328 pages**

GEOTHERMAL ENERGY 1975

by Edward R. Berman

Energy Technology Review No. 4

This book describes in detail the nature of the geothermal resources, their extent, and the currently available technology by which these natural sources of energy can be exploited and utilized to the greatest advantage. There is little pollution resulting from the use of geothermal energy, and with a little care all disturbance of natural ecological systems can be avoided.

Earth heat can be used most practically where hot volcanic rocks are comparatively near the surface, and emerging circulating ground waters act as heat collectors, either by producing steam or by serving as heat transfer media. Suitable sites have been discovered in the Continental U.S. and Hawaii, the U.S.S.R., Japan, New Zealand, Iceland, and Italy. In view of the rise in the price of oil and the intensive search for new sources of energy, geothermal power appears to be in for a period of rapid development.

This Energy Technology Review is based on international studies conducted by industrial and engineering firms or university research teams under the auspices of various governments and governmental agencies.

A partial and condensed table of contents follows here.

1. U.S. RESEARCH & EXPLORATION TECHNIQUES

Historical Background
Recent Explorations
U.S. Agencies Involved
World Survey of Major Geothermal Installations (except USSR)

2. RESEARCH AND PRACTICE IN THE USSR

Soviet Geothermal Electric Power Generation
Paratunka Geothermal Electric Power Station (using Freon®)
Pauzhetka Power Station
Bol'she-Bannaya Station
Makhachkala Station
Additional Sites
Non-Electric Uses of Thermal Waters
Equipment and Instruments

3. DRY GEOTHERMAL RESERVOIRS

Conventional Drilling
Ultra-Deep Drilling
Limitations of Current Technology
Areas of Further Research
Novel Techniques
A Proposed Commercial System
Man-Induced Fracturing of Hot Rocks
Artificial Geothermal Reservoirs

4. THE PLOWSHARE CONCEPT

Geothermal Heat Extraction Using Nuclear Explosives
Site Guidelines & Geology
Heat Source Developments
System Thermodynamics
Seismic Design Considerations

5. EXPERIMENTAL STUDIES

Scale Formation in Simulated Geothermal Brine
Wairakei (New Zealand) Well Analysis
Aquifer Chemistry in Geothermal Systems
Plowshare Geothermal Steam Chemistry

6. GEOTHERMAL RESOURCES OF CALIFORNIA

The Geysers
Imperial Valley
Mono Lake
Desalination of Geothermal Water
Supplemental Water for Reinjection
Land Subsidence and Cave-Ins
System Synthesis and Costs

7. ENERGY RECOVERY FROM NATURAL HOT BRINE

Chemical Composition
Fluid Pressures
Turbine Systems
Corrosion and Scale
Sperry Rand Corp. Method

8. FEASIBILITY STUDIES FOR SPECIFIC U.S. AREAS

Texas
Port Mansfield Site
California
Point Mugu Site
Twenty-Nine Palms Site
China Lake Site
Idaho
Raft River Valley Site

9. PROPOSED RESEARCH

ISBN 0-8155-0563-9 **336 pages**

THERMAL ENERGY FROM THE SEA 1975

by Arthur W. Hagen

Energy Technology Review No. 8
Ocean Technology Review No. 5

Recent advances in heat transfer research and thermal power plant design suggest that sea thermal power can now be made competitive with more conventional generating methods.

Utilization of ocean thermal gradient systems appears relatively attractive from many points of view. The ocean acts as a large solar energy heat reservoir which reduces energy storage requirements and permits the system to be operated the year round, 24 hours per day.

The purpose of this book is to provide a condensed data base to aid in proof-of-concept experiments and continued R & D to prove the technical feasibility and economic viability of generating either electricity or hydrogen by harnessing the temperature gradients in the sea.

The optimum locations for such generating plants appear to be in latitudes from 23°N to 23°S. Transmission and storage problems may limit the amount of power delivered by ocean-based plants. Construction of power production facilities must consider design problems associated with the hazards of marine environment.

Reports of proposed solutions to these problems are also contained in this book which is based on government-sponsored studies by engineering firms and university research teams. A partial and condensed table of contents follows:

1. **AN OVERVIEW OF SEA THERMAL POWER**
   Introductory Material and NASA Reports
   Site Analysis
   Cost
   Systems Analysis

2. **OPEN CYCLE THERMAL GRADIENT OCEANIC POWER PLANT**
   Introductory Material
   NASA and PB Reports
   Preliminary Design Investigation
   Overall System Design
   Spray Evaporator Design
   Turbine Design
   Deaeration Losses
   Condenser Design
   Cold Water Pump Considerations
   Fresh Water Production
   Feasibility Study of a
   100 Megawatt Plant

3. **CARNEGIE MELLON UNIVERSITY DESIGN**
   Introductory Material and PB Reports
   Design Overview
   Water Circulation
   Heat Transfer Fluid Circulation
   Seawater Inlet-Outlet Hydrodynamics
   Boiler Technology Considerations
   Cold Water Pipe vs. Ammonia Pipe
   Vertical Tube Heat Exchangers
   Antifouling Costs
   SSPP (**S**olar **Sea** **Po**wer **P**lant) Topology

4. **UNIVERSITY OF MASSACHUSETTS DESIGN**
   Introductory Material and PB Reports
   Overall Design Concept (Mark I)
   Cold Water Supply & Suction Pipe
   Anchor & Mooring Systems
   Energy Umbilical
   Control of Biofouling
   Heat Exchanger Design (Mark I)
   Variations and Improvements (Mark II)
   Evaluation of Turbomachinery
   Technical & Economic Feasibility Studies

5. **COMPARATIVE STUDY OF CLOSED CYCLE TECHNOLOGIES**
   NASA Reports
   CMU vs. UMASS Designs
   Cost Evaluations

6. **ADDITIONAL TECHNOLOGIES**
   Hydrogen Utilization
   Hydrogen Energy
   NITINOL (Ni-Ti-Naval Ordnance Lab.)
   #55 Alloy Utilization
   Further SSPP Proposals
   OTEC Proposal (Ocean Thermal Energy Conversion Proposal)
   The Danish System
   A Japanese System

ISBN 0-8155-0597-3 **150 pages**

ENERGY IN THE CITY ENVIRONMENT 1973

Edited by Robert N. Rickles

This book is based upon material supplied by Robert Rickles, the Institute for Public Transportation, and the New York Board of Trade.

I: CRISES IN ENERGY SOURCES AND PRODUCTION

1. OVERVIEW (1972-1985)
   - Primary Reliance on Conventional Fuel Sources
   - Problems in Research and Technology
   - Lead Time for Breeder Reactors
   - Fusion
   - Solar Energy
   - Coal Gasification

2. OIL
   - NEPA and Other Environmental Issues
   - Accelerated Government Leasing Programs on Outer Continental Shelf
   - Oil Imports
   - Costs and Risks: Economic, National Security and Environmental Problems
   - Improved Exploration Incentives
   - Product Prices versus Foreign Imports
   - Tax Incentives

3. GAS
   - Accelerated Government Leasing of Outer Continental Shelf
   - Amendments to Natural Gas Act
   - Security of Contract Terms
   - Competitive Pricing
   - LNG/SNG — Imported and Domestic
   - Interstate and Intrastate Marketing Policies

4. ELECTRICITY
   - Generating Plant Licensing and Regulations
   - Nuclear Technology
   - Pumped Storage
   - Availability of Low Sulfur Fuel Oil and Gas
   - Efficiency of Conventional Sources
   - Purchasing Power (Imports)

II: CRISES IN ENERGY DEMANDS

1. OVERVIEW
   - Are Our Priorities and Objectives Compatible and Ready for Action?
   - Competition for Available Supplies: National and International Demands Constrain City Energy Resources
   - Effects of Differing Demand Breakdowns
   - Institutional Alternatives for Energy Management

2. REALISTIC APPRAISAL OF DEMAND GROWTH RATES (through 1985)
   - Population Growth (Inevitable Modest Demand Expansion)
   - Social Side Effects of Energy Constraint Policy
   - Additional Energy of Environmental Protection
   - Air Pollution and Energy

3. CONSERVATION OF ENERGY RESOURCES
   - Rationing and End Use Controls
   - Taxing Certain Energy Uses
   - Changes in Life Style
   - Improved Energy Efficiency
   - New Buildings
   - Mass Transit
   - Appliances

III: ELEMENTS IN THE ENERGY-ENVIRONMENT BALANCE

1. OVERVIEW
   - Need to Develop New Energy Reserves Without Unreasonable Environmental Delays
   - Ten Year Planning Lead
   - New Explorations to Preclude Shortages

2. IMPACT OF INADEQUATE ENERGY SUPPLIES
   - Economy-GNP-Unemployment-Inhibition of Investments
   - Insecurity and Reduced Standard of Living
   - Slowdown and Abandonment of Public Sector Programs
   - National Security
   - Environmental Protection: Air Pollution, Water and Waste Management

3. ENERGY-ENVIRONMENT BALANCING
   - Air Pollution
   - Offshore Drilling
   - Increased Tanker Traffic
   - Reliance on Fuel Imports
   - Oil and LNG
   - Nuclear Power
   - Radiation and Waste Disposal
   - Interim Nuclear Licensing
   - Selection and Nature of Plant Sites
   - Air Quality
   - Thermal Pollution

IV. NYS POWER PLANT SITING ACT

The important **charts** in the **appendix** present a list of energy conservation measures for the short-term (1972-1975), mid-term (1976-1980) and the long-term (beyond 1980) for the Transportation, Residential/Commercial, Industrial and Electric Utility sectors. The charts also indicate estimated maximum attainable energy savings, possible means for implementing each conservation measure, and pros and cons.

ISBN 0-8155-5019-7

173 pages

COAL CONVERSION TECHNOLOGY 1976

by I. Howard-Smith and G. J. Werner

Chemical Process Technology Review No. 66

The industrialized countries of the world are showing renewed and increasing activity in the area of synthetic fuels production from coal. Existing commercial coal conversion processes, such as the Fischer-Tropsch synthesis, the destructive distillation of coal, methanation etc. resulting in liquid and gaseous fuels, are again becoming economically competitive with petroleum.

The many processes and techniques of coal conversion have as a basic concept the transmutation of coal into forms acceptable to our transportation and heating equipments. To accomplish this, high-sulfur coals must be desulfurized, high-ash coals must be demineralized, and, most important of all, solid coal must be depolymerized into liquid and gaseous products that can be ignited and burned with facility.

Over 100 processes, in various stages of development, are available to carry out these procedures and are discussed fully in this book.

Originally prepared as an "in-house" report for the Millmerran Coal Pty. Ltd. of Brisbane, Australia, it is intended to provide readily accessible and concise data on all major activities in coal conversion technology. A special feature are the flow charts which clearly illustrate all the principal processes. A partial and very condensed table of contents follows here. Chapter headings are given, and some of the more important subtitles are also included.

GENERAL PROCESSES (36)
Bergius
Catalytic Coal Liquefaction
Clean Coke Process
COALCON Hydrocarbonization
COALPLEX
COED
COG (Coal-Oil-Gas Refinery)
COGAS
CONSOL Synthetic Fuel
COSTEAM
Extractive Coking Process
EXXON Liquefaction
Fischer-Tropsch Synthesis
GARRETT Coal Pyrolysis
H-Coal
Intermediate Hydrogenation
LEAS Liquefaction
LURGI-RUHRGAS
Methanol Synthesis
Molten Salt Catalysis
Project Lignite—SRL
Projects Seacoal & Seacoke
Solvent Extraction: CRE & UOP
SRC (Solvent Refined Coal) Processes
COMBUSTION ENGINEERING
MITSUI
PAMCO
SYNTHOIL
TOSCOAL
Underground Liquefaction
UNION CARBIDE

HIGH BTU COAL GASIFICATION (28)
Agglomerating Ash
AIR PRODUCTS Recycle Process
ATGAS/PATGAS
BIANCHI
BI-GAS
CHEVRON Gasification
CO_2 Acceptor
Electric Arc Gasification
Electrofluidic Gasification
GEGAS
HYDRANE
HYGAS
KOPPERS-TOTZEK
SYNTHANE
WINKLER

LOW & MEDIUM BTU GASIFICATION (48)
AVCO Arc Coal
BABCOCK & WILCOX
CONSOL Fixed Bed
KELLOGG Fixed Bed
ICI Moving Burden
In Situ Gasification
KERPELY PRODUCER
MARISCHKA
Molten Salt Process
OTTO RUMMEL Slag Bath
PANINDCO
PINTSCH HILLEBRAND
RUHRGAS VORTEX
THYSSEN GALOCSY
UGI Blue Water Gas

ISBN 0-8155-0614-7 **133 pages**